NANOFABRICATION BY ION-BEAM SPUTTERING

edited by
Tapobrata Som
Dinakar Kanjilal

NANOFABRICATION BY ION-BEAM SPUTTERING

Fundamentals and Applications

PAN STANFORD PUBLISHING

Published by

Pan Stanford Publishing Pte. Ltd.
Penthouse Level, Suntec Tower 3
8 Temasek Boulevard
Singapore 038988

Email: editorial@panstanford.com
Web: www.panstanford.com

British Library Cataloguing-in-Publication Data
A catalogue record for this book is available from the British Library.

Nanofabrication by Ion-Beam Sputtering: Fundamentals and Applications

ISBN 978-981-4303-75-0 (Hardcover)
ISBN 978-981-4303-76-7 (eBook)

Printed in the USA

To my wife and parents

—T.S.

To my wife

—D.K.

Contents

Foreword

The irradiation of solid surfaces by energetic particles gives rise to a wide variety of phenomena that are closely correlated with the energy deposition processes of the incoming ions. Over a near-surface region of the solid whose extension depends primarily on the range of the incident particles, the ions transfer energy and momentum to the target atoms, displacing them from their original positions. These processes may cause the ejection of target atoms and molecules from the surface (i.e., sputtering), while displacement cascades result in an accumulation of point defects (vacancies and interstitials) or more extensive defect structures; the concurrent incorporation of the implanted species may also lead to phase changes within that zone. The amount of research done in this field has been quite substantial over the past decades, and the understanding of the pertinent processes appears now to be highly evolved.

Based on that experience, ion beam processing of materials became an important and widely employed technological tool. In microelectronic device fabrication, for example, ion implantation matured into an indispensable method for the highly controlled doping of semiconductors. Other applications included ion-beam-assisted thin film growth, modification of magnetic or optical properties of pertinent materials, or alteration of polymer surfaces for biological applications.

At the surface, ion irradiation may result in substantial morphological changes. At low fluences, isolated defects such as vacancies and adatoms can be created. If these defects are mobile (for example, at elevated sample temperatures), they may annihilate or form adatom and vacancy clusters. For higher bombarding fluences, such structures may cause a coarsening of the surface;

the extent of this roughening will depend again on the specimen temperature. Eventually, prolonged ion bombardment often leads to the development of a very specific surface morphology that can exhibit a high degree of periodicity: features such as "nanodots" or "ripple"-like contours, with sizes in the nm range, were observed. Naturally, these self-organized surface nanostructures caused by ion irradiation attracted considerable interest and they have been studied in some detail. In fact, the various aspects of the evolution of the surface morphology caused by ion bombardment and their potential applications constitute the subject of this book.

First reports on the production of submicrometer-sized patterns on the surfaces of solid targets irradiated by low- and medium-energy ions date back at least to the early 1960s. The periodic ripple morphologies observed in those early studies have been well documented in later experimental observations. The full exploration of such structures became more advanced with the availability of high-resolution surface-imaging methods, in particular scanning probe techniques such as atomic force microscopy and scanning tunneling microscopy. Using these tools, highly regular arrays of dots (each with a linear size of a few ten nanometers) were eventually observed on ion-bombarded surfaces in 1999. However surprising this finding was at that time, it triggered a flow of further investigations in this area.

Theoretical models describing the evolution of surface topography due to ion irradiation have been put forth by Carter, Barber, and others in the 1970s. Renewed interest in such concepts came with the work of Bradley and Harper. This frequently cited description of the morphological effects of ion sputtering was based on Sigmund's classic theory of sputtering and initiated a considerable number of further theoretical studies. Recently, also computer simulations were employed for the elucidation of the mechanisms leading to the development of those surface nanostructures.

Although a substantial degree of self-organization is often observed in the nanostructures produced by ion irradiation, many basic questions still remain open as to the specific ways in which these processes of structural ordering unfold. Nevertheless, the considerable extent of control achieved in experimentally created nanopatterns on a variety of targets has spurred the imagination of researchers to exploit these structures in a range of potential applications for the nanofabrication of materials. While these technological uses can be considered to be still in their infancy, this situation may change rapidly given the enormous interest these procedures have recently attracted and the large number of researchers now active in this field.

The present volume is an attempt to account for this growing interest and to respond to the demand for a well-balanced, yet detailed state-of-the-art presentation of this thriving research area. The contributions in this compilation convey a representative view of its present status and the unresolved issues that drive most current and future efforts. They are authored by a number of highly knowledgeable scientists and their expertise leads to concise and informative reviews of those topics.

Hubert Gnaser
Department of Physics
University of Kaiserslautern
Kaiserslautern, Germany

Preface

The advantages of using an energetic ion beam to modify surfaces are precise control, reproducibility, and cleanliness of the process due to high-vacuum environment. Reports on the production of submicron and nanometric patterns on the surfaces of solid targets eroded by ion irradiation (sputtering) at low and medium energies date back to the 1960s. However, the advent of high-resolution tools for surface and interface characterization helped us develop a deeper understanding of the underlying physical processes and recognize the high potential of this method for efficient production of well-ordered surface patterns. The significance of this method for patterning surfaces is that the technique is fast, simple, and relatively less expensive. The possibility to create patterns on large areas at once makes it even more attractive. Such self-organized morphologies are made up of periodic arrangements of nanometric-sized features (e.g., dots and ripples) that show properties of high technological interest. Over the last decade, a lot of efforts were made toward harnessing this nanofabrication technique. These include intensive and fruitful experimental and theoretical works albeit there is no consensus yet on any particular model or dependence of pattern on experimental parameters. In spite of this, such ion-induced self-organized nanopatterns onto a large variety of targets have led to the possibility of a wide range of applications (e.g., photonics, magnetic, catalytic, and others). Although these technological uses can be considered to be still in a nascent stage, the situation may change rapidly given the enormous interest this nanofabrication tool has recently attracted. As a result, a large number of researchers are now active in this field. This volume is an attempt to account for this growing interest and to respond to the demand for a state-of-the-art presentation of this research area.

The contributions are authored by a number of experts in this field, leading to concise and informative accounts of those topics.

In general, sputtering forms the basis for the evolution of all morphological features observed on ion-bombarded surfaces. This book starts with a specific discussion of sputtering theory in Chapter 1. In this overview by Prof. Peter Sigmund, he outlines pertinent aspects of high-energy cascade dynamics such as sputtering yields, spatial distribution of the deposited energy, and fluctuations. In addition, he discusses the tools required as input for the theoretical treatment (potentials, cross sections, and stopping concepts), characteristics of the emitted flux, compositional changes, and the occurrence of spikes. Experimental aspects of nanostructure formation by ion irradiation are described for semiconductor surfaces by Mollick and Ghose (Chapter 2) and Sarkar (Chapter 3), for ionic insulators by Krok *et al.* (Chapter 4), and for oxide films by Karmakar (Chapter 5). These contributions highlight the dependence of ripple formation and its characteristic features on various ion beam parameters such as energy, angle of incidence, ion fluence, and specimen temperature. In addition, the possible role of the chemical reactivity of the surface is pointed out. These data nicely illustrate the substantial influence various experimental parameters may have and the wide variety of surface morphological features that might evolve. Recently, it has been identified that the presence of contaminants on sputtered surfaces may drastically change the topographical structures that develop. Although this phenomenon is still not understood in detail, it could exhibit a tremendous influence on the applicability of these ion-induced nanostructures, by providing the possibility to tailor them to specific technological requirements. In Chapter 6, Gago *et al.* describe the range of patterns and properties that can be achieved by an intentional incorporation of metallic impurities into the bombarded surface. Simulation studies clearly constitute a very promising method to scrutinize the mechanisms that lead to the formation of self-organized nanostructures on ion-bombarded surfaces. This approach and the results obtained thereby are presented in the contributions of Chan and Chason (Chapter 7), Kree and Yasseri (Chapter 8), and Ódor *et al.* (Chapter 9). Characteristically, these simulations offer the option to explore a wealth of parameters (some not accessible in experiments) and their influence on surface patterning. Finally, Chapter 10 concludes the book with a detailed description, by Ranjan *et al.*, of the use of ion-induced patterned substrates in plasmonics. Arguably, this might be one of the most imminent technological applications of such nanostructures. A range of plasmonic properties (such as optical anisotropy and dielectric functions) of metallic nanostructures, deposited on rippled surfaces, are discussed in this contribution; this provides also a flavor of the many pressing issues that still have to be resolved for their full exploitation.

We have tried to show in the book where the field is now and in which direction it is going to move. We hope that this book will be useful for a broad readership, including young researchers carrying out research in this interesting field.

It has been a great pleasure to work with some of the leading scientists in this area during the preparation of this book. We are extremely grateful to all the contributors for kindly accepting our invitation to contribute to the book and presenting their latest results in this exciting field. Our special thanks are due to Professor Hubert Gnaser, who was very kind to write the foreword and provide valuable ideas in projecting the strong points of the book. We would also like to thank Profs. John E. E. Baglin, Rodolfo Cuerno, and R. Mark Bradley for writing highly encouraging words about the book.

Tapobrata Som
Dinakar Kanjilal

Chapter 1

Elements of Sputtering Theory

Peter Sigmund
Department of Physics and Chemistry and Pharmacy, University of Southern Denmark
DK-5230 Odense M, Denmark
sigmund@sdu.dk

This is a brief introduction into sputtering theory, focusing on aspects which may be of interest in the context of nanopatterning. The subject matter is classified into high- and low-energy processes. Prime quantities are the sputter yield and its dependence on ion type, energy and angle of incidence, the stoichiometry of the sputtered flux from a compound target, and fluctuations. Comments are made on standard input such as interatomic potentials and electronic stopping cross sections.

1.1 Introduction

The term *sputtering* may denote either

- the erosion of a surface by particle bombardment (also called *sputter emission* or *sputter erosion*), or
- the production of a thin film by deposition of sputtered material on a substrate (also called *sputter deposition*).

Nanofabrication by Ion-Beam Sputtering: Fundamentals and Applications
Edited by Tapobrata Som and Dinakar Kanjilal

ISBN 978-981-4303-75-0 (Hardcover), 978-981-4303-76-7 (eBook)
www.panstanford.com

Both processes have been known and utilized for more than 150 years and have been described in numerous articles, reviews, and books. The literature is still growing exponentially and currently amounts to ~3000 items per year.

In this chapter, I try to give an introduction to the theory of sputter emission (denoted *sputtering* in the following), with a focus on aspects that may play a role in nanopatterning by ion bombardment. There is unquestionably some arbitrariness in this selection. The reader who needs information on other aspects is encouraged to consult more general summaries (Behrisch, 1981, 1983; Behrisch and Wittmaack, 1991; Sigmund, 1992; Behrisch and Eckstein, 2007). Numerous more specific references are given throughout the article for a reader who wants to dig deeper.

Fundamental research in sputtering is distinguished by a very tight connection between experiment, theory, and simulation. Measurable quantities are the number of emitted atoms and their distribution in energy, ejection angle, elemental composition, state of aggregation as well as charge and excitation states. Changes in bulk and surface structure may be observed on the target. Technology to observe sputter events in real time is within reach.

Variable parameters are the kind of incident particle, usually an ion, its state of aggregation, energy and angle of incidence, and the material, in particular its elemental composition, bulk and surface structure. Target temperature plays a role mainly in the neighborhood of phase transitions.

1.2 General Aspects

A central aspect in the theory of ion–matter interaction is the distinction between nuclear and electronic processes (Bohr, 1948; Sigmund, 2006). In this field, "nuclear" processes are elastic interactions between a projectile ion or atom and a target atom, while electronic processes comprise excitation, ionization, and charge transfer. The two types of process are coupled in principle, but such coupling is most often ignored in particle penetration and related phenomena. Nuclear processes tend to dominate at low incident energies, whereas electronic processes take over at higher

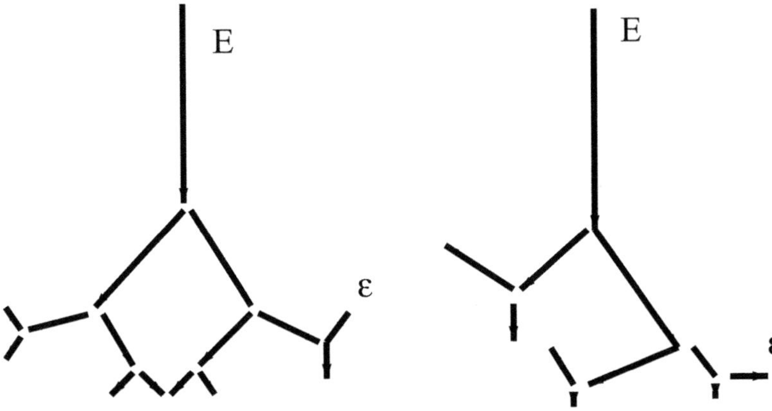

Figure 1.1 Collision cascade. Left: Schematic and somewhat misleading. Right: More realistic.

energies. The cross-over energy depends on the atomic numbers Z_1, Z_2 of projectile and target, respectively (Lindhard *et al.*, 1963b), and increases monotonically with increasing Z_1.

Except for the lightest ions (H, He, Li) beam energies in the keV range and below lie in the nuclear regime. Sputtering in this energy range is a universal phenomenon. Sputtering by electronic processes has been observed for certain classes of materials but will not be discussed here. Summaries by Johnson and Schou (1993) and Assmann *et al.* (2007) may provide a point of entrance into this active and interesting field of research.

A central concept in the theory of sputtering by nuclear processes is the collision cascade, which is illustrated in Fig. 1.1. An ion with some energy E is impinging from the top of the graph and generates a series of primary and higher-order recoiling atoms. This is a fractal structure. In the older literature, you will often find an equipartition between the energies of the two partners emerging from a collision (left graph). It is one of many important contributions by J. Lindhard to this field, to have pointed out a strong asymmetry inherent in the differential cross section, which favors low recoil energies (Lindhard *et al.*, 1963a), as indicated in the right graph.

There are two obvious ways of how cascade multiplication can end:

- Once the energy of a recoil atom has reached its lattice binding energy, it is likely to lose its remaining kinetic energy into heat. In this situation, we talk about a *linear cascade*.
- Alternatively, the cascade may be so dense that the chance of hitting a target atom at rest is small while the atom is still moving. In that case we talk about a *spike*.

1.2.1 *Cascade and Sputtering*

Figure 1.2 shows a simple model of sputtering from a linear collision cascade. An ion with an atomic number Z_1 impinges on a monoatomic material with atomic number Z_2 and a plane surface S. As a result of cascade development, atoms (*in casu* A and B) arrive back at the surface and retain enough energy to overcome a characteristic surface binding energy U.

In a first (quite common) approximation, the target surface S is replaced by a reference plane in an infinite medium. This leads to useful scaling relations regarding angular and energy distributions in the cascade. It also implies that "sputtered" atoms may be scattered back and reenter the surface, an effect that needs to be corrected for. Alternative treatments for a half-infinite medium exist (e.g., Glazov (2000)) but are less transparent. More complex

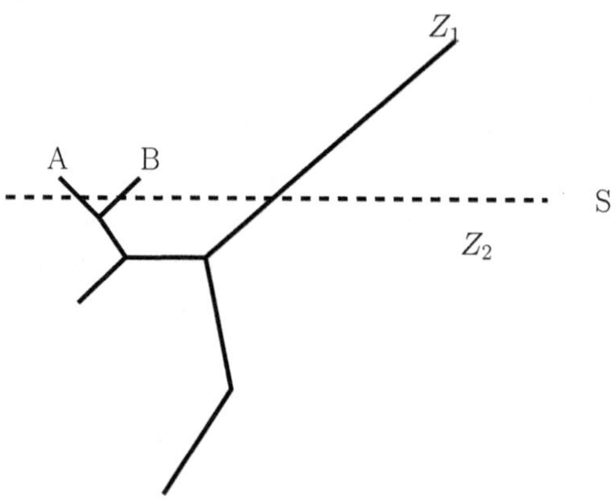

Figure 1.2 Sputtering from a linear collision cascade (schematic).

configurations involving, for instance, a target with a nonuniform composition or a nonplanar surface may be treated by simulational techniques.

1.2.2 *Characteristic Energies and Depths*

A linear cascade is characterized primarily by two representative energies, the beam energy E and the surface binding energy U. A third important energy may be hidden in the scattering cross section.[a] It is a necessary condition for sputtering that $E > U$. In fact, sputter thresholds typically lie at least one order of magnitude above U. Therefore, asymptotic expansion in powers of E/U is a relevant and useful tool in sputter theory.

The overall dimensions of a cascade may be characterized as a first approximation by the penetration depth of the impinging ion and its lateral and longitudinal fluctuation (*straggling*). Figure 1.3 shows a calculated example for the case of $M_1 = M_2$, where M_1 and M_2 are masses of projectile and target atom, respectively. The right part shows the penetration profile of the ion, averaged over many trajectories. Contour lines go in steps of 10%. The length unit is the calculated *average pathlength* of the ion, which depends on Z_1, Z_2, and the beam energy. The left part shows a corresponding profile for the density of deposited energy after dissipation down to some energy $\ll E$.

You may note that both profiles extend beyond the surface at the bottom of the graph. Integration over these parts allows to estimate the reflection coefficient for ions (right) and energy (left). An early result from quantitative transport theory of sputtering concerned the energy reflection coefficient γ, which was predicted to be only weakly dependent on beam energy (Sigmund, 1968). Figure 1.4 shows a successful experimental confirmation.

There are at least two important characteristic depths in sputtering (Sigmund, 1969). First, there must be a certain depth beneath the target surface in which events take place that ultimately lead to sputtering. This depth will generally be a certain fraction of

[a] There is also a bulk binding energy which, however, is of secondary importance. Its main effect is to drain the cascade for part of its energy. Moreover, bulk and surface binding energies are not unrelated.

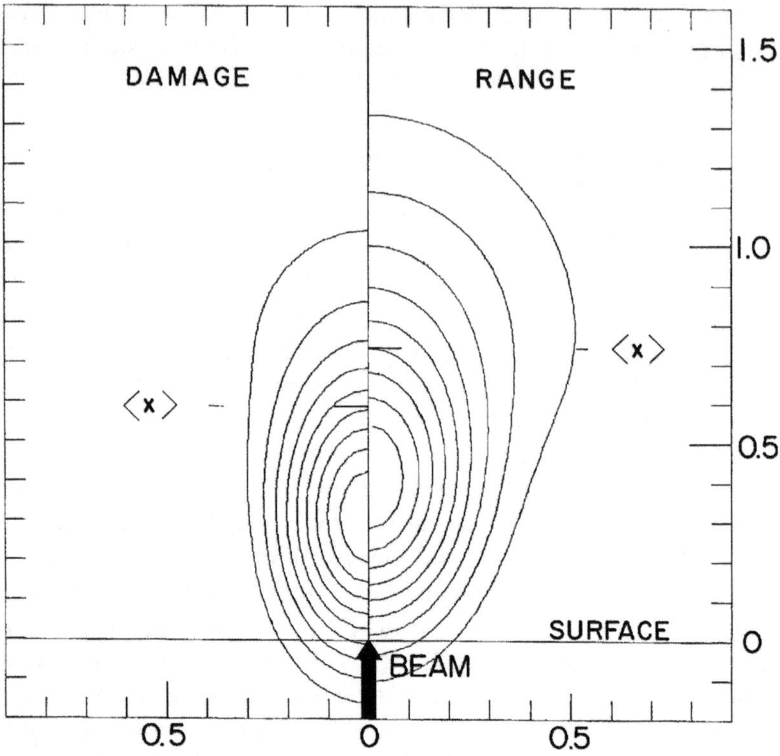

Figure 1.3 Calculated energy deposition (left) and range (right) profile in an infinite medium. Both profiles are averages over many trajectories and cylindrically symmetric. The length unit is the average path length of the incident ion. Contour lines from 0.1 to 0.9 in steps of 0.1. From Winterbon *et al.* (1970), with permission. See text.

the ion range and, hence, will increase with increasing beam energy. Second, there is the depth of origin of sputtered particles. For linear cascades, this quantity is very small and only weakly dependent on bombardment conditions.

1.2.3 *Two Energy Regimes*

The existence of two energies E and U, which differ by one or several orders of magnitude, provides a natural way to divide cascade dynamics into a high-energy and a low-energy regime: High-energy

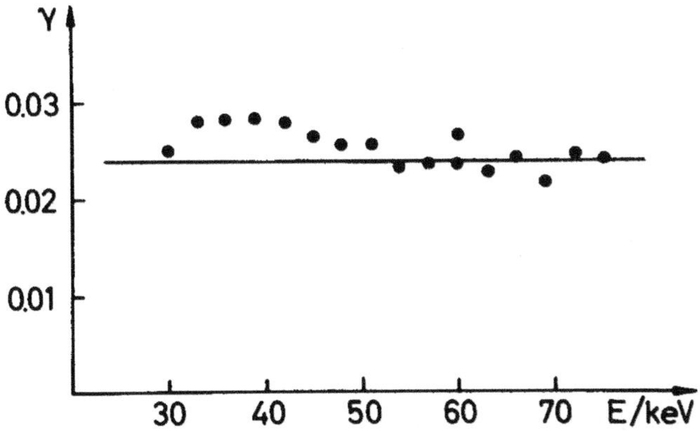

Figure 1.4 Energy reflection coefficient (*sputter efficiency*) γ for Pb on Pb. Theoretical prediction from Sigmund (1968). Measurement and graph from Andersen (1968), with permission.

dynamics governs the penetration depth of the beam, the spatial extension of a cascade, and the distribution of deposited energy. Low-energy dynamics governs rearrangement processes within the cascade, defect formation, and emission of sputtered atoms.

High-energy dynamics is conveniently treated by kinetic equations of linear transport theory or by binary-collision simulation. The accuracy of these tools deteriorates in the low-energy regime because of the many-body nature of the pertinent interactions, neglect of attractive forces, etc. Low-energy dynamics is the domain of molecular-dynamics simulation. There is an upper energy limit of applicability of molecular dynamics, which, however, increases with the development of hardware. The main problem is achieving adequate statistics.

An obvious way out of these limitations would be a combination of these tools. Such studies have only rarely been reported (Urbassek *et al.*, 1995; Vicanek *et al.*, 1989).

1.3 High-Energy Cascade Dynamics

High-energy cascade dynamics in the standard form known today has been developed in the 1960s primarily by Lindhard *et al.*

(1963a), Robinson (1965a), Robinson (1965b); Sigmund and Sanders (1967), and Sanders (1968). Thompson (1968) developed an expression for the energy spectrum of sputtered atoms, and the present author condensed available information into a comprehensive scheme, which led to estimates of sputter yields and their dependence on ion type, energy, and angle of incidence (Sigmund, 1969). For a systematic introduction to cascade and range theory, the reader is referred to Sigmund (1972a,b,c).

1.3.1 *Kinetic Equation*

A fundamental quantity is the particle density $G(E, \epsilon, t)$, where $G(E, \epsilon, t)d\epsilon$ is the average number of atoms moving with an energy $(\epsilon, d\epsilon)$ at time t as a consequence of a primary particle impinging at time $t = 0$ with an energy E. This quantity obeys a kinetic equation,

$$N\upsilon \int d\sigma\,[G(E, \epsilon, t) - G(E - T_n - T_e, \epsilon, t) \\ - G(T_n, \epsilon, t)] = -\frac{\partial}{\partial t} G(E, \epsilon, t), \tag{1.1}$$

where $d\sigma = d\sigma(E, T_n, T_e)$ is the differential cross section, T_n the nuclear and T_e the electronic energy loss, and N the number of target atoms per volume. Equation (1.1) assumes a monoatomic homogeneous medium. In the quoted form, the equation applies to self-bombardment, i.e., $Z_1 = Z_2$. Application to $Z_1 \neq Z_2$ and/or a polyatomic medium is a matter of adding suitable indices and performing appropriate summations.

Equation (1.1) is easily derived by expressing $G(E, \epsilon, t)$ by the weighted sum of the possible outcomes of collision events occurring in an infinitesimal time interval $(0, \delta t)$ (Lindhard *et al.*, 1963a).

Inclusion of directional and spatial variables is straightforward (Sigmund, 1969, 1981). In vector form, the kinetic equation then reads

$$Nv \int d\sigma \left[G(\vec{v}, \vec{w}, \vec{r}, t) - G(\vec{v}', \vec{w}, \vec{r}, t) - G(\vec{v}''\vec{w}, \vec{r}, t)\right] \\ = -\vec{v} \cdot \nabla G(\vec{v}, \vec{w}, \vec{r}, t) - \frac{\partial}{\partial t} G(\vec{v}, \vec{w}, \vec{r}, t), \tag{1.2}$$

where $\vec{v}$ and $\vec{v}'$ represent velocity vectors of a projectile before and after a collision, respectively, $\vec{v}''$ the recoil velocity, and the cross section $d\sigma = d\sigma(\vec{v}, \vec{v}', \vec{v}'')$ now incorporates also directional variables. Finally, $\vec{w}$ represents the velocity of the cascade recoils, the distribution in space, energy, and direction of motion of which we are aiming at. Actual calculations are performed in energy instead of velocity space, so that $\vec{v}$ is replaced by E and a unit vector $\vec{e}$ and $\vec{w}$ by ϵ and a unit vector $\vec{\Omega}$.

The flux per incident ion of atoms with a velocity $(\vec{w}, d^3\vec{w})$ through a plane at $x = 0$ may be expressed by

$$J(\vec{w})d^3\vec{w} = w|\cos\Theta|\, d^3\vec{w} \int_0^\infty dt \times \int_{-\infty}^{\infty} dy \int_{-\infty}^{\infty} dz\, G(\vec{v}, \vec{w}, x = 0, y, z, t), \quad (1.3)$$

where Θ is the angle between $\vec{w}$ and the inward surface normal.

1.3.2 *Asymptotics*

Mathematical tools to solve equations like Eq. (1.2) have been developed in transport theory. Angular dependencies may be expressed by expansion in spherical harmonics, and spatial dependencies may be transcribed into spatial moments. In view of Eq. (1.3), only the integral over time and the lateral coordinates y and z are of interest. These integrations can be performed directly, leaving an initial condition defining the energy and direction of motion of the impinging ion.

Equations for moments can then be solved recursively starting from the zeroth moment. A few simplifications allow explicit analytical integration:

(1) The differential cross section is assumed to have power form (Lindhard *et al.*, 1968),

$$d\sigma = C_m E^{-m} \frac{dT_n}{T_n^{1+m}}, \quad (1.4)$$

where E is the energy of the colliding atom, T_n the recoil energy, m a numerical constant, $0 \leq m \leq 1$, and C_m a well-defined parameter depending on atomic numbers and masses of the collision partners.

(2) Electronic energy loss and bulk binding are neglected.

(3) Moments are expanded in powers of ϵ/E.

The main results are as follows (Sigmund, 1969):

- If only the leading term for $E \gg \epsilon$ is retained, the zeroth moment goes as E/ϵ, while all higher moments become independent of ϵ.
- Moments of order $\upsilon > 0$ depend on energy as $(E^{2m}/NC_m)^{\upsilon}$. Note that also the average pathlength of the projectile is proportional to E^{2m}/NC_m.
- The particle flux can be factorized into a function characterizing the spatially and directionally integrated flux and the density in space of the deposited energy.
- The particle flux is isotropic.

For the sputter yield, i.e., the average number of atoms emitted per incident ion, this implies that

$$Y(E,\theta) = \Lambda F_D(E,\theta,0), \tag{1.5}$$

where $F_D(E,\theta,x)\,dx$ is the average energy deposited in a layer (x,dx) in a cascade initiated by a projectile impinging with an initial energy E at an angle θ to the x-axis (the inward surface normal). The constant Λ is a material constant, including the surface binding energy and a cross section for target atoms colliding with each other at low energies.

The deposited-energy density $F_D(E,\theta,0)$ can be factorized according to

$$F_D(E,\theta,0) \simeq \alpha(M_2/M_1,E,\theta)NS_n(E), \tag{1.6}$$

where $S_n(E) = \int T_n\,d\sigma$ is the nuclear stopping cross section and $\alpha(M_2/M_1,E,\theta)$ a quantity depending primarily on the mass ratio M_2/M_1. The dependence of α on beam energy E is weak, and the angular dependence can be approximated as

$$\propto \cos^{-f}\theta \tag{1.7}$$

with $1 \le f < 2$.

1.3.3 *Sputter Yields*

Equation (1.5) determines the variation of the sputter yield with the atomic number of the ion, its energy, and its angle of incidence

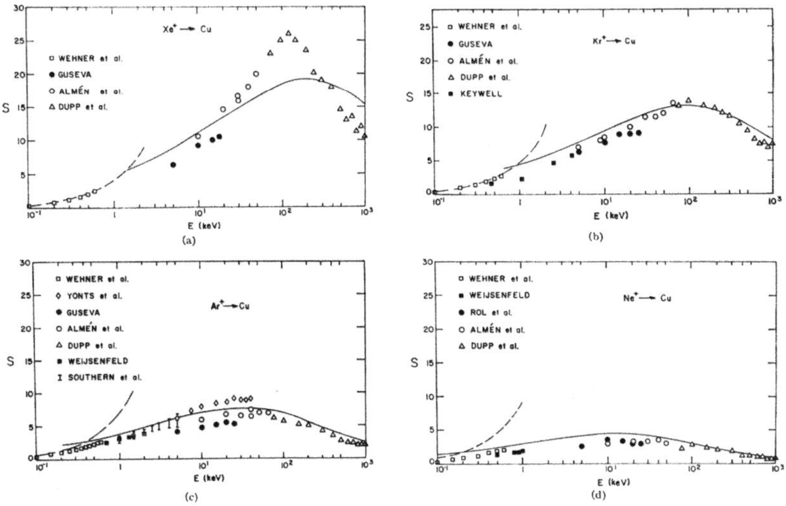

Figure 1.5 Sputter yields vs. energy for noble gas ions on copper. Lines represent two approximations for the deposited-energy function at high resp. low beam energies. From Sigmund (1969), with permission.

within the range of applicability of the asymptotic scheme, i.e., at not too low beam energies. It was a pleasant surprise that a simple expression emerged for Λ,

$$\Lambda \simeq \frac{3}{4\pi^2}\frac{1}{NC_0U} \simeq \frac{0.0420}{NU\,\text{Å}^2}, \tag{1.8}$$

where C_0 is related to the cross section at low energies. This made it possible to calculate absolute sputter yields, which compared favorably with experiment.

Figure 1.5 shows sputter yields for four noble gas ions incident normally on copper (Sigmund, 1969). The agreement with experimental data available at the time is essentially within experimental scatter, with the exception of xenon, where a significant spike contribution may be expected.

A systematic experimental effort to determine sputter parameters of prime importance was carried out by Andersen and Bay in the 1970s and early 1980s. This included studies of the dependence of the sputter yield on the type and energy of the bombarding ion on a number of materials, which were summarized in a review (Andersen and Bay, 1981).

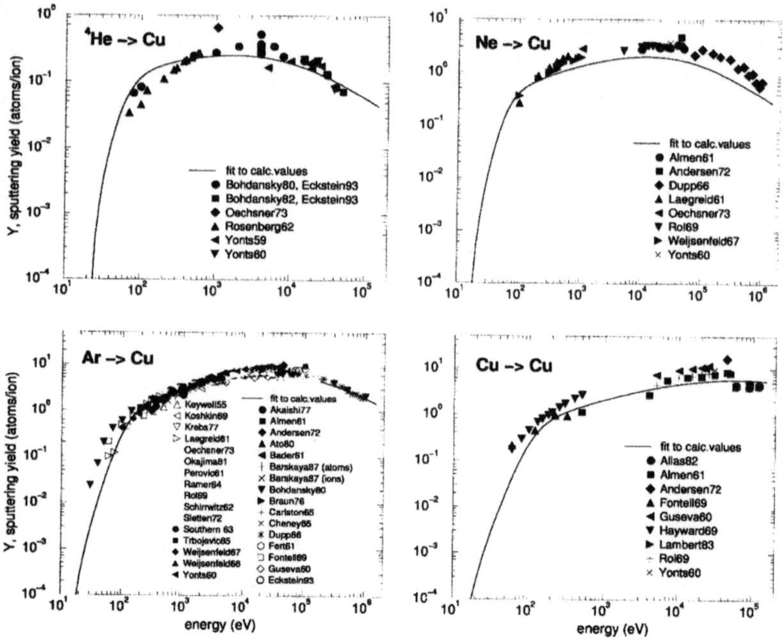

Figure 1.6 Sputter yields versus energy for various ions on copper. Lines represent computations by the TRIM-SP binary-collision code. From Eckstein (2007), with permission.

Figure 1.6 shows a recent graph, which can be compared to Fig. 1.5. There have been more measurements, some of them presumably more reliable, and the calculated curves have been generated by a well-developed Monte Carlo simulation code. Keeping in mind the logarithmic plot, it appears justified to conclude that with regard to absolute sputter yields, progress has been marginal over the past 40 years.

Figure 1.7 shows sputter yields as a function of the angle of incidence. Transport theory for an infinite medium yields a monotonic increase according to Eq. (1.7). Monte Carlo simulation on a semi-infinite medium leads to a maximum at grazing incidence and a subsequent drop. This drop is barely visible in many experimental data, presumably because the assumption of a strictly planar surface is rarely fulfilled in experiments.

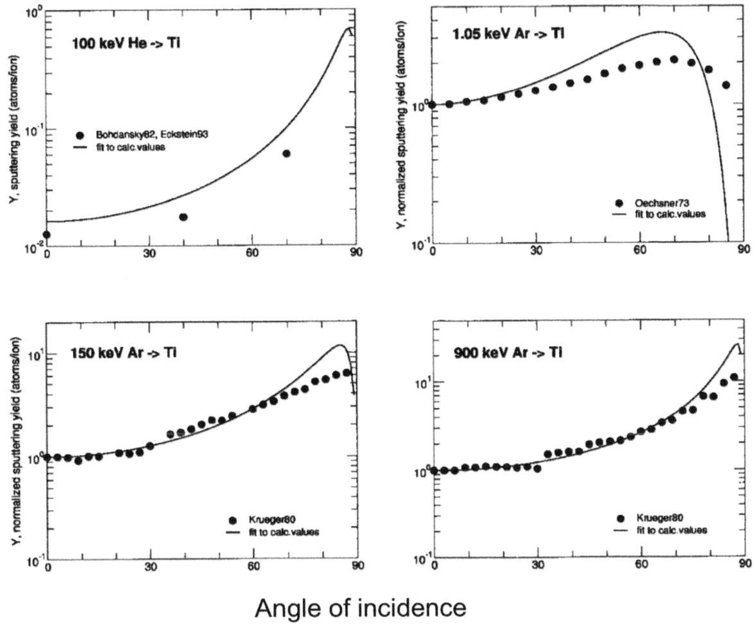

Figure 1.7 Sputter yields vs. angle of incidence for various ions on titanium. Lines represent computations by the TRIM-SP binary collisions code. From Eckstein (2007), with permission.

1.3.4 *Spatial Distribution of the Emitted Flux*

Equation (1.5) relates the number of sputtered atoms per incident ion to the mean energy deposited at the surface. It is tempting to assume this relation to be valid locally on the bombarded surface,

$$dY(\vec{r}) = \Lambda f_D(\vec{r}) d^2\vec{r}, \tag{1.9}$$

where $dY(\vec{r})$ is the mean number of atoms sputtered from an area $d^2\vec{r}$ on the surface and $f_D(\vec{r})$ the average deposited energy per volume in $\vec{r}$. The relation is valid as long as the assumption of an isotropic cascade is justified (Sigmund, 1973). An immediate consequence of Eq. (1.9) is illustrated in Fig. 1.8: For oblique incidence of an ion beam on a plane surface, the point of maximum sputtering moves away from the point of impact of the ion in the downward direction, as the angle of incidence increases. In the absence of a smoothening process, this gives rise to an instability

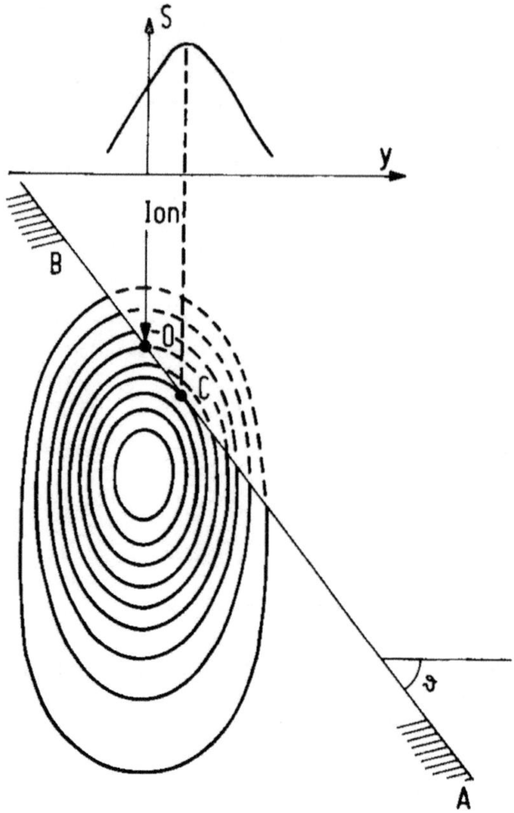

Figure 1.8 Energy deposition profile of an ion incident at an angle θ on a plane surface AB. Contour lines as in Fig. 1.3. The upper part of the graph indicates the variation of the differential sputter yield dY (called S in the graph) across the beam profile. From Sigmund (1973), with permission.

causing roughening of the bombarded surface (Sigmund, 1973). The significance of this phenomenon to surface roughening and pattern formation has been studied extensively after its rediscovery by Bradley and Harper (1988).

1.3.5 *Beyond the Leading Term*

Sanders (1968) pointed out that the leading correction to the E/ϵ term in the zeroth moment goes as $(E/\epsilon)^{1/2}$ and that it is related to the initial momentum of the projectile. This has the obvious consequence that asymptotic expressions for the sputter yield and related quantities can only be expected to be valid at rather large ratios E/ϵ or E/U.

While it is clear that at beam energies below the asymptotic limit, an expression needs to be added to the leading term in the particle flux, finding its proper form is a nontrivial task: Unlike $F_D(E,\theta,x)$, the density of deposited momentum $\vec{F}_P(E,\theta,x)$ is a vector. Moreover, expressions for spatial moments over this quantity contain integrals that diverge for $m = 1/4$, a value that is by no means unrealistic.

As it turned out gradually, there are actually two corrections to the leading term, which both contain divergencies at $m = 1/4$ that cancel when the two corrections are included jointly. While the first correction term is proportional to $\vec{F}_P(E,\theta,x)$, the second one is proportional to the spatial derivative of the deposited-energy density. This term is harder to identify because its zeroth moment vanishes.

I shall not go here into a description of the process that ultimately led to valid results, but just quote the main result for $M_1 = M_2$ in the notation of Sckerl *et al.* (1996).

$$w\int_0^\infty dtG\,(E,\vec{e},\epsilon,\vec{\Omega},x) = \frac{\Gamma_m}{4\pi NS_n(\epsilon)}$$
$$\times\left[\frac{F_D(E,\vec{e},x)}{\epsilon} - K_m\cos\Theta\frac{\partial F_D(E,\vec{e},x)/\partial x}{NS_n(\epsilon)} + 3\frac{\Omega\cdot\vec{F}_P(E,\vec{e},x)}{\sqrt{2M\epsilon}}\right], \tag{1.10}$$

where Γ_m and K_m are well-defined dimensionless constants.

Here the first term in the square brackets represents the asymptotic limit. It is evident that inclusion of the two corrections complicates the scaling properties.

Sckerl *et al.* (1996) also determined corrections for bulk binding and electronic stopping, which will not be mentioned explicitly here.

1.3.6 *Depth Profiles*

While scaling properties of the distributions $F_D(E,\vec{e},x)$ and $\vec{F}_P(E,\vec{e},x)$ may be deduced from the calculated moments, determining explicitly the densities from the moments is a nontrivial mathematical task. For range profiles, Sanders (1968) proposed Hermite polynomial (or Edgeworth) expansions. By taking into account a considerable number of moments, Winterbon *et al.* (1970) determined reliable expressions also for deposited-energy profiles.

For the momentum density, Padé approximants have proven more efficient (Littmark and Sigmund, 1975). For consistency, Sckerl *et al.* (1996) determined also deposited-energy profiles by this method.

Figure 1.9 shows an example. Plotted is the x-component of the deposited-momentum profile, including the gradient of the deposited energy for incidence in the x-direction, $m = 0.2$, $M_1 = M_2$, and E/U ranging from 10 to 1000. Solid curves result from Eq. (1.10) via Padé approximants. Points result from a Monte Carlo simulation. Excellent agreement is seen for $E/U \gtrsim 20$. For $E/U = 10$, a clear discrepancy is observed: While the numerical result is negative for $x < 0$, the analytical curve is positive. Clearly, for

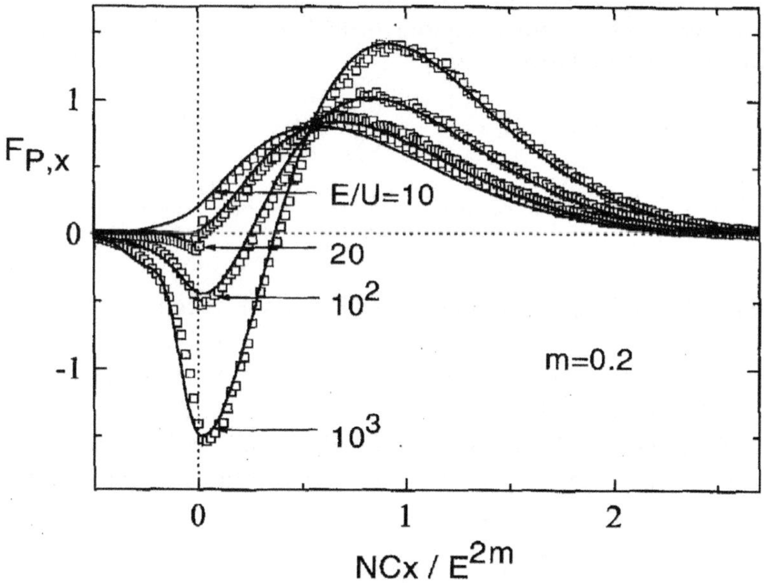

Figure 1.9 Longitudinal component of a deposited-momentum profile. See text. From Sckerl *et al.* (1996), with permission.

that part of the cascade which spreads backward, the net deposited momentum must point into the negative x direction. One may conclude that at least for $m = 0.2$ and $M_1 = M_2$, the range of validity of the scheme discussed in this paragraph extends to $E/U \simeq 20$, i.e., down to much lower beam energies than a solution based only on the leading asymptote.

Profiles shown in Figs. 1.3 and 1.9 have been constructed assuming continuity in value and gradient. This assumption is not necessarily fulfilled, even in an infinite medium (cf. Glazov (1995) and papers quoted there).

1.3.7 *An Application*

Figure 1.10 shows angular distributions of the particle flux for various depths x, mass ratios M_2/M_1, and energies E/ϵ. It is

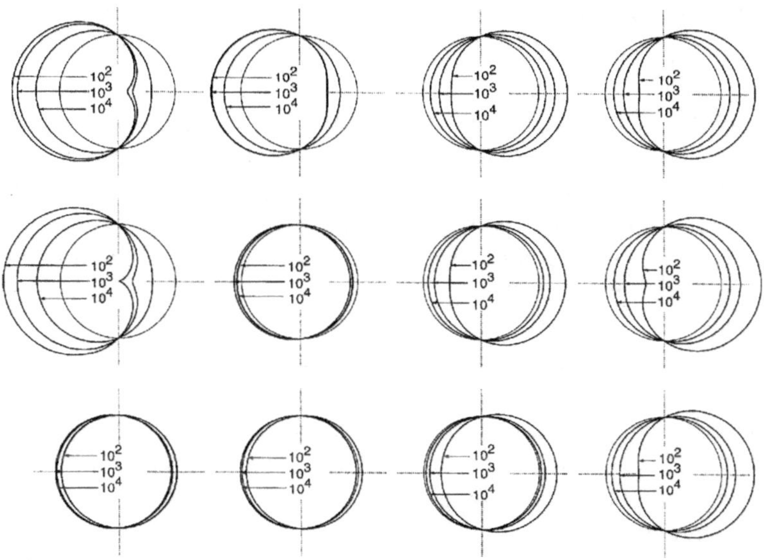

Figure 1.10 Calculated angular dependence of the particle flux for $m = 0.3$ at depths $x = 0, x_D - \Delta x_D, x_D$, and $x_D + \Delta x_D$ (left to right), where x_D is the mean deposition depth and Δx_D the standard deviation. $M_2/M_1 = 4$, 1, and 1/4 (top to bottom). Ratios E/w given in the graph. From Sckerl *et al.* (1996), with permission.

seen that at the average deposition depth and beyond, the flux is forward peaked, while it is more or less backward peaked at $x_D - \Delta x_D$ and $x = 0$, dependent on the mass ratio and the ratio E/ϵ.

An expanded formula has been derived by Sckerl *et al.* (1996) for the sputter yield and other parameters, comparisons with experimental data have not been performed so far.

1.3.8 *Fluctuations*

Figures 1.3–1.10 and the underlying functions all deal with averages over many individual cascades. It is intuitively evident — and has been proven long ago — that the measurable outcome of collision cascades may show pronounced fluctuations.

This aspect has been studied by Westmoreland and Sigmund (1970) within linear transport theory by means of a kinetic equation governing the pair distribution function between ion range and deposited energy,

$$F_{RD}(E, \vec{e}, x_D, x_R) = F_R(E, \vec{e}, x_R)g(E, \vec{e}, x_D; x_R), \qquad (1.11)$$

where $F_R(E, \vec{e}, x_R)$ is the range profile and $g(E, \vec{e}, x_D; x_R)$ the deposited-energy profile for a given penetration depth x_R. The treatment was explorative rather than comprehensive: Only second and third moments were evaluated. Figure 1.11 indicates that fluctuations can be quite pronounced for $M_2/M_1 \gtrsim 1$. The quantity of interest in sputtering is the value $g(E, \vec{e}, x_D = 0; x_R)$ at the surface.

Figure 1.12 illustrates this point on the case of 50 keV Xe on Ni at oblique incidence. According to a binary-collision simulation by Eckstein (1988), the sputter yield is 22 atoms/ion. A Poisson distribution would imply a standard deviation of 5 atoms/ion. This is evidently much smaller than the actual fluctuation that results from the simulation.

The total number of recoil atoms in a cascade is known to obey a sub-Poisson distribution (Leibfried, 1958). Fluctuations shown in Fig. 1.11 are the most obvious cause — albeit not the only one — for the fluctuations shown in Fig. 1.12.

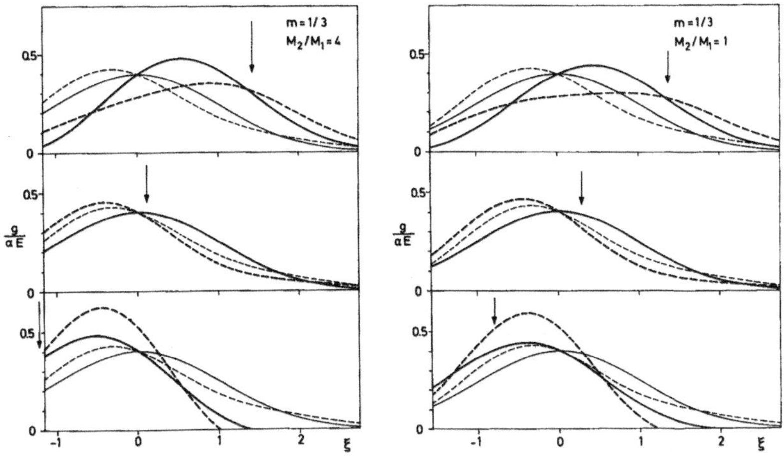

Figure 1.11 Deposited-energy profiles for given ion range, normal incidence. $M_2/M_1 = 1/4$ (left) and 1 (right). The penetration depth of the ion is marked by an arrow. The standard deviation (straggling) in penetration depth is the abscissa unit. Thick lines: correlated profiles; thin lines: average profiles. Solid lines: gaussian; broken lines: corrected for skewness. From Westmoreland and Sigmund (1970), with permission.

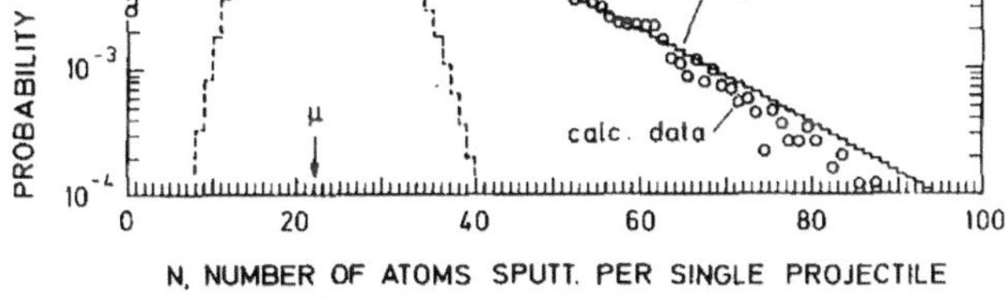

Figure 1.12 Distribution in sputter yield for 50 keV Xe in Ni at 60 deg incidence angle, calculated by Monte Carlo code TRIM-SP. From Eckstein (1988), with permission.

1.4 Input

Quantitative predictions, whether found from transport theory or from simulation, require reliable input regarding nuclear as well as electronic interactions. This section serves to collect some observations regarding the high-energy cascade regime.

1.4.1 *Potentials*

Elastic collisions are characterized by an interatomic potential, which may be written in the form

$$V(R) = \frac{Z_1 Z_2}{R} \Phi\left(\frac{R}{a_{12}}\right), \tag{1.12}$$

where R is the internuclear distance, Φ a screening function most often independent of Z_1 and Z_2, and a_{12} a screening radius specific for the collision partners. Any dependence on charge states is usually neglected.

Most frequent among options for a universal screening function $\Phi(\xi)$ have been

$$\Phi(\xi) = e^{-\xi}, \text{ Bohr (1948)} \tag{1.13}$$

$$\Phi(\xi) = \left(1 + y + 0.3344y^2 + 0.0485y^3 + 0.002647y^4\right) e^{-y}; \tag{1.14}$$

$$y = \sqrt{9.67\xi}, \text{ Lenz (1932); Jensen (1932)} \tag{1.15}$$

$$\Phi(\xi) = 0.35e^{-0.3\xi} + 0.55e^{-0.55\xi} + 0.10e^{-6.0\xi} \text{ (Molière, 1947)} \tag{1.16}$$

$$\Phi(\xi) = 0.02817e^{-0.2016\xi} + 0.28018e^{-0.4029\xi} + 0.50986e^{-0.9423\xi} \tag{1.17}$$

$$+ 0.18179e^{-3.2\xi} \text{ (Ziegler } et\ al.\text{, 1985)} \tag{1.18}$$

in addition to the (numerically given) Thomas–Fermi function (Gombas, 1956). The Lenz–Jensen potential has been found to represent a reasonable approximation to experimental scattering cross sections (Loftager *et al.*, 1979), but deviations from universal

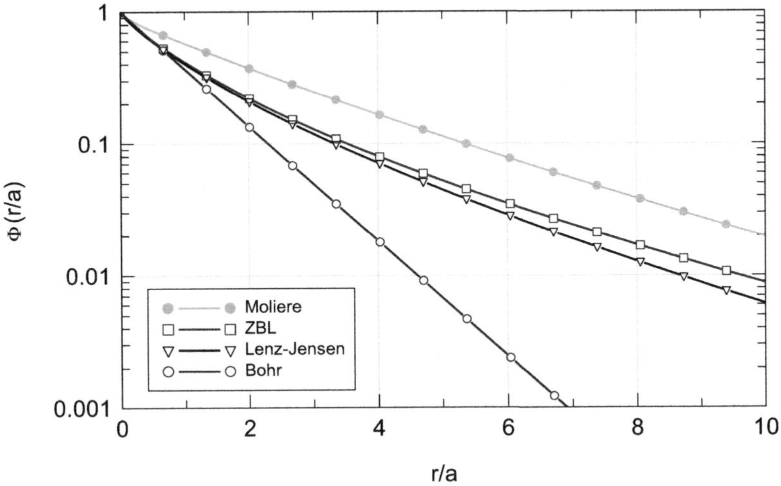

Figure 1.13 Universal screening functions employed in calculations of elastic collisions.

behavior are comparable in magnitude to the difference between Molière and Lenz–Jensen, which is seen in Fig. 1.13.

For the screening radius, the following options have most frequently been adopted:

$$a_{12} = 0.8853 a_0 / \sqrt{Z_1^{2/3} + Z_2^{2/3}}, \quad \text{Bohr (1948); Lindhard } et\ al. \text{ (1963b)} \qquad (1.19)$$

$$a_{12} = 0.8853 a_0 / (\sqrt{Z_1} + \sqrt{Z_2})^{2/3}, \quad \text{Firsov (1957)} \qquad (1.20)$$

$$a_{12} = 0.8853 a_0 / (Z_1^{0.23} + Z_2^{0.23}), \quad \text{Ziegler } et\ al. \text{ (1985)} \qquad (1.21)$$

Figure 1.14 shows ratios of screening radii vs. Z_1/Z_2. It is seen that the Lindhard radius differs from Firsov's by at most 12 %. For the potential of Ziegler *et al.* (1985), the difference is much more pronounced, in particular so for $Z_1 \gg Z_2$. This has consequences.

Figure 1.15 shows that the mean penetration depth of Pb ions in carbon, measured over three orders of magnitude in energy, is in excellent agreement with the prediction of Lindhard *et al.* (1963b) on the basis of Thomas–Fermi interaction. An evaluation by SRIM (Ziegler *et al.*, 1985) shows discrepancies of up to a factor

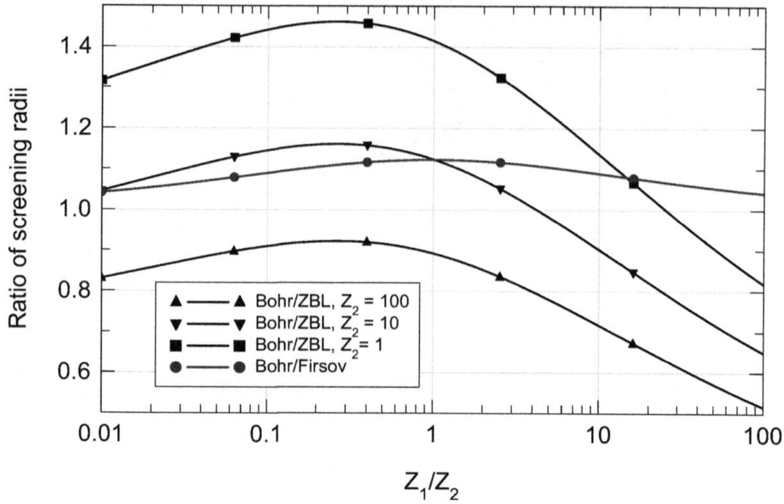

Figure 1.14 Ratio of screening radii for three popular choices. See text.

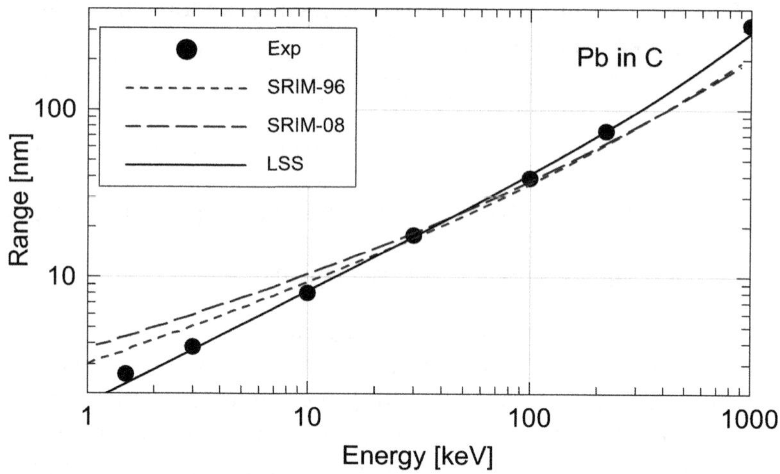

Figure 1.15 Mean range of Pb ions in carbon. Measurements from Friedland *et al.* (1998). Expanded version of an earlier graph from Sigmund (1998).

of two at either end of the energy scale. Note that the discrepancy has become larger for the 2008 version than for the 1996 version of the code. This feature is asserted to be caused by a rather peculiar choice of screening radius and is quite pronounced for all heavier ions.

1.4.2 *Electronic Stopping*

The electronic stopping cross section $S_e = \int T_e \, d\sigma$ has been studied extensively for light ions on numerous materials over a wide energy range. In the context of the present volume, studies with heavier ions from nitrogen upward are of primary interest.

Figure 1.16 illustrates the situation for aluminum, a material used frequently for stopping measurements. Oxygen bombardment is well documented experimentally, while very few data exist for Pb and other heavy ions. The overall behavior is well described by the PASS code due to Sigmund and Schinner (2000, 2002) or the CasP code due to Grande and Schiwietz (2010).

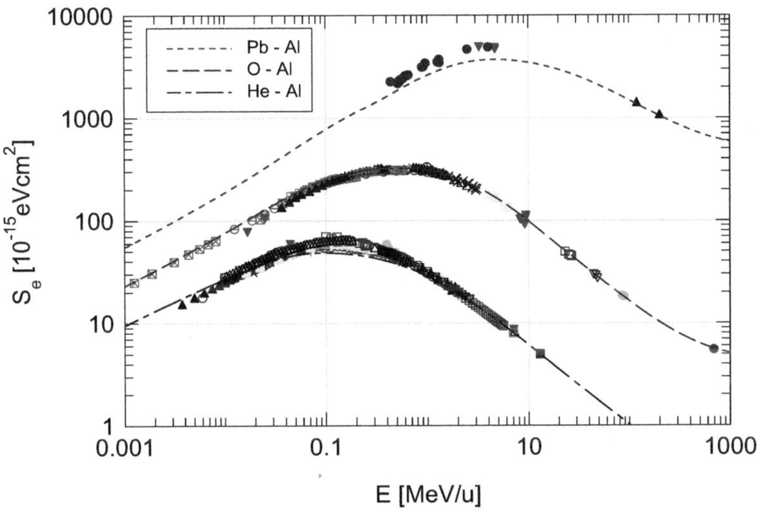

Figure 1.16 Measured stopping cross sections for He, O, and Pb ions in Al. Data collected by Paul (2009). Lines calculated by PASS code (Sigmund and Schinner, 2002).

For sputtering in the nuclear-stopping regime, only the energy range below the maximum is of interest, which is usually described by velocity-proportional cross sections

$$S_e = \xi_e 8\pi a_0^2 \frac{Z_1 Z_2}{Z} m v_0 v, \quad \text{Lindhard and Scharff (1961)} \tag{1.22}$$

$$S_e = 2.3\pi a_0^2 (Z_1 + Z_2) m v_0 v, \quad \text{Firsov (1959).} \tag{1.23}$$

The quantity $\xi_e \simeq Z_1^{1/6}$ was introduced as an empirical parameter to fit measured ranges of fission fragments [Lindhard *et al.* (1963b)]. Figure 1.17 shows that the two estimates differ by up to a factor of two for $1/4 < Z_1/Z_2 < 4$. Experimental values tend to lie between.

The monotonic dependence of the stopping cross section on Z_1 at low velocities predicted by Eqs. (1.22) and (1.23) is only approximate. An oscillatory Z_1-structure has been identified long ago (Ormrod and Duckworth, 1963; Fastrup *et al.*, 1966) and studied extensively.

A problem in the experimental determination of electronic stopping cross sections at low velocities is the dominant role of

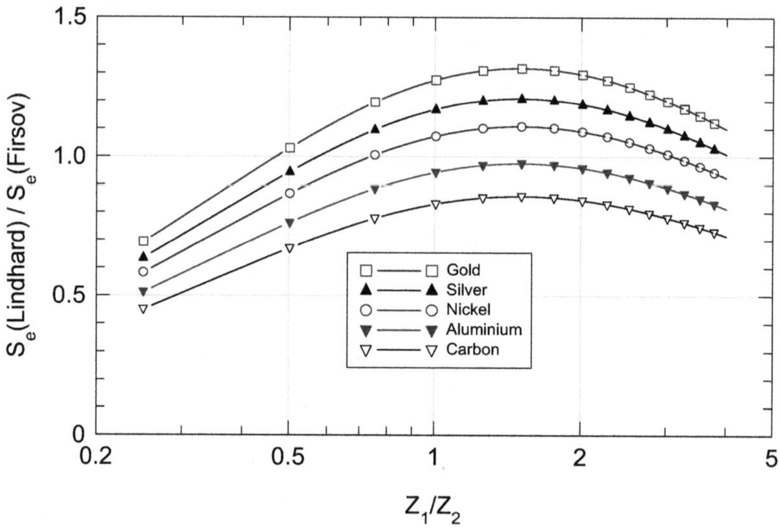

Figure 1.17 Ratio of electronic stopping cross sections due to Lindhard and Scharff (1961) and Firsov (1959) for a number of target elements.

nuclear stopping, in particular for the heaviest ions. A helpful tool in that context is the reversibility rule (Sigmund, 2008), which claims that

$$S_e(v)|_{1\to 2} \simeq S_e(v)|_{2\to 1}\,. \tag{1.24}$$

The rule must be obeyed rigorously for a neutral beam interacting with a dilute cold gas. Low-velocity ions are close to neutral in charge equilibrium and, by comparison with experimental data, neither gas–solid nor insulator–metal differences appear to cause major deviations from reciprocity.

Examples are shown in Fig. 1.18. The top graph shows that reciprocity is well obeyed for C-Si and Si-C, while interpolation codes SRIM (Ziegler *et al.*, 1985) and MSTAR (Paul and Schinner, 2001) show significant deviations from the data which the codes are supposed to interpolate in between. The bottom graph shows the case of Al-Au and Au-Al. No data exist for low-velocity gold in Al. Here the SRIM data are based on range measurements, and S_e is determined from the difference between the calculated range assuming elastic collisions and the measured range. Because of the dominance of nuclear stopping, even a minor error in the interatomic potential will cause a major error in the extracted electronic stopping cross section. As a result, SRIM is likely to overestimate S_e for Au in Al by an order of magnitude.

1.5 Low-Energy Dynamics

Low-energy dynamics determines

- the depth of origin,
- the energy spectrum,
- the angular distribution,
- the elemental and isotopic composition, and
- the excitation/ionization state

of the sputtered flux, as well as the constant Λ determining the magnitude of the sputter yield, Eq. (1.5).

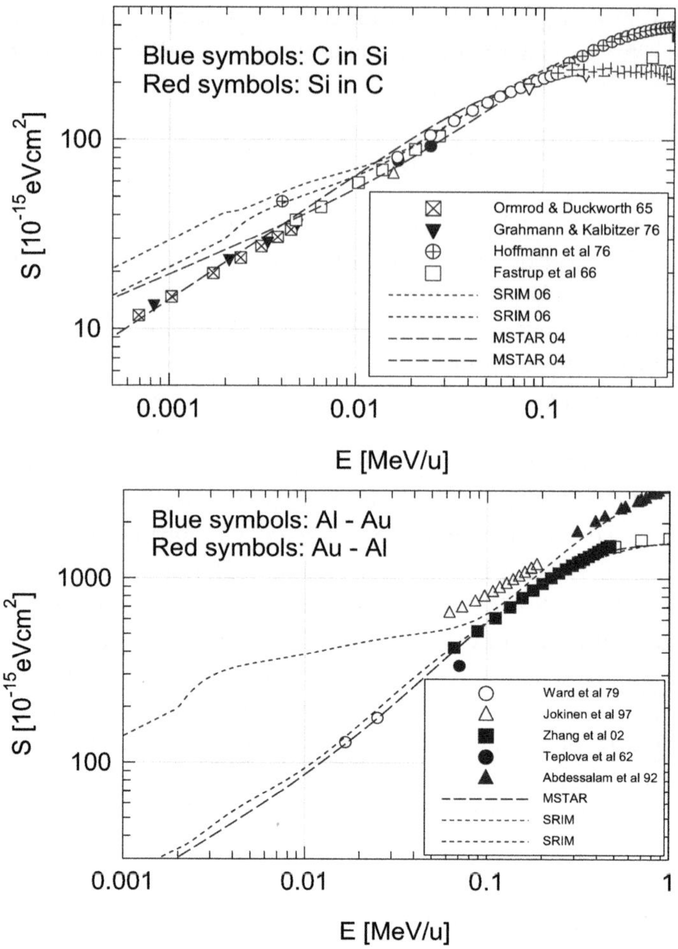

Figure 1.18 Reciprocity of low-velocity stopping cross sections. Top: Si-C and C-Si. Bottom: Al-Au and Au-Al. See text. From Sigmund (2008), with permission.

1.5.1 *Tools*

Low-energy dynamics is the domain of molecular-dynamics simulation (Eckstein and Urbassek, 2007): If a realistic interatomic potential is at hand, bulk and surface binding energies are output rather than input, and the many-body nature of overlapping collisions can be accounted for within a reasonable level of accuracy. Moreover, low-energy cascades involve a limited number of atoms, so that adequate statistics can be achieved even with limited computational capacity.

Also binary-collision codes provide useful information on low-energy dynamics (Eckstein and Urbassek, 2007), in particular when incorporating lattice structure. The many-body nature of low-energy collisions can be incorporated approximately, but in addition to the interatomic potential also bulk and surface binding forces must be entered as independent input. On the other hand, high- and low-energy dynamics are well integrated, because restrictions in accessible beam energy due to computational capacity are less severe than in molecular dynamics.

Transport theory in the form applied in sputtering as well as Monte Carlo simulation codes have the additional weakness that lattice structure is disregarded. Traditionally the free path between subsequent collisions is a random variable, but an alternative is an energy-dependent free path, as applied in the TRIM-SP (Eckstein, 2007) and SRIM (Ziegler *et al.*, 1985) codes and in discrete transport theory (Stepanova and Dew, 2002). On the other hand, adequate statistics can be achieved even with quite moderate computational capacity.

1.5.2 *Operating with Model Cross Sections*

Analytical transport theory requires the use of model cross sections such as Eq. (1.4). On the other hand, its results can be expressed in terms of intuitively relevant physical parameters. I like to illustrate this on the case of Eq. (1.10), which has been quoted above with reference to Sckerl *et al.* (1996) but without a derivation.

Equation (1.10) contains high-energy quantities such as $F_D(E, \vec{e}, x)$ and $\vec{F}_P(E, \vec{e}, x)$ and low-energy quantities such as $S_n(\epsilon)$. Now, $F_D(E, \vec{e}, x)$ is related to the high-energy stopping cross section $S_n(E)$ according to Eq. (1.6). For the power form Eq. (1.4), the stopping cross section becomes $\propto E^{1-2m}$. If this function is to describe a realistic nuclear stopping cross section, the exponent m must depend on energy. But if m is allowed to depend on energy, the kinetic equation cannot be solved.

A way out of this trap is as follows (Sigmund, 1969; Sckerl *et al.*, 1996): First, solve the kinetic equation for a fixed power exponent m. Then express the results by physical parameters that are relevant for dimensional reasons and/or intuitively evident. Stopping cross sections are of particular interest here. They occur either in a form $S_n(E) = NC_m E^{1-2m}/(1-m)$ or $S_n(\epsilon) = NC_m \epsilon^{1-2m}/(1-m)$ (for $M_1 = M_2$), the former in the numerator, the latter in the denominator, as they should.

Once all occurrences of E and ϵ have been identified, the validity of the calculated quantity may be tested by inspection of the remaining dependence on m: The result is considered to be credible if that dependence is weak. Otherwise the result will have to be rejected.

For comprehensive summaries of sputtering by computer simulation, the reader is referred to recent reviews by Eckstein (2007) and Urbassek (2007) for binary-collision and molecular-dynamics simulation, respectively. Here I like to quote a few results that may be of particular interest in the context of this book.

1.5.3 *Depth of Origin*

Theoretical estimates of the depth of origin of sputtered atoms have varied by more than an order of magnitude over the past half century. A first consensus was achieved by a major joint effort (Sigmund *et al.*, 1989). Figure 1.19 shows the distribution in depth of origin of Ge atoms sputtered by Ar ions, simulated by the binary-collision code OXANA (Shulga, 1999). It is seen that the dominating part of sputtered atoms originates in the first two atomic layers, and that the depth distribution is practically independent of the beam energy from 1 to 200 keV.

1.5.4 *Angular Distribution*

At high beam energies, $\gtrsim$10 keV dependent on the ion, angular and energy distributions of sputtered particles are rather insensitive to beam parameters such as energy and angle of incidence. Increasingly pronounced deviations from this behavior are observed with decreasing beam energy. Such behavior can be expected from

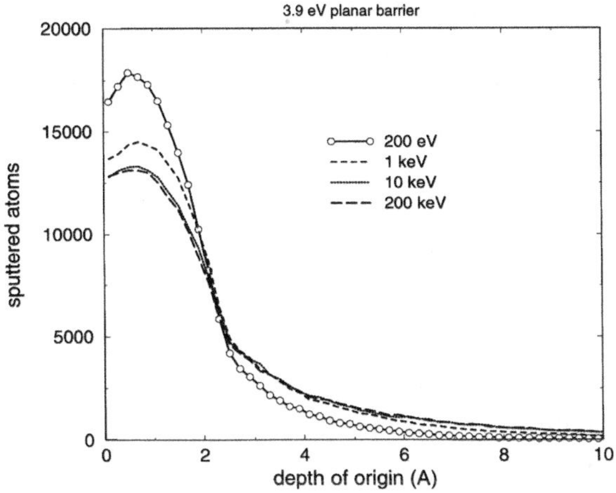

Figure 1.19 Depth of origin of sputtered atoms for Ar on Ge at normal incidence. Beam energy varying from 200 eV to 200 keV. From Shulga (1999), with permission.

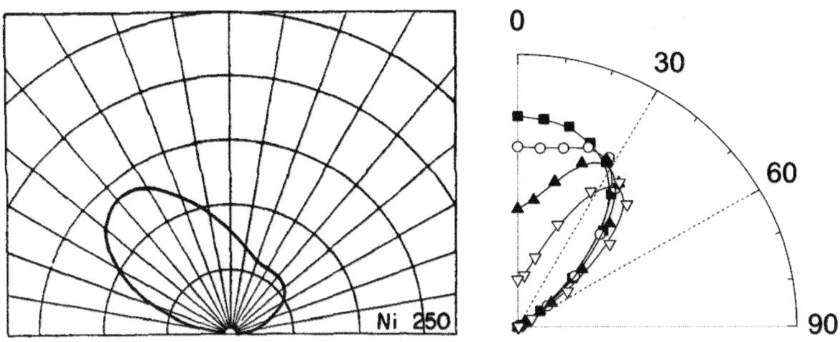

Figure 1.20 Left: Angular distribution of Ni atoms sputtered by 0.25 keV Hg ions at oblique incidence. From Wehner and Rosenberg (1960), with permission. Right: Same for Ar ions of 0.1 keV (empty triangles), 0.2 keV (solid triangles), 0.5 keV (circles), and 1.0 keV (squares) on Ni at normal incidence. From Gnaser (2007), with permission.

Eq. (1.10) but has been well documented from the early days of sputtering, when beam energies up to at most 1 keV were standard (Wehner and Rosenberg, 1960).

Figure 1.20 shows two sets of experimental results, both on nickel and both referring to energies below 1 keV. Angular distributions are clearly skewed toward the forward direction. Equation (1.10) predicts these features to depend on the ratio of E/U. This matches to the observation of Wehner and Rosenberg (1960) of a significant difference between Mo and Ni.

1.5.5 *Elemental Composition*

Sputtering is known to be *preferential*, i.e., the elemental composition of the sputtered flux differs from that of the bombarded target region. This implies that the composition of the target undergoes changes at prolonged bombardment. However, changes in target composition are not only due to preferential sputtering. Ion-induced mixing and segregation are competing processes that, unlike sputtering, may show a significant temperature dependence.

This highly complex problem has been a very active research area, both experimentally and theoretically, from the end 1970s to the end 1980s. Much of this has been summarized by Sigmund and Lam (1993). In that work also a general theoretical scheme was outlined, which incorporates competing processes. General features were explored in a thesis (Sckerl *et al.*, 1998), but much remains to be done here.

Within the scope of this chapter, I shall briefly summarize some results on preferential sputtering.

An early result was found by Andersen and Sigmund (1974) from linear cascade theory in the asymptotic limit $E \gg U$,

$$\frac{Y_1^{(0)}}{Y_2^{(0)}} \simeq \frac{N_1}{N_2}\left(\frac{M_2}{M_1}\right)^{2m}\left(\frac{U_2}{U_1}\right)^{1-2m}, \qquad (1.25)$$

where $Y_i^{(0)}$ denotes the partial sputter yield of the ith species in the isotropic limit and U_i the respective surface binding energy. Here, m is the exponent in the low-energy power cross section Eq. (1.4), a quantity close to 0. Thus, preferential sputtering should be expected to be governed primarily by U_i. This aspect has been studied in detail by Kelly (1978, 1980) on the case of oxides and halides, where U_i is related to the bonding characteristics.

The effect of mass differences can be isolated in measurements on isotope fractionation (Russell *et al.*, 1980; Gnaser and Hutcheon, 1988) and has been analyzed by both molecular-dynamics (Shapiro *et al.*, 1988) and binary-collision (Shulga and Sigmund, 1996) simulation. In view of the smallness of the effect, such simulations have generally been performed by deliberately enhancing mass differences. Measurements and simulations often showed more pronounced isotope fractionation than predicted by Eq. (1.25). Moreover, the effect was also found to depend on the beam energy (Shulga and Sigmund, 1996).

Following the procedure outlined in Section 1.3.5, one may express partial sputtering yields by

$$Y_i = Y_i^{(0)} + Y_i^B + Y_i^C, \tag{1.26}$$

where $Y_i^{(0)}$ represents the asymptotic term, Y_i^B the correction from the momentum term, and Y_I^C the correction from the gradient term. One then finds (Sckerl *et al.*, 1996)

$$\frac{Y_1^B}{Y_2^B} = \frac{N_1}{N_2}\left(\frac{M_1}{M_2}\right)^{1/2-2m} \tag{1.27}$$

$$\frac{Y_1^C}{Y_2^C} = \frac{N_1}{N_2}\left(\frac{M_2}{M_1}\right)^{2m}\left(1+\frac{M_1-M_2}{M_1+M_2}Z(m)\right), \tag{1.28}$$

where $Z(m) \simeq 0.15$ is a well-defined dimensionless constant.

Comparison of these corrections with Eq. (1.25) shows a pronounced mass effect in the momentum term, which leads to preferential sputtering of the heavier species, opposite to the behavior of the leading term, Eq. (1.25).

1.5.6 *Spikes*

A spike is a limited volume with the majority of atoms temporarily in motion (Seitz and Koehler, 1956). Setting the cascade volume to

$$V = N\pi\Delta X_D \rho_D^2, \tag{1.29}$$

where ΔX_D and ρ_D denote the longitudinal and lateral straggling of the deposited-energy profile, we may expect a significant spike component in sputtering if the deposited energy per atom within the cascade volume exceeds the binding energy, i.e.,

$$\frac{E}{N\pi\Delta X_D \rho_D^2} \gtrsim U. \tag{1.30}$$

The dependence on beam energy of the left side is $\propto E^{1-6m}$. Since m runs from close to zero to close to 1 with increasing energy, it is seen that the energy per cascade atom must go through a maximum at some intermediate energy. That maximum lies typically at keV energies, and its magnitude increases strongly with ion and target mass (Sigmund, 1974, 1975).

The magnitude of the spike component to sputtering also depends on the containment time of the spike, i.e., rate of energy transport (Sigmund, 1974).

Attempts to estimate sputter yields from spikes have been based on a number of concepts, all of which have in common that the transport of energy and matter is more of a collective nature than in a linear collision cascade. For an illuminating qualitative survey, the reader is referred to Reimann (1993). More quantitative conclusions may be found in a recent review by Urbassek (2007). Sputtering from spikes has been analyzed in terms of thermal evaporation, various kinds of pressure waves as well as regular flow. The question of which process dominates is to some extent a matter of the magnitude of the sputter yield.

Figure 1.21 shows an extreme case. In view of the low binding energy of solid argon (0.08 eV), an ion with an initial energy of 1 keV may set several thousand atoms in motion. As a result, one finds extensive clouds of emitted gas and a crater that has little similarity with the original target surface.

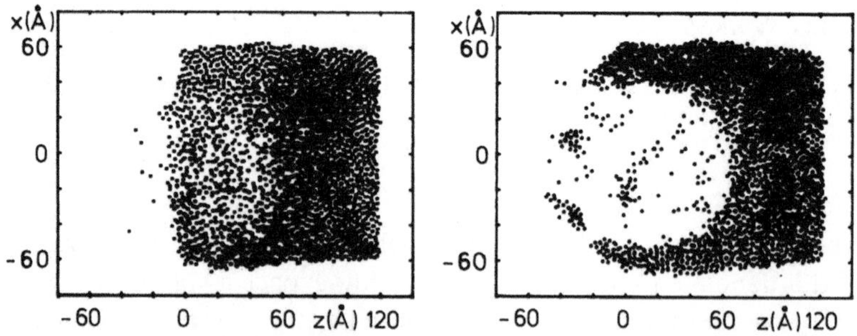

Figure 1.21 Molecular-dynamics simulation of spike induced by 1 keV atomic argon on solid argon at three time steps. From Urbassek and Waldeer (1991), with permission.

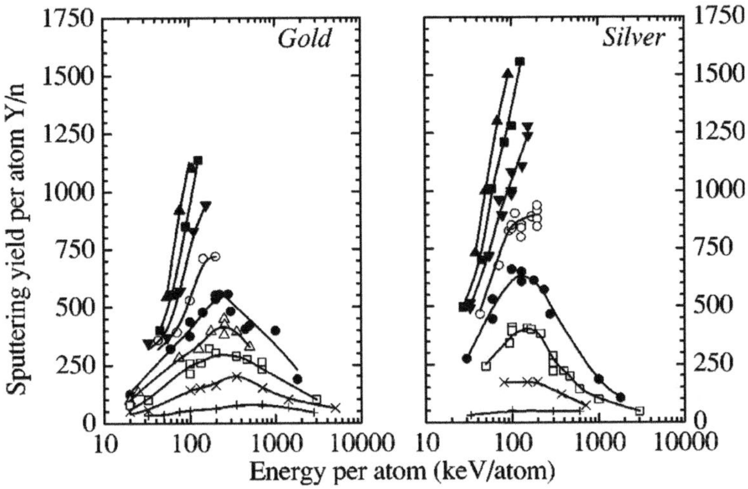

Figure 1.22 Sputter yields *per projectile atom* of gold and silver under bombardment with n-atomic gold clusters. Bottom to top: $n = 1$ to 13. Lines to guide the eye. From Bouneau *et al.* (2002), with permission.

In the opposite end, moderate sputter yields, giving rise to depletion of a couple of monolayers of the target, have been estimated successfully for metallic targets under heavy-ion or cluster bombardment (Sigmund and Claussen, 1981) assuming thermal evaporation, albeit with the use of one adjustable parameter. Dramatically increased sputter yields per projectile atom for cluster bombardment are seen in Fig. 1.22.

A significant fraction of material sputtered from spikes has been found to be emitted in the form of clusters. This feature has received considerable interest during the past decade (Gnaser, 2007), not the least due to applications in mass spectrometry and film deposition.

1.6 Concluding Remarks

This brief introduction has focused deliberately on sputtering by elastic collisions in the linear-cascade regime. This implies that interesting and active research areas have barely been mentioned: Unlike sputtering by elastic collisions, electronic sputtering

(Johnson and Schou, 1993; Assmann *et al.*, 2007) is not a universal phenomenon for all target materials. Unlike sputtering from linear collision cascades, sputtering from spikes (Andersen, 1993; Urbassek, 2007) tends to imply excessively large sputter yields leading to crater formation rather than regular patterns.

I have distinguished between the high- and the low-energy regime, not only because this provides useful insight into the connection between theory and measurable quantities, but also with the aim to encourage the use of combined tools such as transport theory in connection with molecular dynamics, or of binary-collision and molecular-dynamics simulation.

While sputter theory in the asymptotic (high-energy) limit provides insight, actual estimates at energies up to ~10 keV require inclusion of the next term. A useful formalism is available and pertinent references are given (Sigmund and Lam, 1993; Sckerl *et al.*, 1996), but explicit evaluations, including comparisons with experimental findings, await further effort. This includes the important problem of the stoichiometry of the sputtered flux from a compound target.

Fluctuations have been discussed long ago (Westmoreland and Sigmund, 1970), and explicit results have been reported on the distribution of the number of emitted atoms (Eckstein, 1988), but if sputtering is important in nanopatterning at all, fluctuations are likely to be as important as averages and hence deserve increased attention.

Although sputter theory nowadays is heavily dominated by computer simulation, I have given relatively little attention to available codes, their respective benefits and obtained results. This is in part due to my own limited experience but also due to the fact that areas where simulation has proven to be the only available successful tool, such as sputtering of and by molecules and clusters, have been left out for other reasons.

I have discussed pertinent input in the form of interatomic potentials and electronic stopping cross sections with a view to the keV and MeV range, and I expressed some caution concerning the SRIM code (Ziegler *et al.*, 1985).

With regard to input in the low-energy regime — which is central to molecular dynamics — simple scaling relations appear to be of little use. Adequate potentials have been known for many materials for some time (Nieminen, 1993).

Although electronic stopping in the eV regime is a small effect compared to nuclear stopping, it is a central quantity determining the lifetime of a spike. Efforts in this area are directed both at ways of incorporating existing cross sections into molecular-dynamics codes (Duvenbeck *et al.*, 2007) and to develop energy loss rates *ab initio* (Race *et al.*, 2010).

Analytic transport theory in the form described here requires the use of model cross sections. Power cross sections due to Lindhard *et al.* (1968) are particularly useful in this context when utilized with due care.

Acknowledgement

I would like to thank Tapobrata Som for the kind invitation to contribute to this book and to Hubert Gnaser, Michael Aziz, and Hans Henrik Andersen for carefully reading and commenting on the preliminary manuscript. This work has been supported by the Danish Natural Science Research Council (FNU).

References

Andersen, H. H. (1968). The energy efficiency of lead self-sputtering, *Appl. Phys. Lett.*, **13**, 85–86.

Andersen, H. H. (1993). Nonlinear effects in collisional sputtering under cluster impact, *Mat. Fys. Medd. Dan. Vid. Selsk.*, **43**, 127–154.

Andersen, H. H., and Bay, H. L. (1981). Sputtering yield measurements, *Top. Appl. Phys.*, **47**, 145.

Andersen, N., and Sigmund, P. (1974). Energy dissipation by heavy ions in compound targets, *Mat. Fys. Medd. Dan. Vid. Selsk.*, **39(3)**, 1.

Assmann, W., Toulemonde, M., and Trautmann, C. (2007). *Electronic Sputtering with Swift Heavy Ions, Top. Appl. Phys.*, Vol. 110 (Springer, Berlin), 401–450.

Behrisch, R. (ed.) (1981). *Sputtering by Particle Bombardment I, Top. Appl. Phys.*, Vol. 47 (Springer, Heidelbarg).

Behrisch, R. (ed.) (1983). *Sputtering by Particle Bombardment II, Top. Appl. Phys.*, Vol. 52 (Springer, Heidelberg).

Behrisch, R., and Eckstein, W. (eds.) (2007). *Sputtering by Particle Bombardment, Top. Appl. Phys.*, Vol. 110 (Springer, Berlin).

Behrisch, R., and Wittmaack, K. (eds.) (1991). *Sputtering by Particle Bombardment III, Top. Appl. Phys.*, Vol. 64 (Springer, Heidelberg).

Bohr, N. (1948). The penetration of atomic particles through matter, *Mat. Fys. Medd. Dan. Vid. Selsk.*, **18(8)**, 1–144.

Bouneau, S., Brunelle, A., Della-Negra, S., Depauw, J., Jacquet, D., Le Beyec, Y., Pautrat, M., Fallavier, M., Poizat, J. C., and Andersen, H. H. (2002). Very large gold and silver sputtering yields induced by keV to meV energy Au-N clusters ($n = 1$–13), *Phys. Rev. B*, **65**, 144106.

Bradley, R. M., and Harper, J. M. E. (1988). Theory of ripple topography induced by ion bombardment, *J. Vac. Sci. Technol. A*, **6**, 2390–2395.

Duvenbeck, A., Weingart, O., Buss, V., and Wucher, A. (2007). Electron promotion and electronic friction in atomic collision cascades, *New J. Physics* **9**, 38.

Eckstein, W. (1988). Statistics of sputtering, *Nucl. Instrum. Methods B*, **33**, 489–492.

Eckstein, W. (2007). *Sputtering yields, Top. Appl. Phys.*, Vol. 110 (Springer, Berlin), pp. 33–187.

Eckstein, W., and Urbassek, H. M. (2007). *Computer Simulation of the Sputtering Process, Top. Appl. Phys.*, Vol. 110 (Springer, Berlin), pp. 21–32.

Fastrup, B., Hvelplund, P., and Sautter, C. A. (1966). Stopping cross section in carbon of 0.1–1.0 meV atoms with $6 < z_1 < 20$, *Mat. Fys. Medd. Dan. Vid. Selsk.* **35(10)**, 1–28.

Firsov, O. B. (1957). Calculation of the interaction potential of atoms, *Zh. Eksp. Teor. Fiz.* **33**, 696, [Engl. transl. Sov. Phys. JETP 6, 5340–537 (1958)].

Firsov, O. B. (1959). A qualitative interpretation of the mean electron excitation energy in atomic collsions, *Zh. Eksp. Teor. Fiz.*, **36**, 1517–1523, [Engl. transl. Sov. Phys. JETP, **9**, 1076–1080 (1959)].

Friedland, E., Kalbitzer, S., Hayes, M., Klatt, C., Konac, G., and Langpape, C. (1998). Range parameters of gold and lead in carbon and of carbon in gold at reduced energies of $10^{-3} < \epsilon < 1$, *Nucl. Instrum. Methods B*, **136**, 147–152.

Glazov, L. G. (1995). Solution of the kinetic equation for the deposited momentum distribution. II. threshold energy effects, *J. Phys. C*, **7**, 6365–6377.

Glazov, L. G. (2000). Restricted medium: analytical theory of collision cascades and sputtering, *Izv. Akad. Nauk Ser. Fiz.*, **64**, 631–638.

Gnaser, H. (2007). *Energy and angular distributions of sputtered species, Top. Appl. Phys.*, Vol. 110 (Springer, Berlin), 231–328.

Gnaser, H., and Hutcheon, I. D. (1988). Preferential emission of lighter isotopes in the initial-stage of sputtering, *Surf. Sci.*, **195**, 499–512.

Gombas, P. (1956). Statistische Behandlung des Atoms, in S. Flügge (ed.), *Handbuch der Physik*, Vol. 36 (Springer, Berlin), 109–231.

Grande, P. L., and Schiwietz, G. (2010). CasP version 4.1, http://www.helmholtz-berlin.de/people/gregor-schiwietz/casp_e n.html.

Jensen, H. (1932). Die Ladungsverteilung in Ionen und die Gitterkonstante des Rubidiumbromids nach der statistischen Methode, *Z. Physik*, **77**, 722.

Johnson, R. E., and Schou, J. (1993). Sputtering of inorganic insulators, *Mat. Fys. Medd. Dan. Vid. Selsk.*, **43**, 403–494.

Kelly, K. (1980). On the problem of whether mass or chemical bonding is more important to bombardment-induced compositional changes in alloys and oxides, *Surf. Sci.*, **100**, 1, 85–107.

Kelly, R. (1978). Attempt to understand preferential sputtering, *Nucl. Instrum. Methods*, **149**, 553–558.

Leibfried, G. (1958). Statistische Schwankungen der von einem Primärteilchen erzeugten Frenkel-Defekte, *Nukleonik*, **1**, 57–59.

Lenz, W. (1932). Über die Anwendbarkeit der statistischen Methode auf Ionengitter, *Z. Physik*, **77**, 713.

Lindhard, J., Nielsen, V., and Scharff, M. (1968). Approximation method in classical scattering by screened coulomb fields, *Mat. Fys. Medd. Dan. Vid. Selsk.*, **36(10)**, 1–32.

Lindhard, J., Nielsen, V., Scharff, M., and Thomsen, P. V. (1963a). Integral equations governing radiation effects, *Mat. Fys. Medd. Dan. Vid. Selsk.*, **33(10)**, 1.

Lindhard, J., and Scharff, M. (1961). Energy dissipation by ions in the keV region, *Phys. Rev.*, **124**, 128–130.

Lindhard, J., Scharff, M., and Schiøtt, H. E. (1963b). Range concepts and heavy ion ranges, *Mat. Fys. Medd. Dan. Vid. Selsk.*, **33(14)**, 1.

Littmark, U., and Sigmund, P. (1975). Momentum deposition by heavy-ion bombardment and an application to sputtering, *J. Phys. D*, **8**, 241–245.

Loftager, P., Besenbacher, F., Jensen, O. S., and Sørensen, V. S. (1979). Experimental study of effective interaction potentials, *Phys. Rev. A*, **20**, 1443.

Molière, G. (1947). Theorie der Streuung schneller geladener Teilchen i. Einzelstreuung am abgeschirmten Coulomb-Feld, *Z. Naturforsch.*, **2a**, 133–145.

Nieminen, R. M. (1993). Perspectives of molecular dynamics, *Mat. Fys. Medd. Dan. Vid. Selsk.*, **43**, 81–96.

Ormrod, J. H., and Duckworth, H. E. (1963). Stopping cross sections in carbon for low-energy atoms with $z \leq 12$, *Can. J. Phys.*, **41**, 1424–1442.

Paul, H. (2009). Stopping power graphs, www.exphys.uni-linz.ac.at/stopping/.

Paul, H., and Schinner, A. (2001). An empirical approach to the stopping power of solids and gases for ions from ^{3}Li to ^{18}Ar, *Nucl. Instrum. Methods B*, **179**, 299–315.

Race, C. P., Mason, D. R., Finnis, M. W., Foulkes, W. M. C., Horsfield, A. P. and Sutton, A. P. (2010). The treatment of electronic excitations in atomistic models of radiation damage in metals, *Repts. Prog. Phys.*, **73**.

Reimann, C. T. (1993). Theoretical models for sputtering and desorption of bio-organic molecules under collisional and electronic excitation by ion impact, *Mat. Fys. Medd. Dan. Vid. Selsk.*, **43**, 351–402.

Robinson, M. T. (1965a). Energy spectra of atoms slowing down in structureless media, *Philos. Mag.*, **12**, 145–156.

Robinson, M. T. (1965b). Influence of scattering law on radiation damage displacement cascade, *Philos. Mag.*, **12**, 741–765.

Russell, W. A., Papanastassiou, D. A., and Tombrello, T. A. (1980). The fractionation of ca isotopes by sputtering, *Radiat. Eff.*, **52**, 41–52.

Sanders, J. B. (1968). *On penetration depths and collision cascades in solid materials*, Ph.D. thesis, Leiden University.

Sckerl, M. W., Lam, N. Q., and Sigmund, P. (1998). Compositional changes in alloys during ion bombardment at elevated temperatures, *Nucl. Instrum. Methods B*, **140**, 75–90.

Sckerl, M. W., Sigmund, P., and Vicanek, M. (1996). Particle fluxes in atomic collision cascades, *Mat. Fys. Medd. Dan. Vid. Selsk.*, **44**, 3, 1–63.

Seitz, F., and Koehler, J. S. (1956). Displacement of atoms during irradiation, *Solid State Phys.*, **2**, 307–448.

Shapiro, M. H., Tombrello, T. A., and Harrison, D. E. (1988). Simulation of isotopic mass effects in sputtering. 2. *Nucl. Instrum. Methods B*, **30**, 152–158.

Shulga, V. I. (1999). Depth-dependent angular distribution of sputtered atoms, *Nucl. Instrum. Methods B*, **155**, 382–394.

Shulga, V. I., and Sigmund, P. (1996). Analysis of the primary process in isotope sputtering, *Nucl. Instrum. Methods B*, **119**, 359–374.

Sigmund, P. (1968). Sputtering efficiency of amorphous substances, *Can. J. Phys.*, **46**, 6, 731–737.

Sigmund, P. (1969). Theory of sputtering. 1. sputtering yield of amorphous and polycrystalline targets, *Phys. Rev.*, **184**, 383–416.

Sigmund, P. (1972a). Collision theory of displacement damage, ion ranges, and sputtering. 2. *Rev. Roum. Phys.*, **17**, 969–1000.

Sigmund, P. (1972b). Collision theory of displacement damage, ion ranges, and sputtering. 3. *Rev. Roum. Phys.*, **17**, 1079–1106.

Sigmund, P. (1972c). Collision theory of displacement damage, ion ranges, and sputtering, *Rev. Roum. Phys.*, **17**, 823–870.

Sigmund, P. (1973). A mechanism of surface micro-roughening by ion bombardment, *J. Mater. Sci.*, **8**, 1545–1553.

Sigmund, P. (1974). Energy density and time constant of heavy-ion-induced elastic-collision spikes in solids, *Appl. Phys. Lett.*, **25**, 169–171.

Sigmund, P. (1975). Energy density and time constant of heavy-ion-induced elastic-collision spikes in solids (erratum), *Appl. Phys. Lett.*, **27**, 52.

Sigmund, P. (1981). Sputtering by ion bombardment: Theoretical concepts, *Top. Appl. Phys. (Springer)*, **47**, 9–71.

Sigmund, P. (1992). *Fundamental Processes in Sputtering of Atoms and Molecules, Mat. Fys. Medd. Dan. Vid. Selsk.*, Vol. 43 (Royal Danish Academy of Sciences and Letters, Copenhagen).

Sigmund, P. (1998). Stopping power in perspective, *Nucl. Instrum. Methods B*, **135**, 1–15.

Sigmund, P. (2006). *Particle Penetration and Radiation Effects, Springer Series in Solid-State Sciences*, Vol. 151 (Springer, Berlin).

Sigmund, P. (2008). Reciprocity in the electronic stopping of slow ions in matter, *Europ. Phys. J. D*, **47**, 45–54.

Sigmund, P., and Claussen, C. (1981). Sputtering from elastic-collision spikes in heavy-ion-bombarded metals, *J. Appl. Phys.*, **52**, 990.

Sigmund, P., and Lam, N. Q. (1993). Alloy and isotope sputtering, *Mat. Fys. Medd. Dan. Vid. Selsk.*, **43**, 255–351.

Sigmund, P., Robinson, M. T., Baskes, M. I., Hautala, M., Cui, F. Z., Eckstein, W., Yamamura, Y., Hosaka, S., Ishitani, T., Shulga, V. I., Harrison, D. E., Chakarov, I. R., Karpuzov, D. S., Kawatoh, E., Shimizu, R., Valkealahti, S., Nieminen, R. M., Betz, G., Husinsky, W., Shapiro, M. H., Vicanek, M., and

Urbassek, H. M. (1989). Round robin computer-simulation of ejection probability in sputtering, *Nucl. Instrum. Methods B*, **36**, 110.

Sigmund, P., and Sanders, J. B. (1967). Spatial distribution of energy deposited by ion bombardment, in P. Glotin (ed.), *Application of Ion Beams to Semiconductor Technology* (Editions Ophrys), 215–238.

Sigmund, P., and Schinner, A. (2000). Binary stopping theory for swift heavy ions, *Europ. Phys. J. D*, **12**, 425–434.

Sigmund, P., and Schinner, A. (2002). Binary theory of electronic stopping, *Nucl. Instrum. Methods B*, **195**, 64–90.

Stepanova, M., and Dew, S. K. (2002). Discrete-path transport theory of physical sputtering, *J. Appl. Phys.*, **92**, 1699–1708.

Thompson, M. W. (1968). Energy spectrum of ejected atoms during high energy sputtering of gold, *Philos. Mag.*, **18**, 377.

Urbassek, H. M. (2007). *Results of Molecular Dynamics Calculations, Top. Appl. Phys.*, Vol. 110 (Springer, Berlin), pp. 189–230.

Urbassek, H. M., Mayer, G., Gades, H., and Vicanek, M. (1995). Effect of bulk binding forces on energetic-ion-induced collision cascades - a combined simulational and analytical approach, *Nucl. Instrum. Methods B*, **103**, 275–283.

Urbassek, H. M., and Waldeer, K. T. (1991). Spikes in condensed rare gases induced by keV-atom bombardment, *Phys. Rev. Lett.*, **67**, 105–108.

Vicanek, M., Jimenez-Rodriguez, J. J., and Sigmund, P. (1989). Depth of origin and angular spectrum of sputtered atoms, *Nucl. Instrum. Methods B*, **36**, 124.

Wehner, G. K., and Rosenberg, D. (1960). Angular distribution of sputtered material, *J. Appl. Phys.*, **31**, 177–179.

Westmoreland, J. E., and Sigmund, P. (1970). Correlation functions in the theory of atomic collision cascades: Ion location and distribution in depth and size of damage clusters, *Radiat. Eff.*, **6**, 187–197.

Winterbon, K. B., Sigmund, P., and Sanders, J. B. (1970). Spatial distribution of energy deposited by atomic particles in elastic collisions, *Mat. Fys. Medd. Dan. Vid. Selsk.*, **37(14)**, 1–73.

Ziegler, J. F., Biersack, J. P., and Littmark, U. (1985). *The Stopping and Range of Ions in Solids, The Stopping and Ranges of Ions in Matter*, Vol. 1 (Pergamon, New York).

Chapter 2

Sputter-Ripple Formation on Flat and Rough Surfaces: A Case Study with Si

Safiul Alam Mollick and Debabrata Ghose*

Saha Institute of Nuclear Physics,
Sector I, Block AF, Bidhan Nagar, Kolkata 700064, India
*debabrata.ghose@saha.ac.in

The formation of nanoscale ripple patterns on solid surfaces by low-energy ion sputtering has attracted a high level of interest in the last few decades from the fundamental point of view as well as for various technological applications. The interplay between ion sputtering and different surface relaxation processes is thought to be the origin of ripple formation. The size, shape, orientation, and compositional modulation of the pattern can be controlled by manipulating the ion beam parameters and sample temperature. Recently, it has been shown that pre-roughened surface is also a determining factor for the ripple development and its dynamics. This chapter presents a comparative study of ripple formation on flat and rough Si surfaces as a function of angle of ion incidence and fluence.

Nanofabrication by Ion-Beam Sputtering: Fundamentals and Applications
Edited by Tapobrata Som and Dinakar Kanjilal

ISBN 978-981-4303-75-0 (Hardcover), 978-981-4303-76-7 (eBook)
www.panstanford.com

2.1 Introduction

The interaction between keV ion and a solid target gives rise to two main types of collision processes, namely, inelastic and elastic collisions. The inelastic collisions involve interactions with electrons leading to secondary electron emission, emission of characteristic X-rays, and optical photon emission. The second one considers the elastic collision with the lattice leading to displacement of lattice atoms, formation of point defects, and the erosion of the target by sputtering. At low keV energies, the sputtering effect is dominant. Briefly, the phenomenon is as follows: the primary ion transfers momentum to the target atoms, which recoil with enough energy to set other atoms in motion. These may then subsequently set other target atoms moving, and thereby a collision cascade is generated in space and time. Some of the recoils, which lie close to the surface, may overcome the surface binding energy and get sputtered. Sigmund [1] derived a comprehensive theory of sputtering, where it is assumed that the collision cascade is linear, i.e., a small fraction of atoms within the cascade is in motion and all the collisions are between pairs of a moving atom and a stationary atom. It shows that the sputtering yield Y (atoms/ion) is directly proportional to the nuclear stopping power or the energy deposited by the ion at the surface.

The ion beam sputtering not only erodes the surface but also modifies the surface topography. If the surface is flat and free from imperfections, it is expected on preliminary considerations that the sputtered surface would remain flat. However, the real surface often departs from the flatness; also impurities and/or defects can be present on the surface. All these cause differential erosion around the perturbed regions, which may generate microstructures on the surface. It is now well established that sputtering could produce highly correlated arrays of ripples or dots at sub-micrometer and nanometer length scales on different materials including amorphous and crystalline over a large surface area in a short time [2–6]. The simplistic one-step process, high output and pattern ordering make the ion-induced self-organized process remarkably advantageous over conventional optical lithography and costly e-beam lithography techniques.

Since the first observation of sputter ripples on glass surfaces by Navez *et al.* [7], it took nearly two decades to understand the underlying mechanisms when Bradley and Harper [8] put forward the linear continuum theory for pattern formation. Later, the Bradley–Harper model was extended to incorporate nonlinear features of the ripple structures.

In the present chapter, the studies of ripple formation by ion beam sputtering have been carried out with Si, basically for two reasons. Silicon is an important technological material in semiconductor industry and readily available in high purity and quality. Second, at room temperature, Si amorphizes during ion irradiation much early before any structure could develop on the surface. The latter property makes it a suitable material to test the predictability of different continuum models developed for monatomic and isotropic targets. The literature shows that ripple formations were studied mostly with ion fluence, ion energy, angle of ion incidence, and substrate temperature. On the contrary, studies with surface roughness are relatively scarce. The role of surface roughness on ripple nucleation and growth was first noticed in metal film sputtering [9, 10], since metal films intrinsically develop rough surface morphology during deposition process. It appears from experiments that if roughness could be incorporated on the surface by any means before ion bombardment, ripple formation would be accelerated. In the present chapter, ripple topography formations in flat and pre-roughened Si surfaces are investigated at different ion incident angles and fluences.

2.2 Mechanisms of Pattern Formation

Theories of ion-bombardment-induced surface topography are evolved from the pioneering works of Prof. G. Carter and his group [11, 12], who rely mainly on the effects of angular dependence of sputtering yield. A typical variation of sputtering yield $Y(\vartheta)$ as a function of angle of ion incidence ϑ for amorphous solids is shown in Fig. 2.1. In general, the sputtering yield $Y(\vartheta)$ increases with the angle of incidence ϑ between the beam and normal to the surface roughly as sec ϑ up to angles of about 45°, reaching a maximum

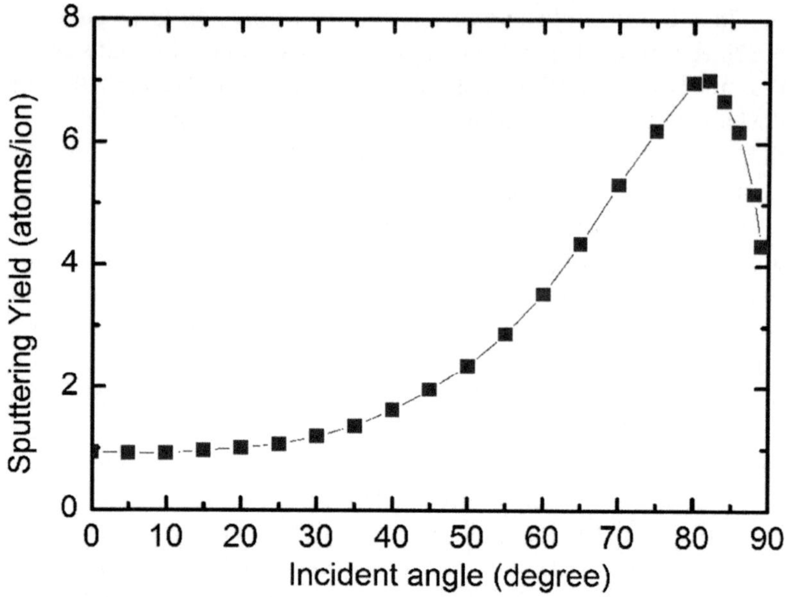

Figure 2.1 Variation of sputtering yield of Si as a function of angle of ion incidence for 8 keV O^+ ion bombardment as calculated by SRIM [20].

around $\vartheta_{max} = 75–80°$ and then decreases rapidly toward zero as ϑ approaches 90°. For simplicity, let us consider the morphological evolution of a two-dimensional irregular surface as illustrated in Fig. 2.2. Let the contour of the unbombarded surface be in the *xoz* plane and the initial mean surface level be parallel to the *x*-axis (Fig. 2.2a). The incoming beam with flux density J makes an angle ϑ with z and lies in the same *xoz* plane. The height of the surface at any time t is measured along the *oz* direction from the mean surface level. An element ds of the surface over the mean surface height at any time t can be represented by the equation: $h(x,t) = bx + h_0$, where $b = \tan\alpha$ is the slope of the surface element and h_0 is the mean surface height. The local surface normal z' forms an angle ψ with the beam direction. Thus $\psi = \vartheta - \alpha$, where ϑ is a fixed angle (the global angle of incidence), but ψ changes from point to point along the surface. Taking into account the variation of current density with respect to the incident beam direction, the erosion speed v normal to the surface element is given by

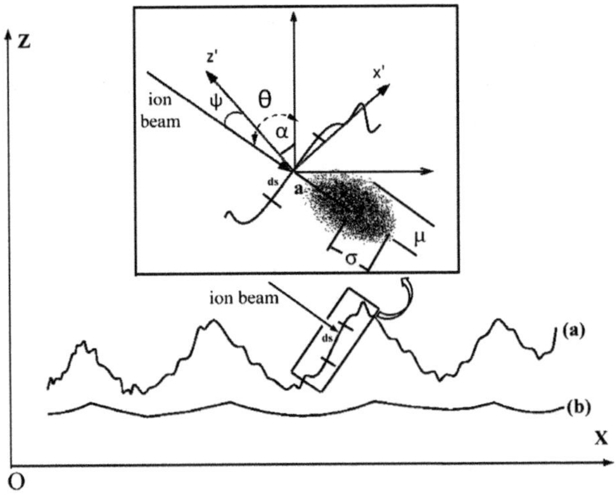

Figure 2.2 Sputtered morphology evolution of an initially rough surface during ion bombardment; (a) represents the gradient-dependent erosion and (b) represents the curvature-dependent erosion after the initial surface becomes nearly flat.

$$v(\psi) = \frac{Y(\psi)J\ \cos\psi}{N}, \tag{2.1}$$

where N is the target atom density. Equation 2.1 shows that v is an explicit function of the surface gradient through the dependence of $Y(\psi)$. The rate of surface height reduction in the $-oz$ direction is then

$$\frac{\partial h}{\partial t} = \frac{v}{\cos\alpha} = \frac{Y(\vartheta-\alpha).J\ \cos(\vartheta-\alpha)}{N\cos\alpha}. \tag{2.2}$$

As the erosion proceeds, the orientation of the surface profile also changes. The motion of the surface element of orientation ψ can be described by a propagating wave equation in the variable ψ [11] such that

$$\frac{\partial\psi}{\partial t} + p(\psi)\frac{\partial\psi}{\partial z} = 0, \quad \text{where } p(\psi) = \frac{J}{N}\frac{\partial y(\psi)}{\partial\psi}\cos^2\psi. \tag{2.3}$$

As a general approach to the profile evolution, Carter *et al.* [13] assumed that the eroding surface as an advancing wave obeying Huygen's principle analogous to that in geometrical optics. The form of Huygen's wavelet at different positions is determined by the normal erosion velocity component and the shape of the $Y(\vartheta)$–ϑ function. The successive sputtered topographies with increasing time can be generated by the so-called characteristics (equivalent to the optical rays) along which points on time-evolving surface move while maintaining constant orientation. It was shown [13] that the peaks erode faster than the troughs, because edges formed by the convergence and intersection of characteristics from either side of the peaks proceed downward with higher velocities. For troughs, the characteristics are divergent and edges are formed with characteristics initially from neighboring troughs and finally from distant troughs. Furthermore as the sputtering proceeds, the wavelets corresponding to deeper troughs extend over those from shallower ones preventing the latter not to be a part of the eroding surface. The surface is ultimately left with irregular shallow depressions slowly widening provided no other roughening or smoothing processes being operative. The final surface is, therefore, not perfectly flat; instead, it appears as irregular uneven surface with small height fluctuations and possessing a broad range of spatial frequencies. All the elemental portions of the surface have vanishingly small gradients with respect to the mean surface direction (e.g., Fig. 2.2b). Under this condition, $\tan\alpha \approx \alpha \approx \frac{\partial h}{\partial x}$, $\cos\alpha \approx 1$, and $Y(\vartheta - \alpha)$ and $\cos(\vartheta - \alpha)$ can be expanded around ϑ by Taylor' series so that Eq. 2.2 becomes

$$\frac{\partial h}{\partial t} = -v(\vartheta) + \frac{\partial v(\vartheta)}{\partial \vartheta}\frac{\partial h}{\partial x}. \tag{2.4}$$

The first term on the right-hand side of Eq. 2.4 represents the velocity of recession of the unperturbed flat surface, while the second term represents the transverse velocity of the modulation. Carter did not consider further the effects of other possible sources of roughening, e.g., the surface curvature that can greatly modify the erosion velocity. It was Prof. P. Sigmund [14] who first recognized the importance of curvature-induced roughening instability that can enhance the growth of small irregularities on a relatively smooth surface when the local radii of curvature are comparable to, or larger than, the dimensions of the collision cascade. He showed that on a length scale of the collision cascade, the crest of any perturbation will erode more slowly than troughs because the maximum energy of deposition is not at the point of impact but further downstream (Fig. 2.2b). Such a process can lead to increasing amplitude instability unless some smoothing process also operating on the atomic scales suppresses it. Considering that the erosion velocity depends both on the local slope and curvature, Bradley and Harper [8] showed that, in the limit $a/R_{x',y'} << 1$, Eq. 2.1 can be approximated as

$$v(\psi, R_{x'}, R_{y'}) = \frac{J}{N}Y(\psi)\left[\cos\psi - \Gamma_x(\psi)\frac{a}{R_{x'}} - \Gamma_y(\psi)\frac{a}{R_{y'}}\right], \quad (2.5)$$

where $R_{x',y'}$ are the local radii of curvature in x'- and y'-directions, respectively, and $\Gamma_{x',y'}$ are the surface tension coefficients, which depend on the shape of the collision cascade, i.e., the penetration depth a, and longitudinal and lateral straggling widths σ and μ, respectively. The roughening instability amplifies the surface noise of all frequencies. On the other hand, the surface diffusion tends to oppose the growth of the noise scale. Bradley and Harper [8] assumed a temperature-dependent self-diffusion process where the diffusing species is associated with the gradient of the chemical potential due to the curvature [15, 16]. Thus the temporal change of surface height due to diffusion is

$$\frac{\partial h}{\partial t} = -\frac{D_s C'\gamma\Omega^2}{k_B T}\nabla^4 h. \quad (2.6)$$

Here D_s, C', γ, Ω, k_B, and T denote the surface diffusion constant, areal density of mobile surface defects, surface free energy, atomic volume, Boltzmann constant, and temperature, respectively. However, there are other sources of diffusion acting on the system. Makeev *et al.* [17] have made a Taylor expansion of the erosion velocity and showed that the fourth-order terms mimic surface diffusion without actual mass transport along the surface. Such a smoothing mechanism, called ion-induced surface diffusion (ISD), is temperature independent and entirely determined by ion beam parameters and the distribution of deposited energy [17]. For Si, which turns out amorphous by ion sputtering, the ion-enhanced viscous flow (IVF) is thought to be more relevant for the relaxation process [18]. For the viscous flow in a thin amorphous layer at the surface, the rate of surface height reduction is given by [18]

$$\frac{\partial h}{\partial t} = -\frac{J\gamma a_z^3}{\eta}\nabla^4 h. \quad (2.7)$$

Here η and a_z ($\approx a$, the ion penetration depth) denote the viscosity and thickness of the liquid layer, respectively.

Transforming the local coordinate system $(x'y'z')$ to the laboratory coordinate system (xyz) and applying the small slope approximations to the erosion velocity (Eq. 2.5), the rate of change of surface height in combination with the diffusion terms reads as

$$\frac{\partial h}{\partial t} = -v_0 + \frac{\partial v_0}{\partial \vartheta}\frac{\partial h}{\partial x} + A_x\frac{\partial^2 h}{\partial x^2} + A_y\frac{\partial^2 h}{\partial y^2} - B\nabla^4 h, \qquad (2.8)$$

where we substitute $1/R_{x'} = -\partial^2 h/\partial x^2$ and $1/R_{y'} = -\partial^2 h/\partial y^2$. Equation 2.8 is commonly known as the Bradley–Harper equation. The values of A_x and A_y can be calculated from the ion beam parameters and the geometrical shape of the deposited energy distribution [8, 17]. B is the effective diffusion coefficient, which accounts all possible smoothing mechanisms described above.

To understand the evolution of the surface topography, it is assumed that each Fourier component of the surface morphology grows with an amplification factor R_q, which controls the growth rate as a function of time t. Let h_q be the amplitude of the Fourier component with wavelength $l = 2\pi/q$, then $h_q(t)$ evolves according to

$$h_q(t) = h_q(0)\exp(R_q t), \qquad (2.9)$$

where

$$R_q = -[A_x(\vartheta)q_x^2 + A_y(\vartheta)q_y^2] - B(q_x^2 + q_y^2)^2. \qquad (2.10)$$

In principle, waves of all surface wave vectors q can develop, but the selection of the dominant wave vector is determined for which the growth rate R_q has a positive maximum. Since the diffusion coefficient B is always positive, the sign of surface coefficient A should be negative in order to make R_q positive. The ripple wavelength l_c is given by

$$l_c = 2\pi\sqrt{\frac{2B}{|A(\vartheta)|}}. \qquad (2.11)$$

Here $|A(\vartheta)|$ is the larger of the two negative surface tension coefficient values $-A_x(\vartheta)$ and $-A_y(\vartheta)$. At moderate incidence angles, the ripple wave vector is directed parallel to the ion beam direction (x-direction ripples) as $A_x < A_y < 0$, while above a certain incidence angle ϑ_c, $(A_y < 0, A_x > 0)$ the ripples are aligned perpendicular to the ion beam plane (y-direction ripples). The value of ϑ_c depends on the shape of the collision cascade. To get an idea of the geometrical shape of the cascade, A_x (ϑ) and A_y (ϑ) are expanded to second order in ϑ for small ϑ as described by Bradley and Harper [8]. It is readily evident that $-A_x$ (ϑ) will be less than $-A_y$ (ϑ) when the following condition is satisfied [8]:

$$(a/\sigma)^2 + (a/\mu)^2 \geq 3. \tag{2.12}$$

Equation 2.12 can give an idea of the type of damage-cascade distribution. The first one is the symmetric distribution of the deposited energy, i.e., $\sigma = \mu$. The above condition is met when $a/\sigma \geq \sqrt{2}$. That means for smaller reduced energy depths a/σ, the x-direction ripples cannot appear at low angles of incidence [17]. However, if one assumes an asymmetric distribution such that $\sigma \geq \sqrt{3}\ \mu$, the ripples will always be directed along the x-direction for small ϑ even for small enough a/σ values, as usually observed in the experiment. One can also infer the shape of the energy distribution by plotting the explicit expressions of surface tension coefficients with the ion incidence angle. For spherically symmetric energy deposition ($a \approx \sigma \approx \mu$) ripples can form above $\vartheta_c \approx 8°$ and only y-direction ripples should develop at all incidence angles (see Fig. 2.3a), which is usually not observed in the experiments. On the other hand, if one considers an asymmetric damage distribution

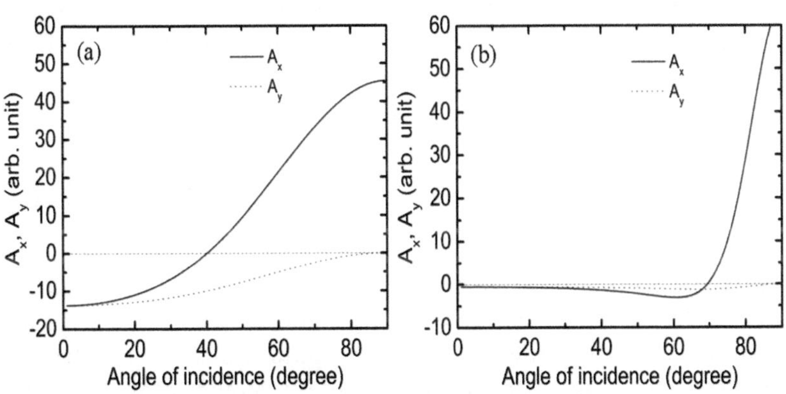

Figure 2.3 Normalized values of surface tension coefficients A_x and A_y versus ion incidence angle calculated for (a) $a = \sigma = \mu$ and (b) $\sigma = a/2$, $\mu = a\ /4$.

typically characterized by $\sigma = a/2$ and $\mu = a/4$ [8], both types of ripples can form in the appropriate angular range, i.e., x-direction ripples at lower oblique angles and y-direction ripples at grazing angles and the ripple rotation occurs at $\vartheta_c \approx 68^\circ$ (Fig. 2.3b) consistent to most of the experimental observations [19]. The SRIM (Stopping and Range of Ions in Matter) simulation [20] also shows that the longitudinal width σ is larger than transverse width μ for many ion-target combinations, i.e., the distribution of energy dissipation is essentially asymmetric.

The Bradley–Harper model requires a noisy starting surface. However, most of the experimental works were carried out with surfaces which were originally flat. Then the question arises, how the ripple topography is nucleated? It is argued that the beam flux possesses local random fluctuations, which are uncorrelated in space and time. These fluctuations can be included in the incoming ion flux as the sum of the average flux J and the noise η with zero average and uncorrelated. This noisy component of the beam is thought to be the driving force for generating random small surface height fluctuations during the early stages of bombardment [21]. However, there are other methods by which the initial roughening process can be initiated. These are: (i) seeding of foreign atoms [22]; (ii) use of multi-component target materials [23] where preferential sputtering causing localized enrichment of one element than others; and (iii) samples with preexisting roughness or deliberate roughening of the surface, say, by chemical etching [24] before bombardment. The latter method is found to be a clean and convenient way to generate the surface noise.

2.2.1 *Extension of Bradley–Harper Theory: The Nonlinear Effects*

Equation 2.8 is a linear deterministic equation, which is successful in predicting many features of ripple formation, e.g., the ripple wavelength, exponential increase in amplitude and ripple rotation. However, it cannot explain other experimental features such as the saturation of the ripple amplitude and the appearance of kinetic roughening at longer sputter time. In order to take all these into account, Cuerno and Barabási [21] developed a more general equation, by adding a tilt-dependent erosion term $\lambda_{x,y}(\partial_{x,y}h)^2$ and a noise term η to Eq. 2.8, known as the noisy anisotropic Kuramoto-Sivashinsky equation. Following this equation, the system initially evolves similar to the Bradley–Harper linear equation, but after a certain cross-over time, the effect of nonlinear terms dominates [25].

The Kuramoto-Sivashinsky equation is further generalized by taking into account other physically relevant fields that also affect the height profile, e.g., the thickness of the mobile surface adatom layer. The so-called coupled two-field equation as developed by Muñoz-García *et al.* [26] specifically predicts the coarsening of ripple wavelength, which is occasionally observed in experiments. The temporal height equation corresponding to this model generates nonlinear terms in the form of: (i) $\sum_{i=x,y}\lambda_i^{(1)}(\partial_i h)^2$ and (ii) $\sum_{i,j=x,y}\lambda_{ij}^{(2)}\partial_i^2(\partial_j h)^2$, where the magnitude and mathematically correct sign of the coefficients $\lambda_i^{(1)}$ and $\lambda_{ij}^{(2)}$ are determined by the additional field, i.e., the amount of redeposited material. The authors [26] showed that the pattern coarsening occurs due to the presence of the coefficient $\lambda_{ij}^{(2)}$ and the relative strength of $\lambda_{ij}^{(2)}$ over $\lambda_i^{(1)}$ determines the coarsening time scale and the exponent value.

2.2.2 *Grazing Incidence Sputtering: The Reflection and Shadowing*

The Bradley–Harper mechanism is based on erosion-induced instability; so it is applicable for incidence angles where sputtering is most effective, i.e., up to the angle $\vartheta_{\max}$, corresponding to the maximum sputtering yield in the $Y(\vartheta) - \vartheta$ curve (Fig. 2.1). At $\vartheta > \vartheta_{\max}$, the incident ions will mostly be reflected with a limited ability to sputter the surface. The trajectories of reflected ions can well be approximated by the concept of "planar channeling" in a (screened) effective potential of surface atoms [27–29]. By analogy with channeling, one can define a critical angle ϑ_R for the specular reflection as [29]

$$\vartheta_R = 90^\circ - 178.2\left[\frac{NZ_1Z_2}{Z_1^{2/3}+Z_2^{2/3}E}\right]^{1/3}, \qquad (2.13)$$

where Z_1, Z_2 are the atomic numbers of the incident ion and the target atom, respectively, ϑ_R is in degrees, E is the ion energy in eV, and N is the target atoms/(Å)^3. At low keV energies $\vartheta_R \sim 80°$, for which the atoms may be sputtered by a few gentle collisions at the surface as the transverse component of the beam energy is insufficient to penetrate inside the lattice. At this condition, the Bradley–Harper mechanism fails to operate in generating the instability because of the incomplete development of the collision cascade. However, if the virgin surface is rough containing spikes or asperities, violent collisions are still possible as the local incidence angles on the column walls can be relatively small. The sharp height modulations on the surface can induce gradient-dependent erosion more dominantly than the curvature-dependent erosion. At the same time, there is also the effect of ion beam shadowing where the taller protrusions may cause hindrance of the incoming ion flux in reaching the smaller protrusions and valleys. The degree of shadowing depends on the aspect ratio (height/width) and the angle of ion incidence. It was shown by Carter [30] that, for a sinusoidal surface, the shadowing effect will be operative when the following inequality is satisfied.

$$\frac{h}{d} \gtrsim \frac{1}{2\pi} \tan(\pi/2 - \vartheta), \tag{2.14}$$

where h is the vertical height and d is the mean spacing of the protrusion peaks ($\approx$ feature width assuming the protrusions are closely spaced). It is thus apparent that for protrusions of a given aspect ratio (h/d), the ion beam shadowing effect will be most prominent when the ion incidence angle is approaching to the grazing angles. Because of the shadowing and local-angle-dependent enhanced sputtering, the protrusions of higher aspect ratio are selectively eroded away and the surface tends to become smoother. As discussed in Section 2.2, such gradient-dependent sputtering mechanism is not an efficient process to make the surface atomically flat. Therefore, at the later stages of bombardment, the residual small surface irregularities can again initiate instability due to local curvature-dependent erosion, which ultimately leads to the formation of ripple structure.

Recently, Toma *et al.* [31] suggested that shadowing instability may also select the initial pattern wavelength that could be larger than the characteristics Bradley–Harper wavelength l_c corresponding to the flat surface. This is not unreasonable since the Bradley–Harper theory allows the activation of a band of long-wavelength modes $\geq l_c$, for which R_q is positive. Since the l_c wave mode has the highest growth rate, it gradually develops with the increasing bombardment time, thereby shifting the peak of the initial wavelength distribution toward the lower value.

2.3 Sample Preparation and Analysis

Commercially available mirror-polished Si (100) wafers, after appropriate degreasing and cleaning, was chemically etched in a solution of 5 mol/L HNO_3 and 12 mol/L HF at room temperature in order to prepare rough surface morphology of different degree by varying the etching time. A number of etched samples with different *rms* roughness, *w* (2–35 nm) were prepared. The *rms* roughness of the as-received Si (100) samples was ~0.1 nm. The sputtering of the samples was carried out in a low energy ion beam set-up [32] under 16.7 keV O_2^+ bombardment at different incidence angles, ϑ ranging from 0–85° and different fluences. The O_2^+ ions were treated as O^+ ions of twice the measured ion current and half the energy of the actual O_2^+ species. The projectile flux was typically around 10^{13} O atoms cm^{-2} s^{-1} The topography of the samples was examined by an MMSPM NanoScope IV (Veeco, USA) in contact mode at ambient conditions. Figures 2.4a and 2.8a show, respectively, the virgin topography of the as-received sample and a chemically roughened sample. The latter one is characterized by grain-like structures with different sizes. The surface roughness *w* is defined as the root mean square deviation of the local height with respect to its mean value. For a rippled surface, the roughness value is also a measure of the ripple amplitude. The wavelength of the ripples is determined either from autocorrelation function or from fast Fourier transformations (FFTs) of the atomic force microscopic (AFM) images.

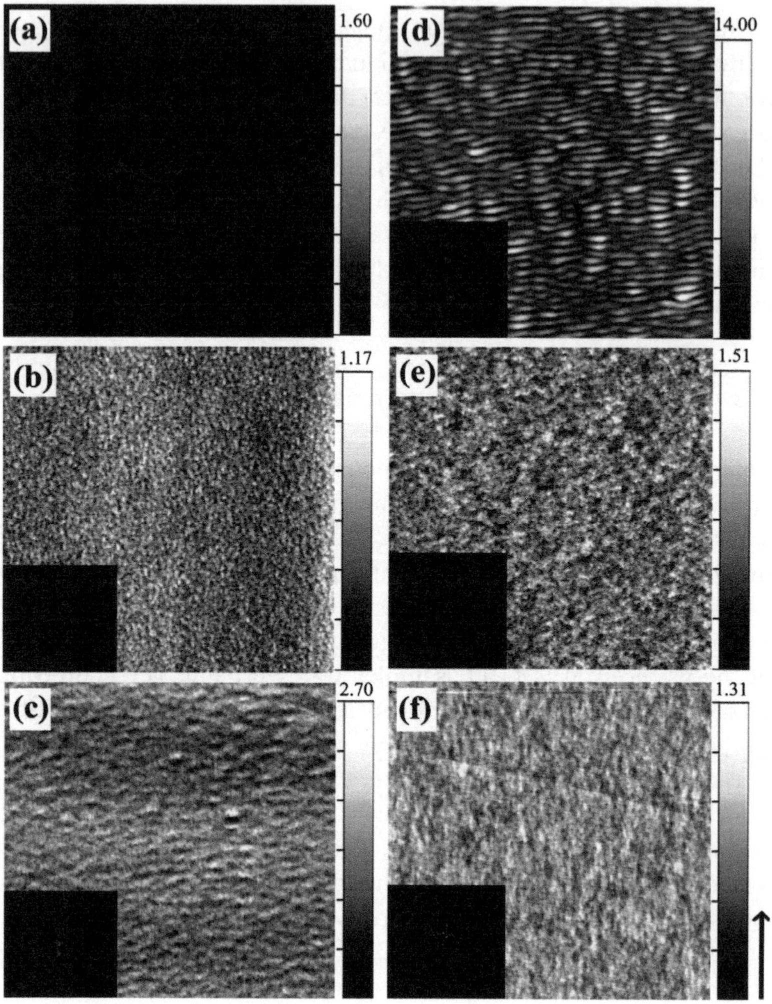

Figure 2.4 AFM images (8 μm × 8 μm) showing the evolution of topographic pattern as a function of angle of ion incidence for the flat Si (100) surface (a), sputtered at 30° (b), 45° (c), 60° (d), 64° (e), and 80° (f). The bombarding fluence is 5×10^{17} O atoms cm^{-2}. The insets are 2D-FFTs. The arrow indicates the beam direction. The vertical scale is in nm. See also Color Insert.

2.4 Results and Discussion

2.4.1 *Pattern Formation as a Function of Angle of Ion Incidence*

The evolution of surface topography of the as-received flat Si under bombardment of 16.7 keV O_2^+ ion as a function of angle of ion incidence is shown in Fig. 2.4. In contrast to the predictions of Bradly–Harper theory, only parallel mode ripples is developed in a limited range of incidence angles $\vartheta \sim 45$–60°. Most importantly, no clear ripple pattern is formed at grazing angles of incidence. At near normal and grazing incidence angles the surface remains more or less flat as the virgin surface. This is quantitatively illustrated in Fig. 2.5, where the *rms* surface roughness *w* is plotted as a function of ϑ. The roughening is found to be maximum around 60° where the ion energy deposition is closer to the surface without significant reflection of the projectile ions.

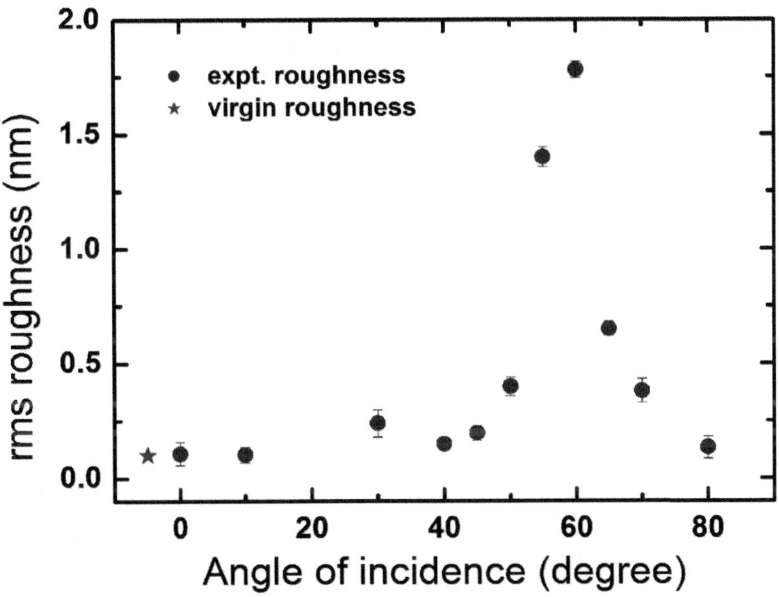

Figure 2.5 Angular variations of *rms* roughness for a flat Si surface bombarded by 16.7 keV O_2^+ ion.

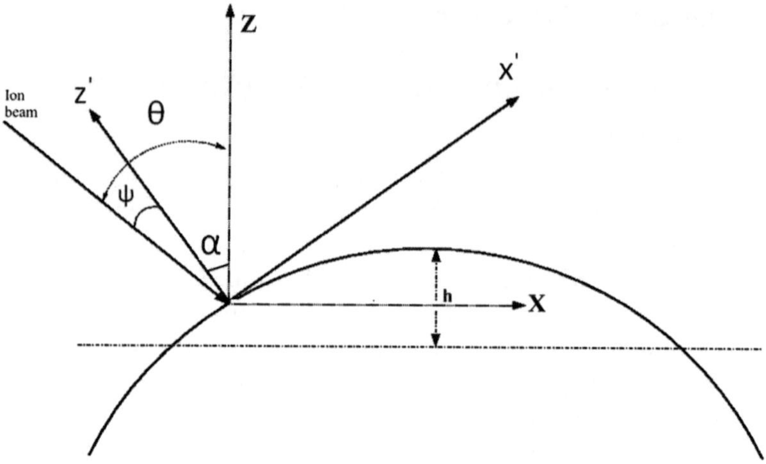

Figure 2.6 Sketch of the ion-induced smoothing mechanism.

The observation of little roughening or smoothing at near normal and grazing incidence leads to the conclusion that there exists an additional smoothing process that is different from other relaxation processes discussed earlier. This extra smoothing process was identified as the ballistic smoothing, which involves ion-impact-induced lateral mass redistribution [33]. As shown schematically in Fig. 2.6, the ions hitting the surface with local slope α, the recoil atoms preferentially move parallel to the ion beam direction. The anisotropic mass flow causes the average surface height $h(x, t)$ to decrease. The x-component of the drifted atoms can be expressed as [33] J cos $(\vartheta - \alpha)$ $f(E)$ δ_x sin $(\vartheta - \alpha)$, where $f(E)$ is the mean number of recoils generated by the impinging ion and δ_x is the mean recoil displacement along the x-direction. At small slope, the gradient of this flux generates a smoothing term $J/Nf(E)$ $\delta_x \cos(2\vartheta)\, \partial^2 h/\partial x^2$. Similarly, the y-component of the flux gives rise to the smoothing term [34–36] as $J/N\ f(E)\ \delta_y \cos^2(\vartheta)\, \partial^2 h/\partial y^2$, where δ_y is the mean recoil displacement along the y-direction. The values of δ_x and δ_y can be extracted from the SRIM simulated full-damaged cascade as prescribed by Zhou *et al.* [36]. This curvature-dependent ballistic smoothing can weaken or even cancel the Bradley–Harper roughening. The results are shown in

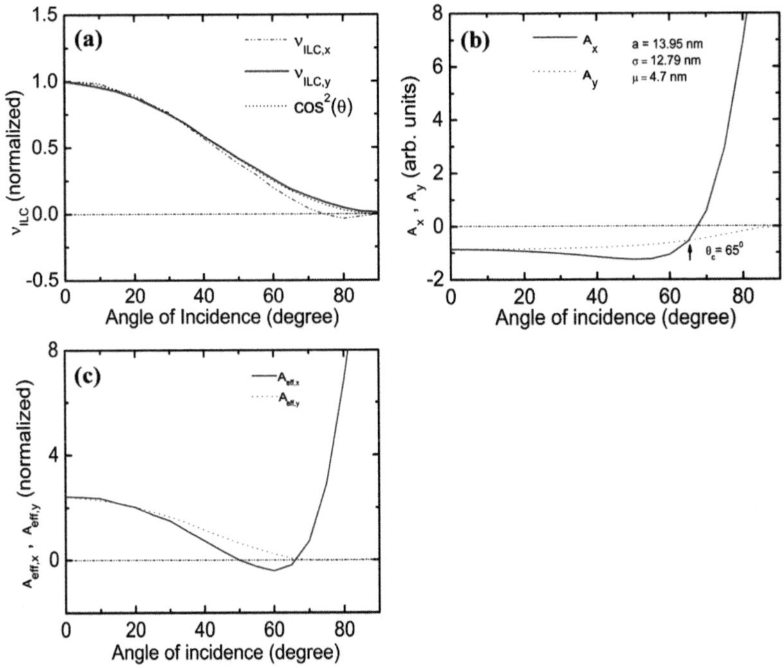

Figure 2.7 Angular variations of (a) ion-induced drift atom currents, (b) Bradley–Harper surface tension coefficients without drift currents, and (c) Bradley–Harper surface tension coefficients with drift currents.

Fig. 2.7, where the Bradley–Harper surface tension coefficients $A_{x,y}$ and the effective surface tension coefficients $A_{\text{eff},x,y}$, i.e., adding with the ballistic contributions with $A_{x,y}$ are plotted as a function of angle of ion incidence. For the present experimental conditions, $A_{\text{eff},x}$ is positive up to an angle $\approx 49°$, but becomes negative around the angular range 49–65°. After that it diverges again toward the positive (stabilized) region. On the other hand, $A_{\text{eff},y}$ is elevated to the positive side at all angles except the near grazing angles where it has slight negative values. All these indicate that the ripples should develop in a narrow angle window as observed experimentally.

The situation is dramatically changed for the pre-roughened surface, where both parallel and perpendicular mode ripples develop over a wide angular range under similar bombardment conditions (Fig. 2.8). In addition, one can identify the threshold

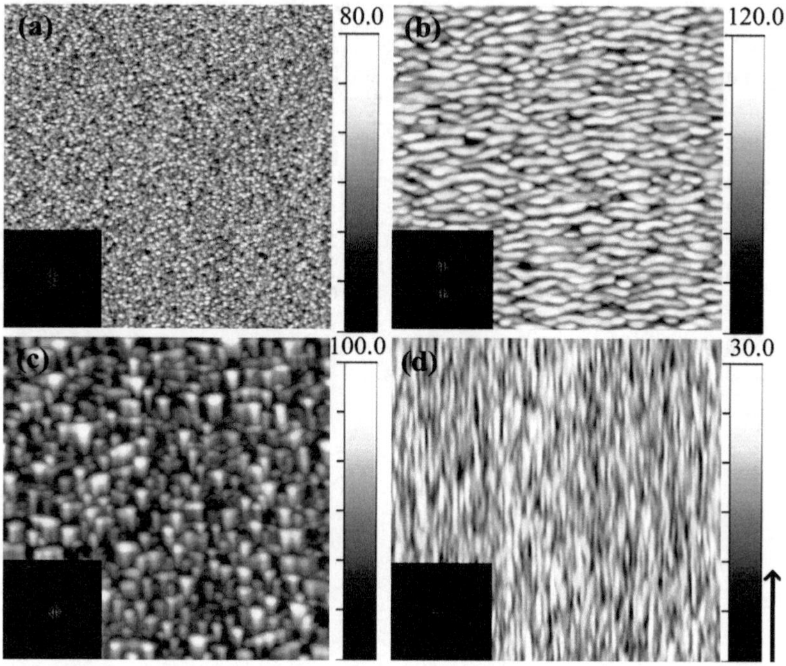

Figure 2.8 (a) A typical chemically roughened Si surface of *rms* roughness 9 nm. (b), (c) and (d) are respectively, the sputtered surface at 60°, 65°, and 80°. The scan area of each image is 8 × 8 μm². The arrow indicates the beam direction. The vertical scale is in nm. See also Color Insert.

angle $\vartheta_c = 64°$, above which the ripple pattern rotates by 90°. At this angle, blunt hill-like surface structures form indicating the suppression of anisotropic sputtering (cf., Fig. 2.8c). The FFTs shown in the inset of the respective AFM images are in conformity with the ordering and the orientations of the pattern with respect to the beam direction at different incident angles. Also for the pre-roughened surface, the angular dependence of the *rms* roughness of the developed topography (Fig. 2.9) is found to be qualitatively similar to that obtained for the flat surface.

In passing, we should mention that if ripples are produced at 60° on flat Si surface (where the ripple wave vector is in the *x*-direction) and subsequently bombarded at grazing angle (80°), the original ripple structure is found to be destroyed and a new ripple pattern

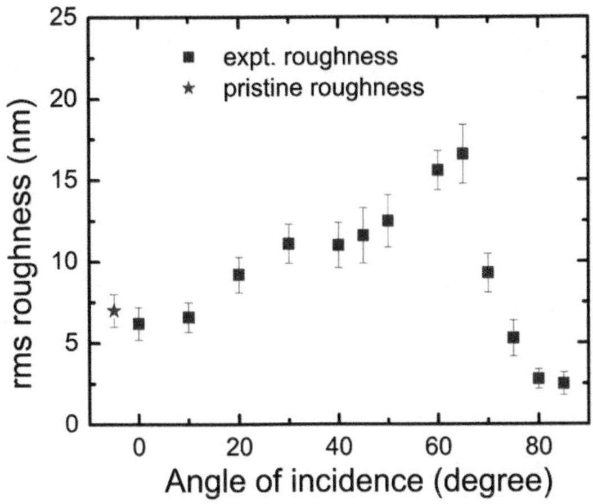

Figure 2.9 Angular dependence of the *rms* roughness for topographic evolution in chemically etched Si surface due to bombardment of 16.7 keV O_2^+ ion.

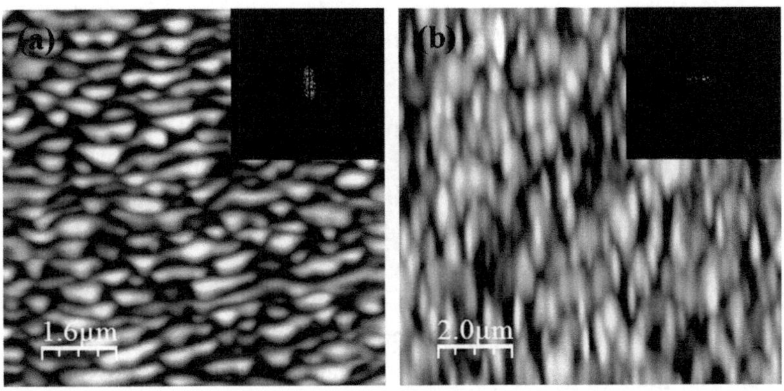

Figure 2.10 Formation of *y*-direction ripples on a flat Si surface by two steps sputtering in sequence. (a) In the first step *x*-direction ripple is formed by bombarding at 60°. (b) In the second step, the same surface is bombarded at 80° keeping the azimuthal angle fixed.

is developed in the *y*-direction (Fig. 2.10). This indicates that the formation of ripples does not depend critically on the details of the initial morphological structures.

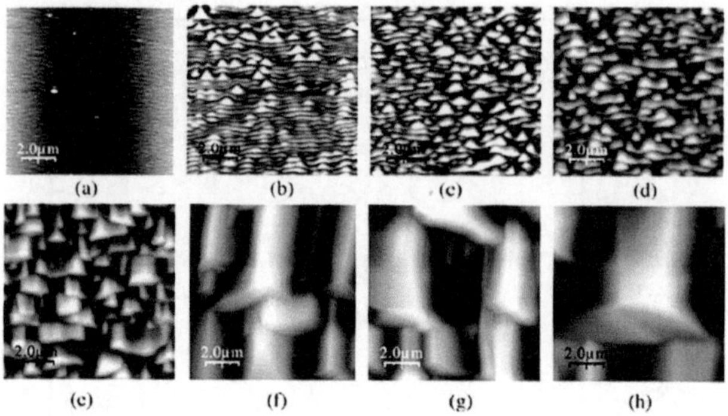

Figure 2.11 AFM images corresponding to flat Si showing the evolution of the surface topography with increasing fluence for 16.7 keV O_2^+bombardment at $\vartheta = 60°$; (a) 2×10^{17} atoms cm^{-2}, (b) 3×10^{17} atoms cm^{-2}, (c) 5×10^{17} atoms cm^{-2}, (d) 8.8×10^{17} atoms cm^{-2}, (e) 1.4×10^{18} atoms cm^{-2}, (f) 4×10^{18} atoms cm^{-2}, (g) 7.5×10^{18} atoms cm^{-2}, and (h) 1.1×10^{19} atoms cm^{-2}; the ion beam direction is from the bottom to the top.

2.4.2 *Evolution of* x-*Direction Ripples on Flat Si with Bombarding Fluence*

Figure 2.11 shows the evolution of the topography for flat Si at bombarding angle 60° as a function of the ion fluence [37]. The ripple topography appears at lower fluences ($2 \times 10^{17} - 8 \times 10^{17}$ atoms cm^2), which is periodic in the direction parallel to the ion beam. At larger fluences, the ripples disintegrate and isolated faceted structures develop. The size of the facets increases with increasing ion fluence. As shown in Fig. 2.12, the ripple amplitude ($\propto w$) grows fast exponentially in the early stages followed by a slower rate of growth at larger fluences. The dashed line represents the least square fit to the data in the linear regime by an exponential function $w \sim \exp(\phi/\phi_o)$ with $\phi_o \approx 4.7 \times 10^{17}$ atoms cm^{-2} from which the growth rate of the ripples is found to be about 2.3×10^{-4} s^{-1}. The solid line represents a power law fitting to the data in the nonlinear regime: $w \sim \phi^\beta$, where $\beta = 0.38 \pm 0.04$ is the growth exponent. As shown theoretically by Cuerno *et al.* [21], a behavior similar to $w = \phi^\beta$ can be a consequence of various nonlinear terms which are neglected in the Bradley–Harper model. In the early stage

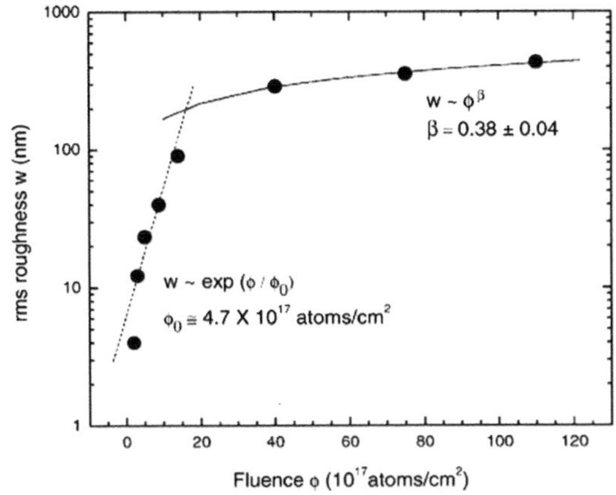

Figure 2.12 Showing the *rms* roughness w as a function of ion fluence ϕ for 16.7 keV bombardment of flat Si at $\vartheta = 60^{\circ}$. The dashed line represents the fitting by an exponential function $w \sim \exp(\phi/\phi_o)$ and the solid line by a power law $w \sim \phi^{\beta}$.

of sputtering, the nonlinear term is not effective; the regular ripples can be observed and the amplitude increases exponentially with fluence. At longer time, a slope asymmetry in the ripple geometry develops, which activates the nonlinear effect. Therefore, in the late stage of the growth, the structure of ripples is not preserved and the surface roughness tends to increase at a slower rate. We have also measured the roughness exponent α by plotting the *rms* roughness w of the sputtered surface as a function of the scan length L for different bombarding fluences ϕ (Fig. 2.13). The result shows that the local width increases as $w \sim L^{\alpha}$ before it saturates. The value of α is found to be fluence invariant with an average value of 1.33 ± 0.08. The scaling exponents α and β are the measure of the extent of nonlinearity developed in the system.

2.4.3 *Fluence Dependence of the x-Direction Ripples Formed on Rough Si Surface*

Figure 2.14 shows the evolution of the topography for the chemically etched surface at bombarding angle 60° as a function of the ion fluence. The most notable thing is that the ripples form at much

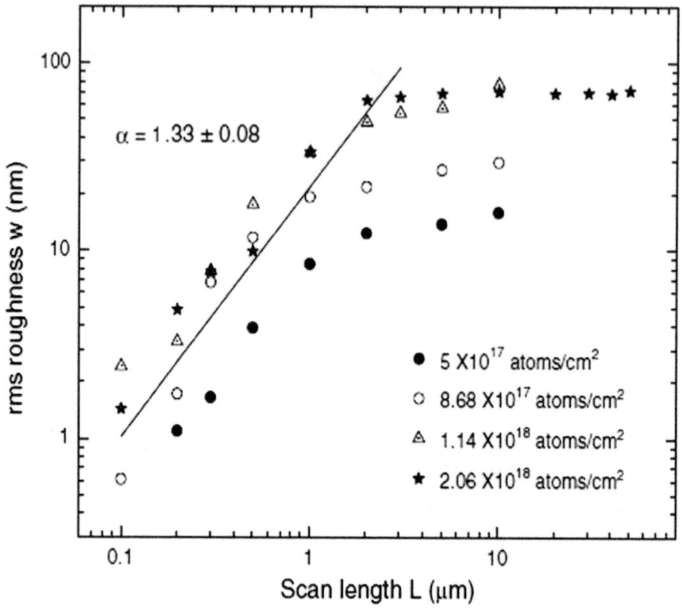

Figure 2.13 Showing the *rms* roughness w versus scan length L at different bombarding fluences for 16.7 keV bombardment of flat Si at $\vartheta = 60°$. The solid line represents the best power law fit $w \sim L^{\alpha}$ with $\alpha = 1.33 \pm 0.08$.

lower fluence (3×10^{16} O atoms cm^{-2}) than that for the flat surface (2×10^{17} O atoms cm^{-2}). Also due to the shadowing instability, the initial wavelength appears to be slightly larger (Fig. 2.15a). The coarsening of the ripple wavelength at the later stages of bombardment is due to the nonlinear effects [26]. The temporal evolution of the ripple amplitude as depicted in Fig. 2.15b is, however, similar as that obtained for the flat surface (cf., Fig. 2.12).

2.4.4 *Fluence Dependence of the y-Direction Ripples Formed on Rough Si Surface*

As we have seen that the y-direction ripples are difficult to form on a flat surface, only a shallow anisotropic structure is found to develop at glancing angle ion sputtering ($\vartheta = 80°$). However, it is already shown that well-defined y-direction ripples can be induced

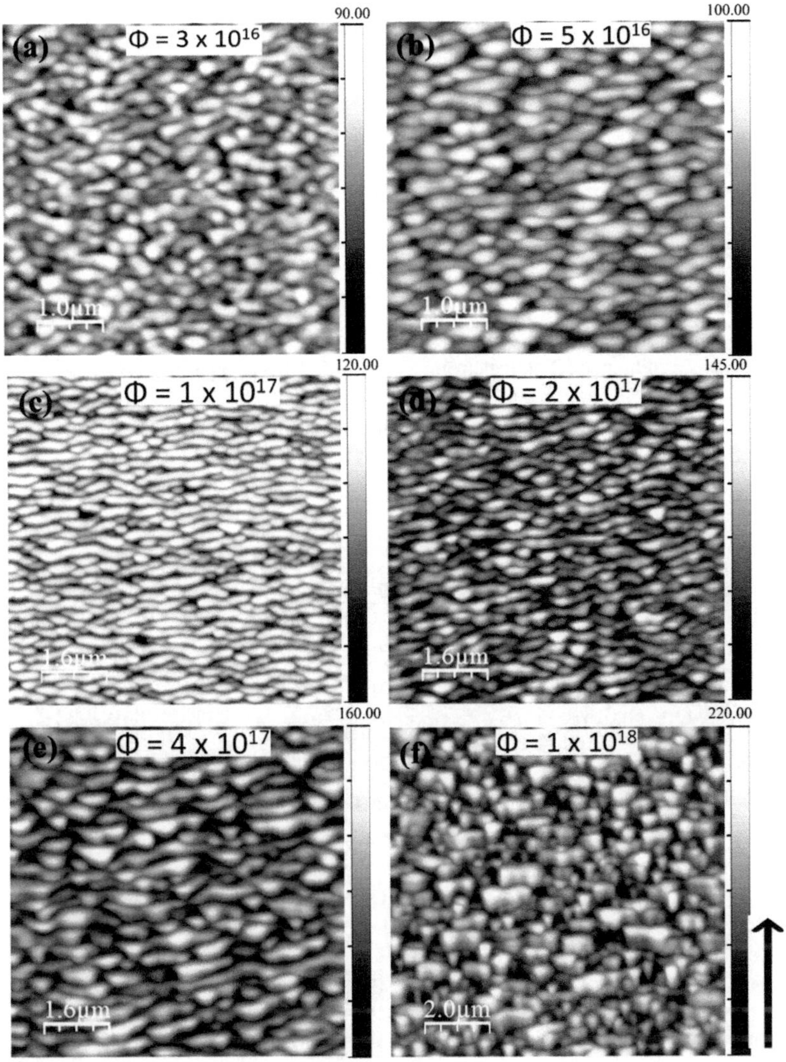

Figure 2.14 The evolution of the topography for etched Si surface (*rms* roughness ~8 nm) bombarded at 60° as a function of the ion fluence. The arrow indicates the beam direction. The vertical scale is in nm.

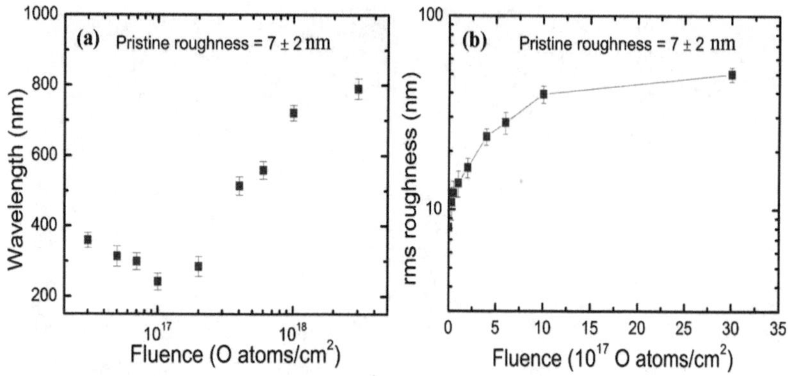

Figure 2.15 Variation of wavelength (a) and amplitude (b) with fluence for the etched Si surface.

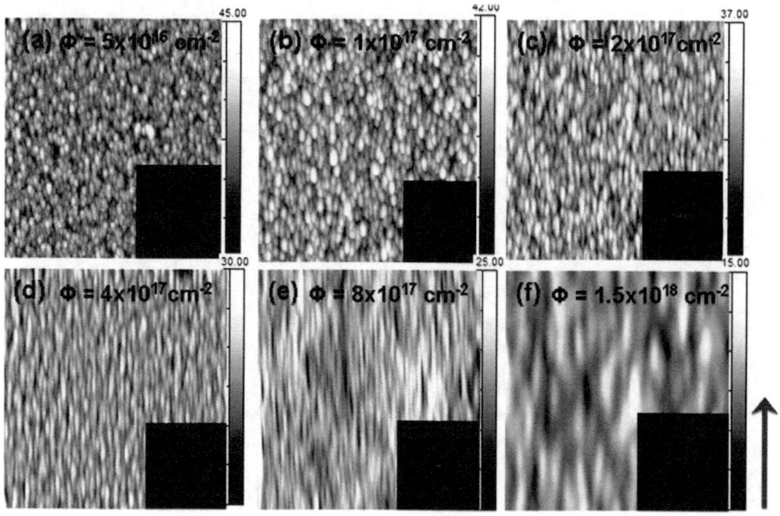

Figure 2.16 (a)–(f) Evolution of the topographic pattern of the etched Si sample (*rms* roughness of 9 nm) sputtered at $\vartheta = 80^\circ$ by 16.7 keV O_2^+ ions. The scan area of each image is $10 \times 10\ \mu m^2$. The arrow indicates the beam direction. The vertical scale is in nm. See also Color Insert.

if the Si surface is pre-roughened. Figure 2.16 shows the evolution of ripple topography on a roughened Si surface as a function of fluence ranging from 5×10^{16} to 1.5×10^{18} O atoms cm^{-2}. At low fluence (Fig. 2.16a, $\phi = 5\times 10^{16}$ O atoms cm^{-2}), the surface exhibits tiny beads which are slightly elongated and lined up along the beam direction; at increasing fluence, these structures overlap and finally a wavy pattern appears. The functional dependence of the lateral periodicity (wavelength) and the vertical height fluctuations (amplitude) of the developed morphology on the fluence are summarized in Fig. 17a,b. Initially, the wavelength appears to be higher due to the effect of shadowing instability and subsequently it decreases with fluence. After reaching a certain minimum value, the wavelength coarsens with the fluence. One can also note that the shadowing instability on the wavelength selection is stronger than for the 60° sputtering case. The amplitude of the pattern reduces with the fluence consistent with the expectations from ion beam polishing effect at grazing incidence sputtering. In another experiment, the development of ripple structure as a function of pristine surface roughness was investigated for the incident angle held at 80° and fluence 4×10^{17} O atoms cm^{-2}, where the ripple wavelength is found to increase with the roughness of the samples (Fig. 2.17c). There is also a significant influence of the degree of roughness on the dynamics of the pattern evolution consistent to our earlier observations for the x-direction ripples [24]. On the other hand, the critical angle ϑ_c is found not to be sensitive to the initial roughness.

The observation of the decrease in ripple amplitude with sputter time is quite interesting, because it contradicts the Bradley–Harper prediction. However, the results are not unlikely because the vertical growth of the topography is suppressed by the ballistic smoothing which is of comparable strength as of the sputter roughening at glancing angles (cf., Fig. 2.7). The Initial roughness reduction is, however, due to strong shadowing effect and involves gradient-dependent sputtering. Once the ripples are formed, the amplification rate as given by Bradley and Harper is getting dampened by the ballistic diffusion term Cq^2 [33] so that the effective growth rate is given by

$$R_{q_y} = -Aq_y^2 - Bq_y^4 - C\,q_y^2. \tag{2.15}$$

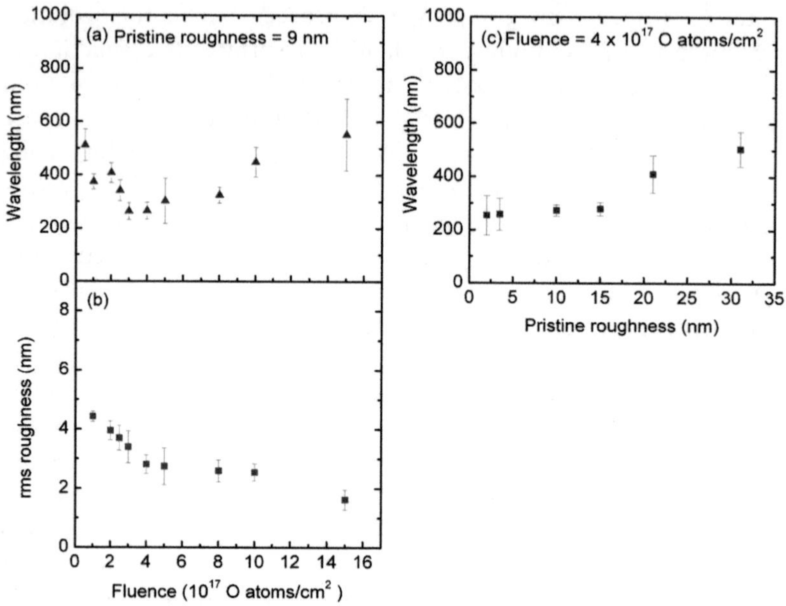

Figure 2.17 (a)–(b) show, respectively, the plots of ripple wavelength and *rms* roughness vs projectile fluence and (c), the plot of the ripple wavelength as a function of pristine surface roughness of the samples. The samples were sputtered at $\vartheta = 80^\circ$ by 16.7 keV O_2^+ ions.

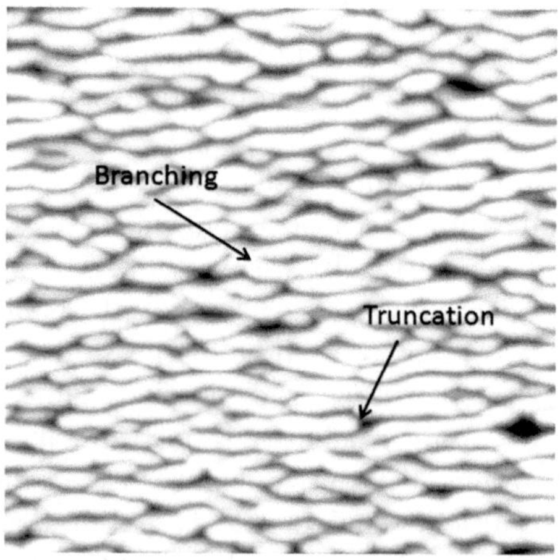

Figure 2.18 AFM image of a rippled surface showing various types of topological defects.

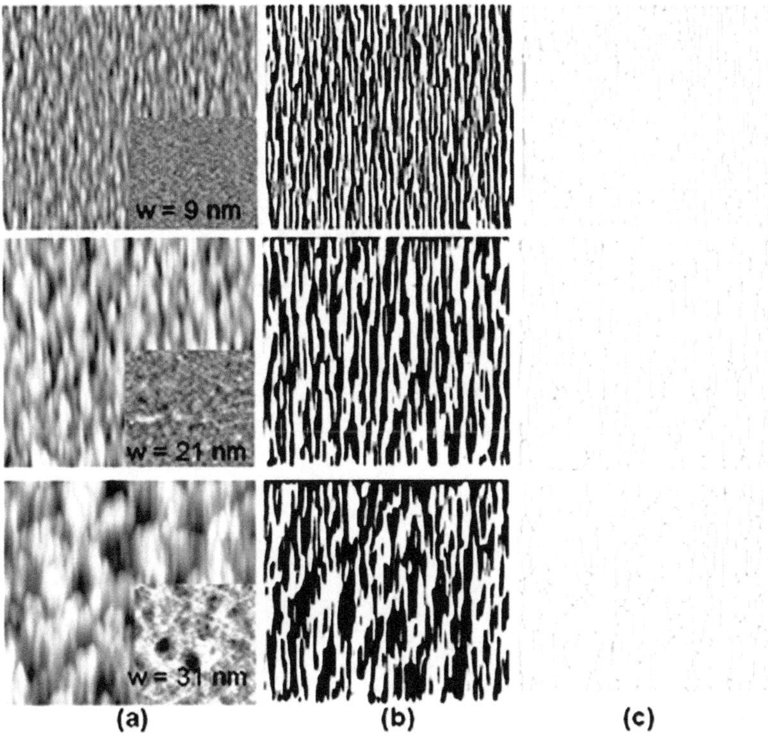

Figure 2.19 10 × 10 μm^2 etched Si samples bombarded at $\vartheta = 80°$ and fluence 4 × 10^{17} O atoms cm^{-2} by 16.7 keV O_2^+ ions. (a) AFM topographies of the rippled surface; the topography of the pristine surface with *rms* roughness is shown in the inset; (b) thresholded images; and (c) the thinned versions. See also Color Insert.

This means that the amplitude of the ripples is unable to increase due to ballistic smoothening of comparable strength.

2.4.5 *Analysis of Topographical Defects of the Rippled Surface*

The ripple topography is generally associated with defects in the form of broken, merged or bended lines as illustrated in Fig. 2.18. Such defects are thought to be associated with the randomness of the initial surface morphology and also can arise due to the stochastic

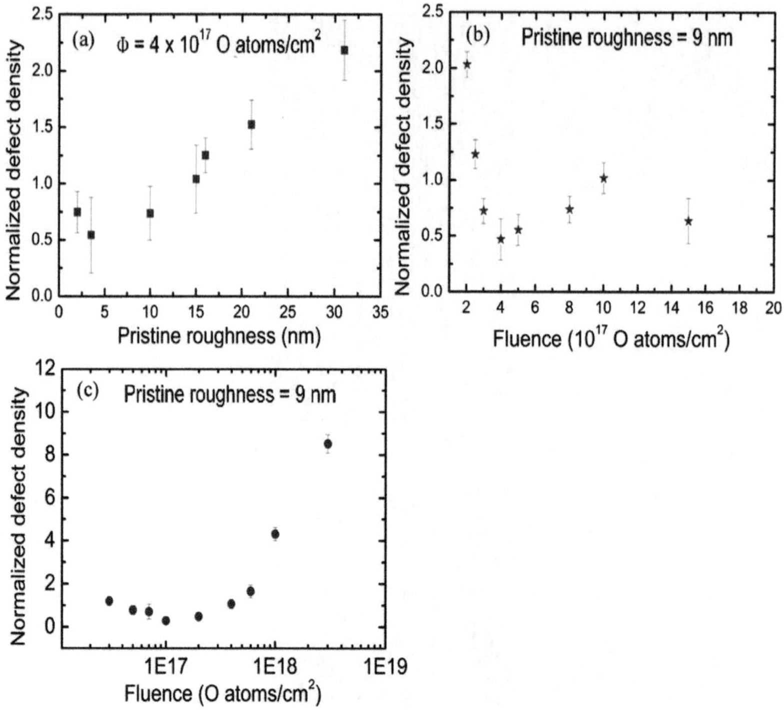

Figure 2.20 (a) and (b) are, respectively, the plots of normalized defect density for ripple morphology vs pristine roughness and projectile fluence. The samples were sputtered at $\vartheta = 80°$ by 16.7 keV O_2^+ ions. (c) The normalized defect density for x-direction ripples as a function of fluence.

nature of the formation process. One can quantitatively estimate the topological defects from the AFM images following the procedures described in references [38–40]. Briefly, a flattened AFM topograph is first reduced to a binary image using a suitable threshold algorithm and then thinned to line traces of single pixel width. The defects are counted where a single line has a branching or truncation at its ends. In order to decouple from the wavelength change, the defect number is normalized by multiplying with the square of wavelength and dividing with the image area. A few representative threshold binary images and their thinned versions corresponding to the AFM ripple topographies on different pre-roughened samples are shown in Fig. 2.19. Figure 20a,b shows the dependence of the normalized defect density (N_D) as a function of substrate roughness and projectile fluence for the *y*-direction ripples. N_D generally increases with roughness for a particular fluence. However, for a given pristine roughness, N_D decreases sharply with fluence and the ripple topography evolves toward more ordered structure until N_D reaches a minimum level around 4×10^{17} O atoms cm^{-2}; after that N_D tends to increase but with lesser degree with fluence. For the *x*-direction ripples, the variation of N_D with fluence is similar, except at higher fluences where more defects are produced (Fig. 2.20c) because of the breaking of ripple structures. It appears from experimental results that the samples of 3–10 nm *rms* roughness when bombarded with fluence around 4×10^{17} O atoms cm^{-2} produces good quality ripples with less topological defects.

2.5 Conclusion

The present study shows that pre-roughening of the Si surface is beneficial for ripple formation particularly at grazing incidence sputtering. The threshold fluence for the development of ripples for rough surfaces can be much lower than that for the flat surface. At the early stages of bombardment, the gradient-dependent sputtering diminishes the aspect ratio of the surface asperities to the extent that triggers the curvature-dependent mechanism leading to the fine scale pattern formation. The geometrical shadowing arising due to surface roughness is found to be stronger at grazing angles as evident from the initial larger wavelength selection and also the suppression of the pattern amplitude. For low oblique angles and higher degree of roughness, the gradient-dependent erosion instability is more dominant than curvature-dependent erosion instability leading to the formation of faceted topography. On the other hand, at glancing angles, the addition of strong shadowing instability makes curvature-dependent erosion more efficient, which is reflected in the formation of ripples with better quality, ordering, and regularity compared to that obtained at lower oblique angle sputtering.

Acknowledgement

Thanks are due to Dr. P. Karmakar for his involvement in the experiment.

References

1. P. Sigmund (1969) *Phys. Rev.*, **189**, 383.
2. J. Muñoz-García, L. Vázquez, R. Cuerno, J. A. Sánchez-García, M. Castro, and R. Gago (2009), in *Lecture Notes on Nanoscale Science and Technology 5* (Z. Wang, ed.), Springer, New York.
3. W. L. Chan and E. Chason (2007) *J. Appl. Phys.*, **101**, 121301.
4. U. Valbusa, C. Boragno, and F. Buatier de Mongeot (2002) *J. Phys.: Condens. Matter*, **14**, 8153.
5. G. Carter (2001) *J. Phys. D: Appl. Phys.*, **34**, R1.
6. D. Ghose (2009) *J. Phys.: Condens. Matter*, **21**, 224007.
7. M. Navez, C. Sella, and D. Chaperot (1962) *C. R. Acad. Sci. Paris*, **254**, 240.
8. R. M. Bradley and J. M. E. Harper J (1988) *Vac. Sci. Technol. A*, **6**, 2390.
9. P. Mishra and D. Ghose (2006) *Phys. Rev. B*, **74**, 155427.
10. P. Mishra and D. Ghose (2009) *J. Appl. Phys.*, **105**, 014304.
11. M. J. Nobes, J. S. Colligon, and G. Carter (1969) *J. Mater. Sci.*, **4**, 730.
12. G. Carter, J. S. Colligon, and M. J. Nobes (1977) *Rad. Effects*, **31**, 65.
13. G. Carter, M. J. Nobes, and I. V. Katardjiev (1992) *Phil. Mag. B*, **66**, 419.
14. P. Sigmund (1973) *J. Mater. Sci.*, **8**, 1545.
15. C. Herring (1950) *J. Appl. Phys.*, **21**, 301.
16. W. W. Mullins (1957) *J. Appl. Phys.*, **28**, 333.
17. M. A. Makeev, R. Cuerno, and A.-L. Barabási (2002) *Nucl. Instrum. and Meth. B*, **197**, 185.
18. C. C. Umbach, R. L. Headrick, and K.-C. Chang (2001) *Phys. Rev. Lett.*, **87**, 246104; S. E. Orchard (1962) *Appl. Sci. Res. A*, **11**, 451.
19. P. Mishra and D. Ghose (2008) *J. Appl. Phys.*, **104**, 094305.
20. J. F. Ziegler, IBM Research, SRIM-2000.40 (PC version), Yorktown Heights, NY, 1999.
21. R. Cuerno and A.-L. Barabási (1995) *Phys. Rev. Lett.*, **74**, 4746.
22. G. Ozaydin, A. S. Özcan, Y. Wang, K. F. Ludwig, H. Zhou, R. L. Headrick, and D. P. Siddons (2005) *Appl. Phys. Lett.*, **87**, 163104.

23. F. Frost, A. Schindler, and F. Bigl (2000) *Phys. Rev. Lett.*, **85**, 4116.
24. P. Karmakar, S. A. Mollick, D. Ghose, and A. Chakrabarti (2008) *Appl. Phys. Lett.*, **93**, 103102.
25. S. Park, B. Kahng, H. Jeong, and A.-L. Barabási (1999) *Phys. Rev. Lett.*, **83**, 3486.
26. J. Muñoz-García, M. Castro, and R. Cuerno (2006) *Phys. Rev. Lett.*, **96**, 086101.
27. A. D. G. Stewart and M. W. Thompson (1969) *J. Mater. Sci.*, **4**, 56.
28. M. J. Witcomb (1974) *J. Mater. Sci.*, **9**, 1227.
29. D. Ghose (1979) *Jpn. J. Appl. Phys.*, **18**, 1847.
30. G. Carter (1999) *J. Appl. Phys.*, **85**, 455.
31. A. Toma, D. Chiappe, B. Šetina Batiè, M. Godec, M. Jenko, and F. Buatier de Mongeot (2008) *Phys. Rev. B*, **78**, 153406.
32. P. Karmakar and D. Ghose (2003) *Nucl. Instrum. and Meth. B*, **212**, 358.
33. G. Carter and V. Vishnyakov (1996) *Phys. Rev. B*, **54**, 17647.
34. M. Moseler, P. Gumbsch, C. Casiraghi, A. C. Ferrari, and J. Robertson (2005) *Science*, **309**, 1545.
35. B. Davidovitch, M. J. Aziz, and M. P. Brenner (2007) *Phys. Rev. B*, **76**, 205420.
36. H. Zhou, L. Zhou, G. Özaydin, K. F. Ludwig, Jr. and R. L. Headrick (2008) *Phys. Rev. B*, **78**, 165404.
37. P. Karmakar and D. Ghose (2005) *Nucl. Instrum. and Meth. B*, **230**, 539.
38. S. A. Mollick and D. Ghose (2009) *J. Appl. Phys.*, **106**, 044309.
39. A. Keller, S. Facsko, and W. Möller (2008) *New J. Phys.*, **10**, 063004.
40. A. Cuenat, H. B. George, K. C.-Chang, J. M. Blakely, and M. J. Aziz (2005) *Adv. Mater.*, **17**, 2845.

Chapter 3

Low-Energy Ion-Induced Pattern Formation in Si-Ge Alloy

Subhendu Sarkar
Department of Physics, Indian Institute of Technology Ropar, Nangal Road, Rupnagar, Punjab 140001, India
sarkar@iitrpr.ac.in

Patterning of surfaces using ion beams is being widely studied for the last few years. Although a lot of work has been done in this direction, yet there remains quite a number of unsolved problems pertaining to this self-organization process. Interestingly, most of the experimental studies have been done on elemental samples. There exist a few studies on compound materials, but among them studies on compound semiconductors are extremely rare. It is to be borne in mind that sputtering in case of compound targets is pretty much different from their elemental counterparts since preferential sputtering plays a major role in such situations. The present article is an attempt to review nanopatterning studies on SiGe surfaces using low-energy ion beams. It is found that pattern formation on such surfaces is greatly influenced by a number of factors like sputter yield effects, initial surface morphologies, constitution of the ion beam, and other effects. Discussion of such effects provides a

Nanofabrication by Ion-Beam Sputtering: Fundamentals and Applications
Edited by Tapobrata Som and Dinakar Kanjilal

ISBN 978-981-4303-75-0 (Hardcover), 978-981-4303-76-7 (eBook)
www.panstanford.com

thorough insight into the sputter patterning phenomenon of such compound semiconductors. This will further help in addressing the open problems that exist in this area of research.

3.1 Introduction

Methods of micro- and nanofabrication of three-dimensional (3D) structures are of increasing interest owing to their importance for the creation of compact 3D devices and their assembly into functional 3D systems. The available methods like photolithography, single-particle manipulation, self-organization, etc. generally suffer from one or more shortcomings like limitations to further size reduction, throughput considerations, limitations in control, etc. Moreover, it is important not only to produce the nanometer-scale objects, but also to have them suitably distributed on an appropriate substrate, which seems to be a major challenge. For this reason, many other fabrication methods currently being investigated rely on first creating and then utilizing some sort of template that guides and dictates the spatial distribution of the material of interest upon its deposition [1]. While all these methods have demonstrated their viability for some specific purposes, the main weakness of these bottom-up approaches is the highly specific combination of factors or components required for their occurrence, as they involve very special substrates or processing techniques that can be difficult to integrate with the technologies currently used in industry.

On the other hand, ion beam sputtering (IBS) offers a controlled and cost-effective method for the production of nanostructures at the surface. Principally, this is a self-organization mechanism where the growth kinetics is determined by etching instead of growth. Ever since their first observation [2], the production of nanostructures in the form of ripples, dots, cones, etc. have been reported by innumerable researchers [3–8] with well-defined vertical roughness, lateral periodicity, controlled step size and orientation. The most common forms of nanostructures observed are in the form of ripples exhibited due to IBS in a wide class of substrates from amorphous or amorphizable [3, 9–11] to metallic targets [12, 13]. The production of semiconductor nanostructures has attracted the interest of many research groups because of the important applications in optoelectronic and quantum devices.

3.2 Theoretical Aspects

The first successful theory of ripple formation was developed by Bradley and Harper [14] based on Sigmund's linear cascade approximation of sputtering processes [15, 16]. Their calculations showed that etching of a surface by ion sputtering is fundamentally unstable because the erosion rate depends on the surface curvature implying a protrusion eroding more slowly than a depression. The generalized sputter-induced surface evolution equation for a surface is given by [17, 18]

$$\partial_t h = \nu \nabla^2 h - D \nabla^4 h - K \nabla^4 h + \frac{\lambda}{2} (\nabla h)^2 + \eta(x, y, t) \qquad (3.1)$$

In the above equation, ν is the roughening prefactor dependent on the angle of incidence of the ions. It is also known as the effective surface tension generated by the erosion process and usually has a negative value leading to a primary (linear) surface instability. D is the relaxation constant for ion-induced effective surface diffusion (ESD). K is the relaxation rate due to thermal surface diffusion. The nonlinear term λ gives the slope-dependent erosion rate, ∇h denoting the actual slope. This term incorporates the lateral growth or erosion and is proportional to the ratio between flux and the penetration depth. It brings forth the saturation of the surface roughness in time. η is a white noise term reflecting the random fluctuations in the growth process and is a set of uncorrelated random numbers with zero configurational average. It is to be noted that all the above terms have x and y components, which can be evaluated independently [18]. Calculations made from Eq. 3.1 clearly separates two regimes in the sputter erosion process. For short erosion times (before a characteristic time $t < t_c$), only the linear terms determine the outcome of the growth equation. This is commonly known as the linear regime where a competition between the unstable negative sputter erosion term $(-|\nu|\partial^2 h)$ trying to roughen the surface and the positive surface diffusion terms (D and K) trying to smoothen it eventually gives rise to ripples with wavelength $l_i = 2\pi\sqrt{D_{tot}/|\nu_i|}$, where i indicates the direction (x or y) along which the corresponding $|\nu_i|$ is the largest. D_{tot} is the sum of the surface diffusion terms K and D and its value depends on the independent strength of these two terms. While successful in predicting the ripple wavelength and orientation [19], this linear theory cannot explain a number of experimental features, such as the saturation of the ripple amplitude [20–22], the observation of rotated ripples [23], and the appearance of kinetic roughening [24, 25]. Consequently, nonlinear terms had to be brought into consideration, which comes into effect after the characteristic time $t_c \sim \left(D_{tot}/\nu^2\right) ln(\nu/\lambda)$ [26] and the surface morphology is determined by either destroying the ripples or generating a new rotated ripple structure. While all predictions of the linear theory can be calculated analytically, the discussion of the nonlinear effects requires a combination of analytical and numerical tools [18], even with which the understanding of nonlinear effects is far less complete than that of the linear theory. Moreover, many experimental findings remain unexplained at low temperatures both in the linear and nonlinear regimes, as well as regarding the nonlinear regime at high temperatures.

The numerical values of ν_i, D_{ij}, λ_i can be found out from the equations given by Makeev *et al.* [18]. In order to do the calculations, one needs to have the values of the energy deposition depth a, longitudinal straggling α, and lateral straggling β. For this purpose, one generally uses Bolse's approach [27, 28] for normal ion incidence. The depth profile of the damage distribution function (target displacements and replacements) is obtained from SRIM (Stopping and Range of Ions in Matter) code [29] with full cascade option and the resulting plot is fitted with the deposited energy density function F_D given by [15, 16]

$$F_D = \frac{E}{(2\pi)^{3/2}\alpha\beta^2} \exp\left(-\frac{(z-h_0-a)^2}{2\alpha^2} - \frac{x^2+y^2}{2\beta^2}\right) \tag{3.2}$$

The growth equation have already incorporated effects like ion-induced effective surface diffusion (ESD) [30], ballistic diffusion [31, 32], viscous relaxation [9, 33], and Ehrlich–Schwoebel barrier diffusion [12] over the last few years. Recent studies [34] based on redeposition models explain ripple formation in a more complete

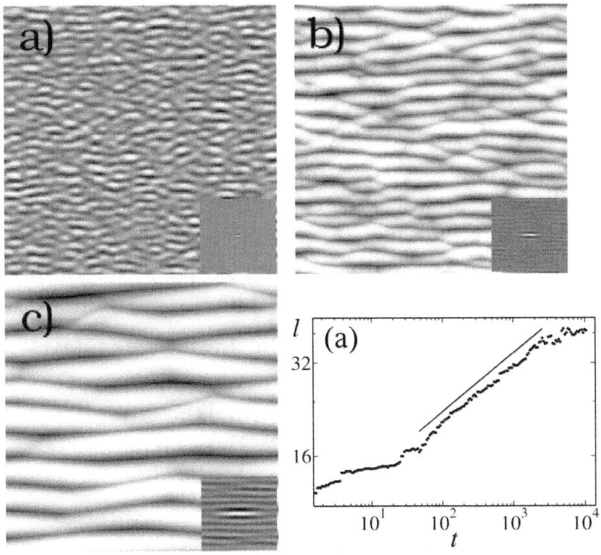

Figure 3.1 Simulation showing ripple coarsening for different times, $t_{(a)} < t_{(b)} < t_{(c)}$. (Ref. [34])

way. Cuerno and his group [34] explained the ripple coarsening phenomenon (Fig. 3.1) by a "hydrodynamic" model, which is a suitable generalization of continuum descriptions of IBS ripples formed under oblique incidence ions. It is proposed that the near surface atoms receiving enough energy and momentum to break their bonds are in principle sputtered away, although they may join the current of surface adatoms that are available to other relaxation mechanisms, such as surface diffusion, before incorporating back to the bulk solid. In addition to the time evolution of surface height, this model also considers the time evolution of thickness of the mobile surface adatoms layer, $R(\vec{r}, t)$. The latter is a function of the fraction $(1 - \phi)$ of eroded material that redeposits onto the growing surface, ϕ being the fraction of atoms that are sputtered away from the surface. The surface evolution is then described by the equation

$$\partial_t h = \sum_{i=x,y}\left[-\nu_i \partial_i^2 h + \lambda_i^{(1)} (\partial_i h)^2\right] + \sum_{i,j=x,y}\left[-K_{ij}\partial_i^2\partial_j^2 h + \lambda_{ij}^{(2)}\partial_i^2 (\partial_j h)^2\right] \tag{3.3}$$

For simplicity, the zeroth and first-order terms as well as the terms that are not responsible for pattern formation itself are omitted from the above equation. ν_i and K_{ij} have the same meaning of sputter erosion and surface diffusion as in Eq. 3.5. However, the relaxation term K_{ij} and the nonlinear terms $\lambda_i^{(1)}$ and $\lambda_{ij}^{(2)}$ depend now on the redeposition parameter ϕ, angle of ion incidence θ, thermal diffusion, and a nonzero average fraction of mobile atoms that also contribute to smoothing. Under the nonlinear regime, the ripple characteristics depend on the relative strength of the terms $\lambda_i^{(1)}$ and $\lambda_{ij}^{(2)}$. When the values of the coefficients $\lambda_{ij}^{(2)}$ increase relative to $\lambda_i^{(1)}$, coarsening stops later, and the amplitude and wavelength of the pattern also increase. The coarsening exponent $(1/z)$ will take an effective value that will be larger when the later coarsening stops and may depend on simulation parameter values.

3.3 Patterning of Alloy Surfaces

Ion beam patterning of alloy targets needs a more careful consideration. Sputtering of alloy surfaces can spontaneously lead to kinetic surface decomposition of the alloy at the peaks and the valleys of the ripples. While the relative phase between the ripple height and composition is found to depend on the ratios of the yields and diffusivities of the alloy components, the magnitude of the ion flux can be altered to control the degree of alloy decomposition. Sputtering can, therefore, be used as a tool to simultaneously achieve nanoscale patterning of both the surface topography and composition. Equations can be derived for the evolution of the height and composition modulations on sputtered alloy surfaces and study the instabilities resulting from the coupling between these modulations. A recent work by Chason [35] showed that for alloy surfaces, the differences in sputter yields and surface diffusivities of the alloy components leads to spontaneous modulations in composition that can be in or out of phase with the ripple topography. The degree of this kinetic alloy decomposition can be altered by varying the flux of the ion beam. In the high temperature and low-flux regime, the degree of decomposition scales linearly with the ion flux, but it scales inversely with the ion flux in the low-

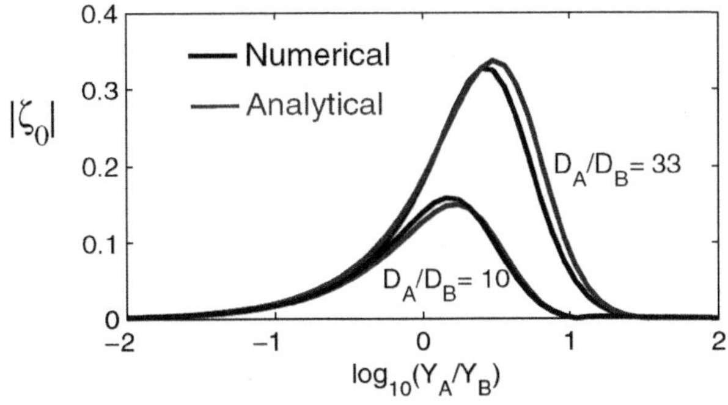

Figure 3.2 Compositional variation of an SiGe alloy for different ratios of diffusion constants of D_A (species A) and D_B (species B). (Ref. [35]). See also Color Insert.

temperature, high-flux regime. According to their study, for an AB alloy, the resulting surface would be enriched in the A species if the sputtering yield of B (Y_B) is larger compared to that of A (Y_A). For the perturbed surface, we finally have a diffusive mass current given by

$$J_i = -D_i \rho_s \nabla \zeta_i + \frac{D_i c_i \rho_s \Omega \gamma}{kT} \nabla \kappa \quad (3.4)$$

where D_i denotes the diffusion constant of species i, ρ_s is the areal density of mobile atoms, ∇ denotes the gradient on the surface, ζ is the composition, Ω is the atomic volume, γ is the surface energy, and $\kappa = \nabla^2 h$ is the curvature. Figure 3.2 shows the resulting compositional variation for such an alloy film.

A similar work by Shipman [36] showed the effect of alloy surface patterning by ion beams. Their theory shows that pattern formation occurs when the initially flat surface of a binary compound is subjected to normal incidence ion bombardment. The study further says that there is a narrow band of unstable wavelengths and that nanodot arrays with short range hexagonal order emerge spontaneously for a certain range of the parameters. Since the hexagonal ordering in the surface height is mirrored in the variations of composition at the surface, normal incidence ion bombardment could be used as a tool to simultaneously achieve nanoscale patterning of the surface topography and composition. Moreover, hysteretic switching between the flat and hexagonally ordered states can occur as the sample temperature is varied. The theory also makes an exciting prediction that normal incidence bombardment will produce surface ripples for a certain range of the parameters. One of the basic differences of this work to the earlier one [35] lies with the fact that in the present one, nonlinear terms had to be introduced in order to get the hexagonal ordering. The coupling of the surface height to the surface composition can lead to a narrow band of linearly unstable wavelengths and so to the formation of nanodot arrays with short-range hexagonal ordering.

However, composition-related differential sputtering yields of binary alloys is still controversial. It has recently been suggested that the surface binding energy of an individual atom, atomic mass, atomic number, and even atomic density are important to explain the observed nonlinear phenomenon [37, 38]. Sputter yields for ion bombardment of $Si_{1-x}Ge_x$ alloys have been computed using molecular dynamics simulations with the target material modeled using the Stillinger–Weber empirical potential [38]. Results show that Si atoms are preferentially sputtered relative to Ge atoms for all x (Fig. 3.3). Altering the parameters of the Stillinger–Weber potential showed the relative importance of atomic mass, bonding energy, atomic number, and atomic density leading to the observed differential sputter yields. The sputter preferentiality showed the

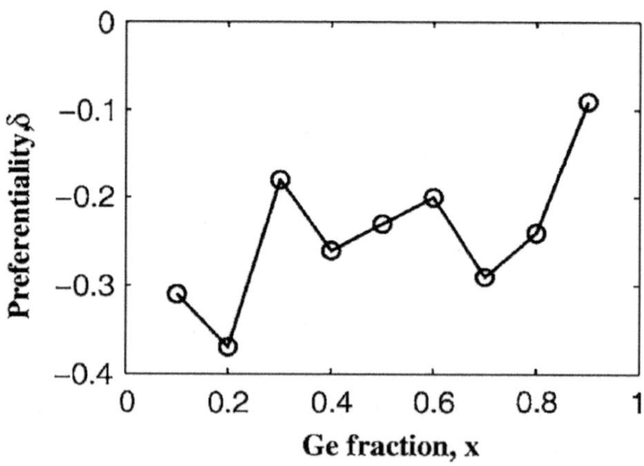

Figure 3.3 Sputter preferentiality as a function of Ge content. (Ref. [38])

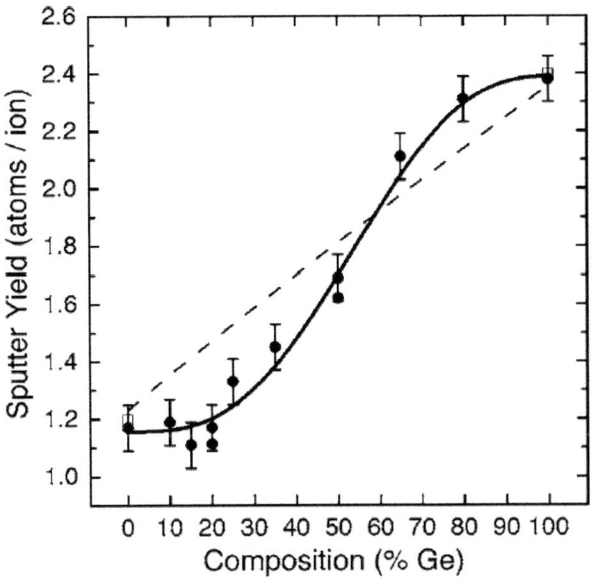

Figure 3.4 Composition dependence of the sputter yield for $Si_{1-x}Ge_x$ bombarded with 3 keV Ar^+ ions. (Ref. [37])

preference of Si sputtering over Ge. This is attributed to the higher sputter yield sensitivity of Si. Furthermore, the individual effects show atomic density to be the most significant parameter for Ge and mass or bonding energy for Si. The combined mass and atomic number for both Si and Ge shows a significant effect on sputter yields compared to either bonding energy or mass alone. The results explain an experimentally observed nonlinear dependence of total sputter yields on composition in $Si_{1-x}Ge_x$ [37]. In this study, sputtering yields were measured for unstrained $Si_{1-x}Ge_x$ ($x = 0 - 1$) alloys when bombarded with Ar^+ ions within the linear cascade regime. Nonlinear S-shape dependence of the sputter yield as a function of the alloy composition was revealed (Fig. 3.4). The dependence was analyzed within the frameworks of the cascade theory conventionally accepted to be the most systematic to date theoretical approach in sputtering. In view of a linear composition dependence predicted for the sputter yield by the cascade theory adapted for polyatomic substrates, the nonlinearity observed in the experiments were shown to be related to the alloying effect on the surface binding energies of the alloy components. Based on this analysis, an interpretation has been proposed for the experimentally observed nonlinear composition dependence of $Si_{1-x}Ge_x$ sputter yield. The yield is expressed by an equation derived from the cascade theory with additional terms of the composition parameter x. The form of the equation implies that for a polyatomic substrate, the surface binding energy of an individual atom is determined not only by its own chemical identity but also, to a considerable degree, by the identities of its neighbors.

3.4 Studies on SiGe Surfaces

One can find studies on ripple formation using compound targets in the literature. As far as semiconductor surfaces are concerned, most of the work have also been done on elemental substrates, although there are only a few studies done on compound semiconductors [11, 39–42]. However, such studies on SiGe surfaces, which are of increasing importance in the omnipresent Si technology are scarce. In what follows, the chapter cites a few examples, mostly from the authors work, that illustrates and discusses the type of pattern formation on SiGe surfaces.

3.4.1 *Sputter Yield Effects*

As discussed above, sputter yield of constituent elements might alter the ripple formation process to a considerable extent. A recent study [43] shows the morphological change at the surface of a strained $Si_{0.8}Ge_{0.2}$ epilayer by bombardment with 500 eV Ar^+ ions at varying incidence angle and tries to understand the impact on strain and ripple patterns, and the corresponding photoluminescence properties. Crystalline SiGe rippled surfaces were formed by aligning the ion beam to 69° from the surface normal. Si atoms are preferentially sputtered from the SiGe layers promoting Ge enrichment at the surface (Fig. 3.5). The ion-beam-induced compositional variations in the SiGe layers were evaluated by carrying out μ-Raman and photoluminescence measurements.

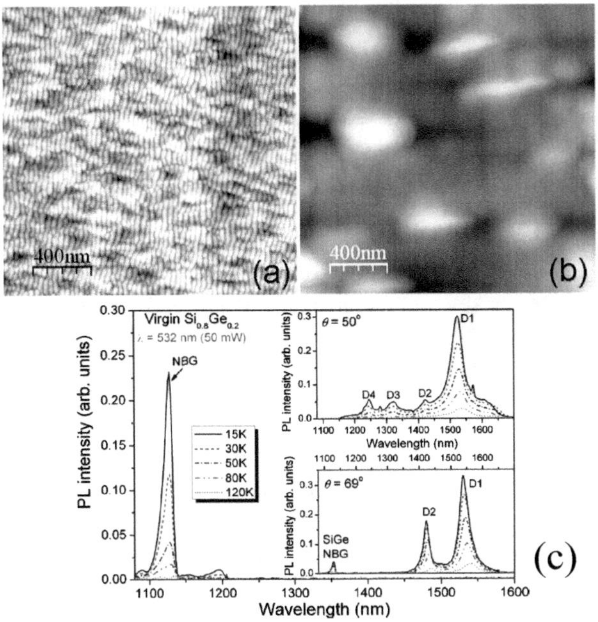

Figure 3.5 AFM images of 500 eV Ar^+ ion bombarded $Si_{0.8}Ge_{0.2}$ epilayer at an angle of (a) 69° and (b) 50° from the surface normal, showing ripple patterns and Ge islands on the surface, respectively; (c) Temperature-dependent PL spectra of the virgin SiGe/Si heterostructure (main frame), and the ion eroded samples with an incidence angle of 69° (lower inset) and 50° (upper inset). (Ref. [43]). See also Color Insert.

When sputtering reaches the SiGe/Si interface, the surplus Ge will accumulate at the surface. In such a situation, the heavy Ge atoms appear to migrate locally to form scattered Ge islands. The crystalline ripples transform to elongated Ge islands by aligning the beam from 69° to 50° from the surface normal.

Photoluminescence investigation reveals two dislocation bands D1 and D2 at ~1540 and 1422 nm, respectively, by suppressing the near-band gap emission of the $Si_{0.8}Ge_{0.2}$ layers in the rippled sample. However, the near-band gap photoluminescence disappears completely in 50° sputtered sample, instead the dislocation bands D1, D2, D3, and D4 are observed at ~1524, 1422, 1322, and 1245 nm, respectively. It is concluded that the near surface

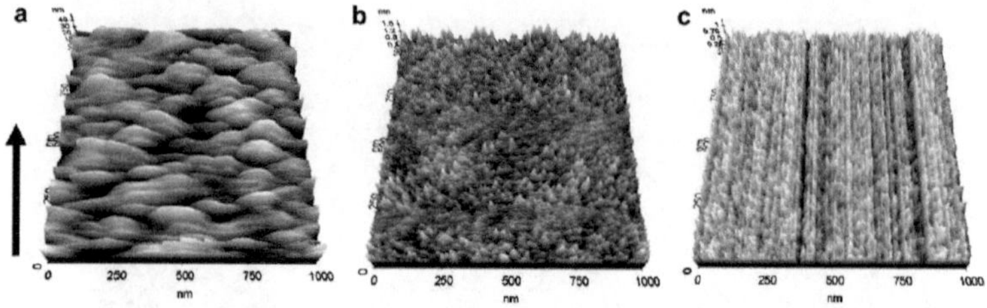

Figure 3.6 AFM images of (a) HOPG, (b) diamond, and (c) ta-C film irradiated with 5 keV Xe^+ ions to a fluence of 2×10^{17} Xe^+ cm^{-2} at an incident angle of 30° with respect to the surface normal at room temperature. An arrow indicates the ion beam direction. (Ref. [44]). See also Color Insert.

dislocations in the rippled sample play the key role for the D1 and D2 bands, whereas the D1-D4 bands in 50° sputtered sample are associated with the ion-beam-induced dislocations in the Si substrate due to the complete sputtering of the SiGe layer.

3.4.2 *Surface Morphology Effects*

The existing theories describing rippling phenomenon essentially does not take into account any contributions from the underlying substrate or the morphology of the substrate surface (having a large surface curvature), both of which decide the fate of the ripples to a large extent. A recent work on allotropes of carbon surfaces [44] indicates that rippling depends on the nature of the underlying carbon materials (Fig. 3.6). In this study, three types of carbon materials — highly oriented pyrolytic graphite (HOPG), single crystalline diamond, and tetrahedral amorphous carbon (ta-C) films — were irradiated with 5 keV Xe^+ at an incident angle of 30°. The irradiation fluence was 2×10^{17} cm^{-2} for all irradiations. No ripples were formed on diamond and ta-C surfaces, while ripples with a wavelength of ~ 100 nm occurred on HOPG. Electron energy loss spectroscopy revealed that the surfaces of all carbon materials became amorphous with similar densities upon irradiation. Works by Aziz *et al.* [45] shows the influence of patterned boundaries on semiconductor substrates, in guiding self-organized topographic ripples spontaneously appearing during uniform irradiation with

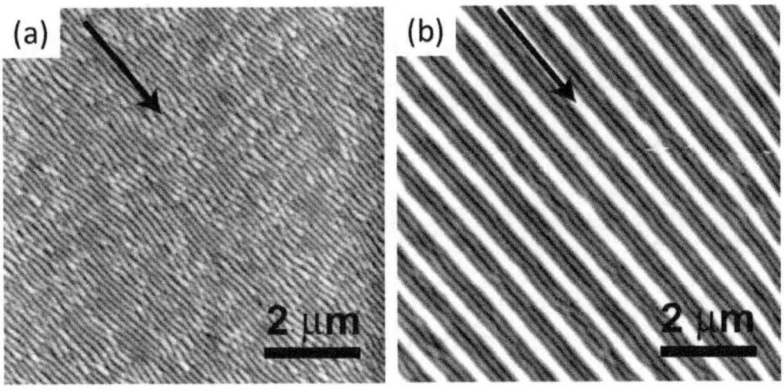

Figure 3.7 a) Self-organized ripples formed on a regular Si surface; (b) long guided ripples in the presence of a lithographically defined template with 430 nm valley on Si. (Ref. [45])

low energy Ar^+ ions (Fig. 3.7). The long-range order of the features can be greatly enhanced by this lateral-templating approach. The emerging pattern can be manipulated by changing the boundary spacing and misorientation with respect to the projected ion beam direction. A dimensionless topological defect density characterizes the degree of order of the pattern. At small boundary separation, greatest order is observed when the separation is near an integer multiple of the spontaneously arising feature size. The defect density is exceedingly low up to a critical misorientation angle, beyond which topological defects develop in proportion to the incremental misorientation. These results suggest the potential utility of lateral templating ion-irradiation-induced topographic patterns for high throughput fabrication at sub-lithographic length scales. A theoretical understanding of the observed phenomenon would, however, require the growth equation to be solved under large slope approximations.

A recent work by Sarkar *et al.* [46] has shown that nanostructures formed on ion-sputtered SiGe surfaces do depend on the initial morphology of the substrate. Investigations were done on ripple formations due to low energy bombardment at oblique incidence on various $Si_{1-x}Ge_x$ surfaces, which differed by crystallinity and also the topography of the substrate surface. Studies were done on epitaxial (strained and relaxed) and polycrystalline SiGe samples. The relaxed samples exhibited cross-hatch patterns on the surface, which

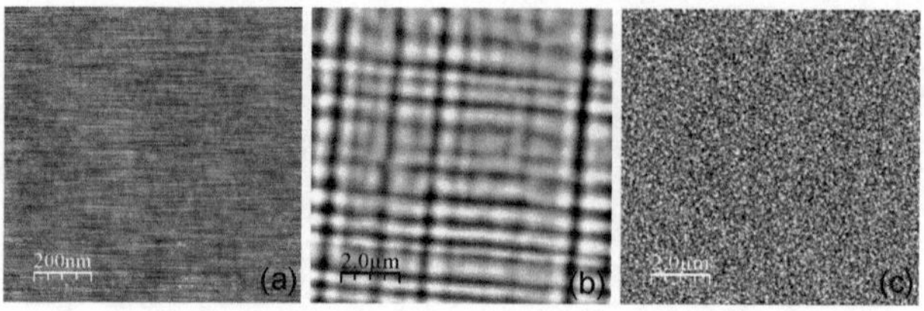

Figure 3.8 AFM images of unbombarded sample surfaces for (a) strained SiGe, (b) relaxed SiGe, and (c) poly SiGe.

generally form owing to a strain relaxation process [47, 48] in the SiGe overlayer. These are often referred to as dislocation lines (DL) in the literature. Therefore, in contrast to the work of Aziz [45], the ion beam in this case interacts with a surface having a sinusoidal boundary rather than a sharp one. Ripples formed on the surfaces after bombarding them with an oxygen beam of energy 1 keV, 18 nA current aligned at 48°. The ion fluence used was 4.3×10^{17} ions cm^{-2}.

3.4.2.1 Currugated SiGe surfaces

Figure 3.8 shows atomic force microscopy (AFM) images of the unbombarded surfaces of strained, relaxed, and poly samples. The *rms* roughnesses of these surfaces were 0.22, 9.32, and 9.42 nm, respectively. The average periodicity of the dislocation lines for the relaxed sample in the two perpendicular directions are 792 and 743 nm. The height was around 30 nm. It is to be noted that the surface of the strained wafer is almost as smooth as that of pure Si (having an *rms* value of $\sim$ 0.12 nm). On the other hand, the case is just the opposite for the relaxed and the poly sample. Although, they have a similar overall roughness, under shorter length scales the relaxed sample has a much smaller roughness, but the poly sample continues to have large roughness values due to its granular nature. It is to be noted here that comparable overall roughness values for these samples arise from two different aspects: random morphology for the poly sample and patterned one for the relaxed sample.

For corrugated surfaces subjected to ion bombardment at an angle away from the normal, one has to look for the limiting condition for shadowing not to occur, which is given by [49]

$\tan\left(\frac{\pi}{2}-\theta\right) \geq \frac{2\pi h_0}{l}$, which places an upper limit on h_0/l for any ion incidence angle θ. Here h_0 is the amplitude of the sinusoid and l is the periodicity. If this ratio tends to be exceeded, the valleys of the sinusoid will not be eroded, but the peaks will be planarized thus changing the sinusoid into a sawtooth-like wave form. Calculations revealed that for the relaxed samples under consideration, there was no shadowing effect even for the maximum fluence condition.

3.4.2.2 Ripple characteristics

Figure 3.9 shows AFM images of the sputtered craters of all the samples, including Si, after a bombardment time of 40 and 100 min shown in the left and right columns, respectively. The strained and poly SiGe layers are eroded after a bombardment time of about 50 and 80 min, respectively. For all the samples, the ripples tend to align themselves perpendicular to the ion beam direction, i.e., the ripple wavelength is parallel to the ion beam direction. It is evident from the figures that ripples evolve much earlier for SiGe samples than that for pure Si even when accounting for the slightly higher sputter yield of SiGe compared to Si [50]. Comparisons between the SiGe samples also reveal certain important features. In particular, for the relaxed case the ripples seem to be aligned with the direction of the dislocation lines. It can be seen that the dislocation lines act as a guiding boundary for the ripples leading to unbroken ripples, which are much longer when compared to those for the strained case. This is exemplified in Fig. 3.10, which shows a histogram of the lengths of the ripples formed for the different films. It is clear that in the relaxed case, extremely long ripples (up to 8 μ) can be found, whereas in the strained case, they hardly exceed 2 μ. Moreover, in the strained case, the length distribution is fairly uniform across the crater surface, whereas in the relaxed case, a shortening of the (long) ripples occurs when moving away from the dislocation line toward the valley in between two dislocation lines. Even further away from the edge, the well-formed ripples eventually break up into small islands. The scenario is altogether very different for the poly sample. Here the ripples are not as well formed as those for the single crystalline samples. It is evident from Table 3.1 that the relaxed sample finally ends up with the surface

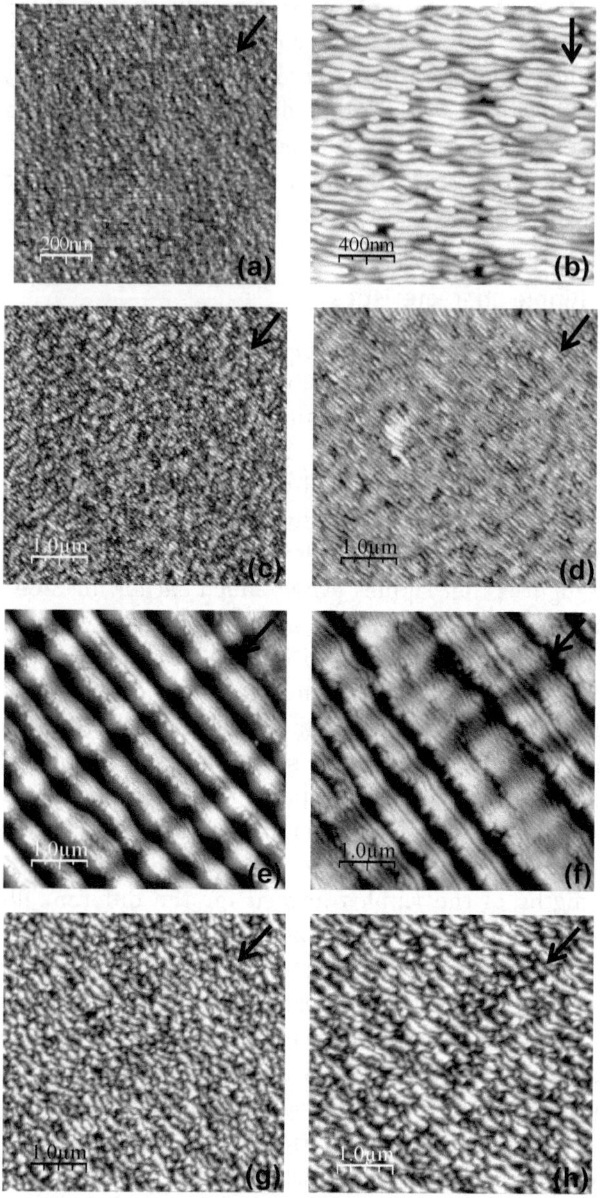

Figure 3.9 AFM images of ion-bombarded surfaces of Si (a) 40 min, (b) 100 min; strained SiGe (c) 40 min, (d) 100 min; relaxed SiGe (e) 40 min, (f) 100 min; and poly SiGe (g) 40 min, (h) 100 min. Arrows in each of the images indicate the ion beam direction. See also Color Insert.

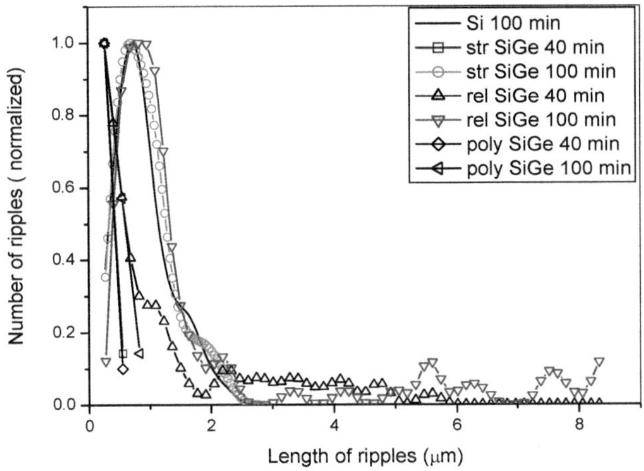

Figure 3.10 Ripple length distribution for the different samples sputtered for the times indicated.

Table 3.1 Roughness (w), ripple wavelength (l), and height (h) of all samples for 40 and 100 min. (All values are in nm)

Parameter	Time (mins)	Si	Strained SiGe	Relaxed SiGe	Poly SiGe
w	0	0.12	0.22	9.42	9.32
	40	0.18	2.83	16.18	9.78
	100	4.38	3.99†	21.88	13.5†
l	40	42	71	53	196
	100	70	88†	60	250†
h	40	–	6	6	36
	100	11	8†	9	52†

† Layer eroded.

having the maximum roughness despite the fact that the relaxed and the poly samples start from the same, initial roughness. Crystalline Si exhibits a rapid change in surface roughness from 40 to 100 min, whereas the strained one shows a more gradual increase although both of them start with almost the same roughness values. So far as the ripple wavelengths are concerned, they are greater for the strained sample than those for pure Si and the relaxed sample. However, the ripple wavelength for the poly sample is by far the largest compared to all the other cases. These differences suggest more than one factor, which could be responsible for the ongoing processes. The surface curvature (due to the dislocation lines) surely plays a major role in deciding the graded ripple pattern for the relaxed sample. Otherwise, the ripple pattern should have looked similar to the strained case. For the fate of the poly sample, the overall granular nature plays a major role. It is probably for this that we get very different values for the poly sample when compared to the other two. The other factor existing at such low energies could be crystallinity. Earlier studies on different allotropes of carbon have already hinted upon such a possibility [44]. For semiconductors bombarded at high energies (which is not our case), this effect might be very low since the ion beam completely amorphizes the topmost layer of the sample [51]. Even under such situations, the surface can dynamically recrystallize and prevent amorphization if the temperature is high enough. Surface patterning determined by effects of surface crystallinity is, of course, quite common for metallic surfaces and have been investigated quite extensively [12]. Differences also exist in the ripple heights of the samples. For the strained and relaxed samples, they are almost similar indicating that small length scale morphologies could be the deciding factor. However, the ripple heights for the poly sample are much greater when compared to any of the other cases. These differences again seem to be a result of the surface curvature and overall morphology of the pristine samples. The ripple characteristics evaluated from the AFM images (Table 3.1) for the 100 min case of strained and poly samples indicate that the underneath Si surface of these samples tend to adhere to the values obtained with the corresponding layers. Therefore, it is clear that the history of the sample surface is remembered for quite some time even after the original layer is eroded (e.g., for almost double the time for the strained case). This suggests a possible route to get ordered nanostructures of a certain material by first creating them on an overlayer for which the exact conditions are known.

3.4.2.3 Estimations from theoretical calculations

One can calculate the different quantities of Eq. (3.1) by putting in appropriate values of a and α for SiGe and Si. The β values have

Table 3.2 Theoretically estimated values of ripple characteristics for SiGe and Si for 1 keV O_2^+ at 48° incidence

Sample	ν_x nm^{2}s^{-1} ($\times 10^{-2}$)	ν_y nm^{2}s^{-1} ($\times 10^{-2}$)	λ_x nm^{2}s^{-1} ($\times 10^{-2}$)	λ_y nm^{2}s^{-1} ($\times 10^{-2}$)	D_x nm^{4}s^{-1} ($\times 10^{-2}$)	l_x nm	Growth rate s^{-1} ($\times 10^{-3}$)
SiGe	−1.34	−1.29	−0.98	−3.92	1.23	8.5	3.67
Si	−1.52	−1.06	−2.25	−0.12	1.16	7.8	4.99

to be obtained from Winterbon tables [52]. Substitution of these values predicts the ripples to be parallel to the ion beam for SiGe, thus contradicting the experiments. This discrepancy seems to be cropping up from a wrong β value, which could be due to the fact that the value obtained from ref. [52] is for an elemental target, although the experiment deals with a compound one. For Si, one gets the correct directionality of the ripples using the β value from the Winterbon tables. Results obtained following the above procedure are tabulated in Table 3.2. It is worthwhile to mention here that the early onset of ripples observed for the SiGe samples compared to Si could be due to the corresponding material parameters that govern the competition between the roughening and relaxation terms and the initial surface roughness, which is nearly impossible to quantify. Moreover, information about initial surface changes is generally not available under such situations. Recent molecular dynamics simulations by Johnson *et al.* [53] have shown that surface corrugations can evolve out of a flat surface just after about 50 ion impacts. It is seen that for both Si and SiGe, $|\nu_x| > |\nu_y|$. Thus the ripples will be aligned perpendicular to the ion beam direction as observed from the AFM images. The characteristic or crossover time (t_c) obtained for SiGe is $\sim$20 s. It is, therefore, evident that the nonlinear terms dominate the surface morphology just after a few moments of sputtering. The morphology, therefore, depends on the relative signs of the nonlinear terms λ_x and λ_y. For both SiGe and Si, $\lambda_x \cdot \lambda_y > 0$, thus signifying an eventual kinetic roughening of the surface. The scaling properties of the surface under such situations are then governed by the isotropic Kardar–Parisi–Zhang (KPZ) equation given by [54]

$$\partial_t h = \nu \nabla^2 h + \frac{\lambda}{2} (\nabla h)^2 + \eta(x, y, t) \tag{3.5}$$

According to the scaling theory [55, 56], the roughness exponent (m) and the growth exponent (n) varies as $w(l) \sim l^m$ and $w(L, t) \sim t^n$, respectively, where w is the surface width. The surface develops coarser and coarser features characterized by the length l parallel to the substrate and $l \sim t^{1/z}$, where z is defined as the dynamical exponent. For the present case, $n \simeq 0.25$ theoretically. The thermal relaxation factor K is usually very low at room temperatures. This is estimated to be $\sim 3 \times 10^{-11}$ nm^4/s, which in turn gives an estimated ripple wavelength to be about 4×10^{-4} nm. It is evident that this value is nowhere close to the experimentally observed values. Here, ion-induced ESD plays a key role in the ripple formation. The calculated value (Table 3.2) is smaller by almost a factor of 7-10 from the experimental values. The maximum deviation is, however, found for the poly SiGe sample whose ripple wavelength differs by almost a factor of 30 from the theoretical value considering a bombardment time of 100 min. The deviations are principally caused by the choice of the bombardment parameters, which in practice could be slightly different from the chosen theoretical values. Moreover, major differences in the starting surface morphology and inherent property (e.g., crystallinity) of the samples can also account for such deviations and differences between the samples. It is to be borne in mind that such differences were not taken care of in the theoretical estimation of the ripple parameters. All the theoretical models expand the morphology in powers of the surface slope, which is treated as a small parameter. Recent theoretical calculations [57] suggest that for a particular material, the resulting surface morphology can vary depending on the surface roughness owing to the difference in sputter yields.

3.4.2.4 Temporal behavior

It is interesting to study the temporal evolution of ripples for relaxed SiGe surfaces. The resulting craters of the ion-bombarded surfaces were analyzed using AFM. Figure 3.11 shows the AFM images obtained. Since, the sample surface already had dislocation lines on it, it would be interesting to see the fate of these dislocation lines with time. Figure 3.12 plots the height, periodicity, and slopes of the dislocation lines with the sputtering time. The heights of

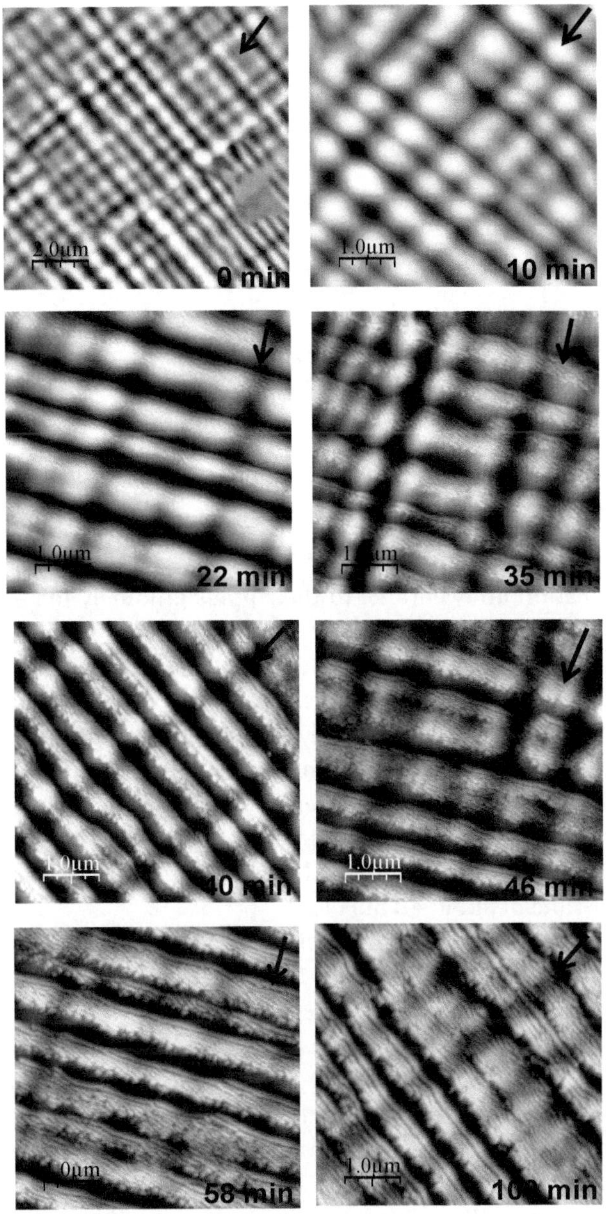

Figure 3.11 AFM images of ion-bombarded surfaces of relaxed SiGe sputtered for different times. Arrows in each of the images indicate the ion beam direction.

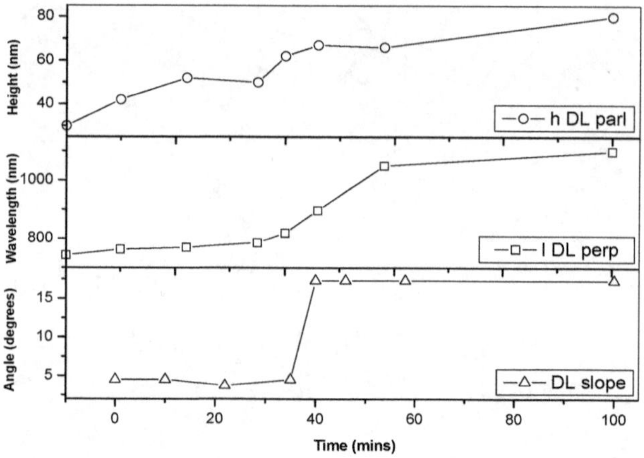

Figure 3.12 Variation of the height (upper panel), periodicity (middle panel), and slope (bottom panel) of the dislocation lines with sputtering time.

the dislocation lines parallel to the ripples gradually increase with time and finally tend to saturate. This further proves that there is no shadowing of the incident beam due to the dislocation lines. The average periodicity of the dislocation lines lying perpendicular to the ripple direction gradually increases suggesting a gradual increase of the correlation length in this direction. In other words, the dislocation lines lying perpendicular to the ripples gradually tend to die out with sputtering time. Another interesting feature of the dislocation lines is that their slopes undergo a sharp change after a sputter time of about 40 min.

As discussed in an earlier section, the lengths of the individual ripples vary to a large extent in the direction of the ion beam for the relaxed SiGe sample. This is a bit similar to what has been observed previously [45] where a template with a sharp vertical boundary served as the driving force for the orientation of the ripples. In comparison to that study, we have in the present case a sinusoidal boundary. Moreover, in contrast to the earlier case, the ion beam direction in the present case is not aligned with the parallel dislocation lines. It is important to note here that a defect-free ripple can actually evolve even in the absence of a sharp boundary although the maximum range of the lateral templating effect might

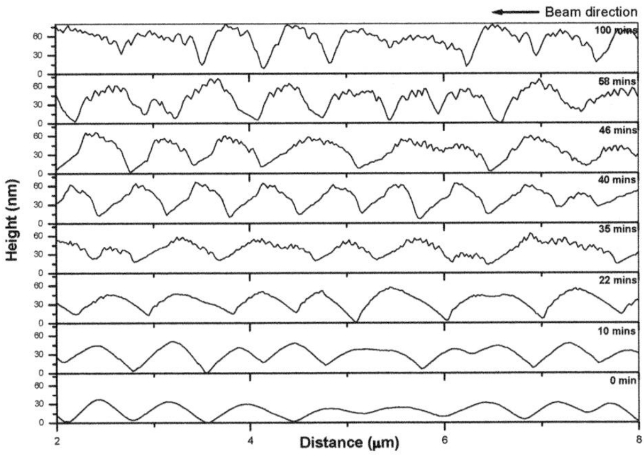

Figure 3.13 Time evolution of the ripple profiles with the sputtering time.

be lower. Currently, there is no theory that takes care of such a situation in ripple formation. In order to account for such an effect, one needs to consider the periodic features of the starting surface (dislocation lines for our case) in the growth equation and carry out calculations beyond the small slope approximation. The fact that shadowing by the dislocation lines is not a cause for ripple shortening is strengthened by the fact that calculations done for the dislocation lines at the end of sputtering (100 min) rule out any such possibility ($\frac{2\pi h_0}{l} = 0.721$).

The gradual formation and propagation of the ripples over the dislocation lines are depicted in the line profile diagrams of Fig. 3.13. The large humps are the dislocation lines that are parallel to the ripples. The small fluctuations on the top of the large humps denote the actual ripples. The ripples initially start to appear at the leading slope of a dislocation ridge. The first ripple that forms at the start of the leading slope is always found to be the longest. Subsequent ripples follow the first one. The entire group of ripples for a particular dislocation ridge is then seen to migrate to the top of a dislocation ridge. Finally, the top of the ridge flattens giving way to the ripple pattern. Ripples are observed as early as 10 min but are quite distinct from about 22 min. Estimates from AFM images reveal that defects in the ripples start appearing after about 2l distance from the edge. Eventually, the ripples become shorter finally ending

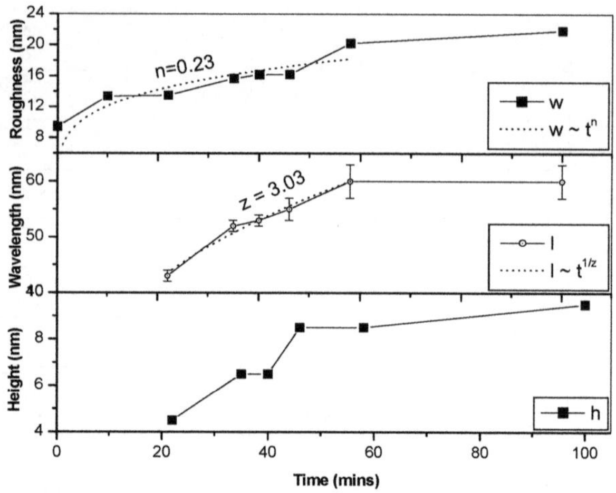

Figure 3.14 Variation of roughness (upper panel), periodicity (middle panel), and height (bottom panel) of the ripples formed with the sputtering time.

up into small islands or mounds over a distance of about 0.4 μ. The driving force behind the long ripple formation is so strong that the ripples are almost unperturbed by the dislocation lines that run perpendicular to them. It is clear that the strength of this factor falls off with the distance from the edge of the dislocation ridge. Figure 3.14 shows the variation of the roughness, periodicity, and height of the ripples formed on the relaxed SiGe surface due to sputtering. The growth rate obtained by an exponential fit to the roughness values gives 1.8×10^{-4} s^{-1}. The calculated value (Table 3.2) is greater than this by an order of magnitude. The growth exponent n is found to be 0.23, which is quite close to what is predicted theoretically. But on the other hand, the dynamical exponent $z = 3.03$. Although this value is almost double of what has been predicted theoretically, yet other studies on semiconductor surfaces have also predicted values close to this [58]. It is, therefore, evident that the nonlinear regime of ripple formation is dominating. The variation of the average height of the ripples also follows a similar nature as the above two. All these parameters tend to saturate after a bombarding time of around 60 min. This is also found to be true for the fate of the dislocation lines, in general.

So far as the ripple coarsening is concerned, the nonlinear $\lambda_{ij}^{(2)}$ terms (Eq. 3.3) play a major role. The exact form is of course unknown at present due to lack of quantified information of the redeposited and mobile atoms. Subsequent to coarsening, a saturation is observed. This is actually due to geometrical shadowing estimated from the shadowing condition of ref. [49], which sets in at around 60 min for our case. But this shadowing is due to the ripples formed on top of the dislocation line ridges. The amplitude to height ratio of the ripples attains such a value that it blocks the incoming beam and hinders the further growth of amplitude resulting in eventual saturation that we observe here. Probably this also explains the flattening of the dislocation ridges (Fig. 3.13) at later stages.

3.4.3 *Ion Beam Effects*

In order to estimate the ripple wavelength, one generally takes help of Eq. 3.1 where the momentum and hence the mass of the bombarding ion determine the value of the ion-induced ESD. In order to calculate the coefficient of thermal surface diffusion, one takes into account the surface mobility of the target atoms [59]. However, the mobility of the bombarding species after they reach the surface is not considered in any of the existing theories till date. In a recent study [60] using a mixed beam experiment with isobaric ions (Cs and Xe ions have almost identical masses, 132.9 and 131.3 amu, respectively), it was demonstrated that differences do exist in the ripple characteristics depending on the Cs/Xe ratio even though the thermal surface diffusion and the momentum transfer imparted by Cs and Xe individually to the target atoms are almost identical.

Figure 3.15 shows the variation of ripple wavelengths for different Cs/Xe ratios in the ion beam. It is evident from the figure that the higher the ratio of the cesium ions, higher is the wavelength of the ripples. But according to the existing theories, the wavelength should not vary if the Cs/Xe ratio is changed (isobaric beams!). This clearly suggests that there is at least another factor that governs the wavelength change under such situations. For the present case the primary energy for both Cs and Xe was 0.5 keV. At such a low energy, the penetration depth of the incoming primary ions into the

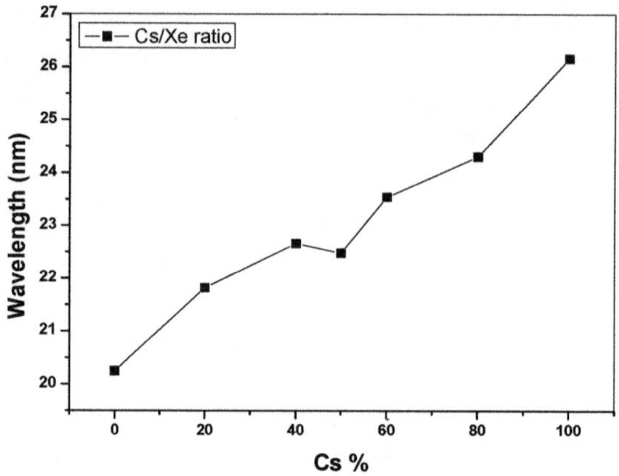

Figure 3.15 Experimental ripple wavelengths under varying Cs/Xe ratio. Primary energy = 0.5 keV.

target is extremely low and ion retention fairly high. Therefore, a large fraction of the primary ions reside either on the surface or just at the sub-surface level. If the change in wavelength is due to the amount of Cs residing at the surface compared to Xe, this effect should diminish with the decrease in the number of primary ions residing at the surface. This can be achieved by increasing the energy of the primary beam as primary ion retention will decrease with the increased sputter yield at higher energies [61].

Theoretical calculations using existing models [12, 18] show that bombarding with Cs or Xe should give almost identical ripple wavelengths under varying energies (as shown in Fig. 3.16(a)). This is in contradiction to what is found experimentally as shown in Fig. 3.16(b). Over a broad energy range, the Xe results have similar energy dependence as the theoretical calculations (decreasing wavelength with decreasing energy). Comparing the values for Cs versus Xe bombardment, the ripple wavelengths appear to deviate from each other with decreasing bombarding energy whereby Cs leads to values exceeding the theoretical calculations. At higher energies (> 1.5 keV), the ripple wavelengths for Cs and Xe are almost equal and follow the theoretical trend. It is to be noted, however, that the theoretical and experimental values of the wavelengths vary from each other by almost a factor of 2. Such a discrepancy

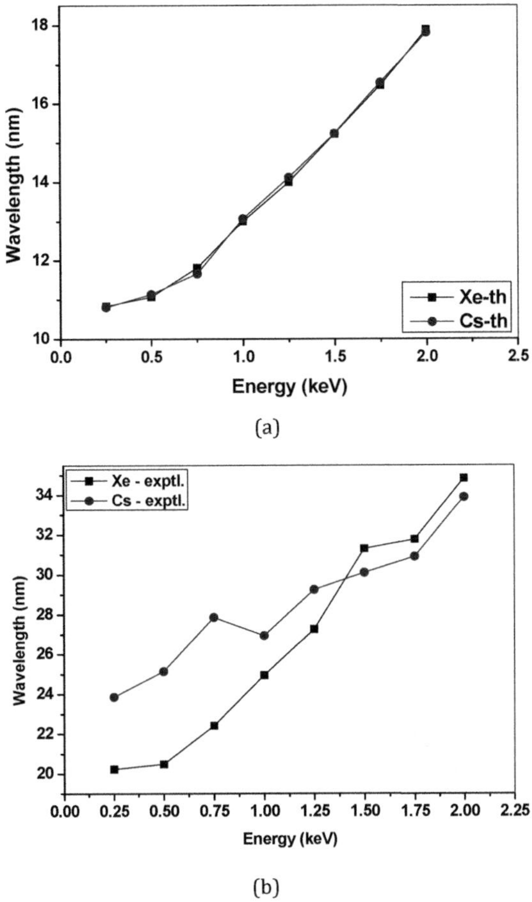

Figure 3.16 (a) Theoretical ripple wavelengths under varying bombarding energies for pure Xe and Cs. (b) Experimental ripple wavelengths for pure Cs and Xe sputtering at different bombarding energies.

also observed in earlier studies [62] indicate that many other assumptions pertaining to such nonequilibrium processes need to go into the existing growth models. But presently, the important thing is that the existing models do not give unrealistic answers even with the uncertainties.

As surface diffusion plays an important role in the KS equation, the larger value of ripple wavelength is ascribed to the high

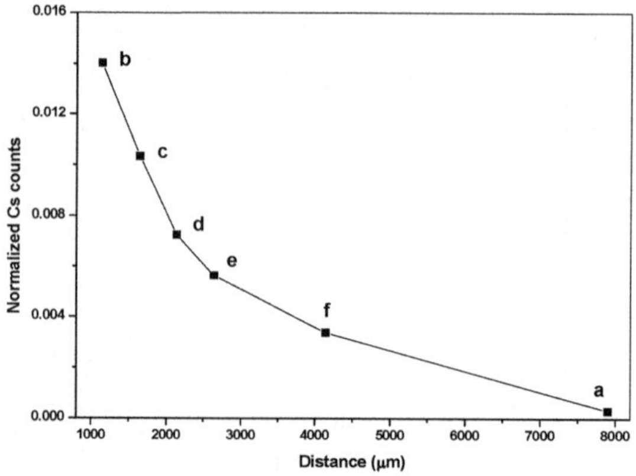

Figure 3.17 Variation of normalized Cs count with distance from the centre of SIMS crater. Inset: Experimental arrangement.

diffusivity of surface cesium concentration across the sample surface [63]. Figure 3.17 shows an exponential decay of the normalized cesium counts plotted as a function of distance from the centre of a SIMS crater made by a 0.25 keV Cs beam. The cesium crater had a size of 500 × 500 μm. Areas b, c, d, e, and f were measured using ion imaging. It is thus evident that Cs atoms migrate over a large distance across the surface. Hence, Cs exhibits very large surface diffusivity on the wafer surface. This experiment is in line with previous observations of Cs surface migration. On the other hand, the migration of Xe atoms is very low compared to that of Cs. In fact, several studies have even shown that noble gases rather tend to form bubbles inside materials under bombardment at low energies [64, 65]. Under such a situation, Xe atoms may get trapped and thus will not contribute to any further enhanced diffusion. A study by Akhter and Venables [63] has clearly shown that Cs overlayers on material surfaces have a high diffusion coefficient. This surface diffusivity is, however, also Cs-concentration dependent as at sub-monolayer coverages Cs atoms may form bonds with the underlying substrate atoms and thus remain rather immobile. The latter would be the case for higher bombardment energies where Cs retention and its surface concentration become quite low. With subsequent increase of coverage, Cs weakly interacts with the substrate atoms, but rather with other Cs atoms thus giving a much smaller value of the cesium sticking coefficient [66] or binding energy [63]. This adds to the increased mobility of Cs atoms at higher coverages. This will be the case for low bombarding energies where the expected surface concentration can become quite high [67].

The ripple wavelength is given by the relation

$$l = 2\pi\sqrt{D_{tot}/\mid \nu \mid} \tag{3.6}$$

where ν is the roughening prefactor dependent on the angle of incidence of the ions and D_{tot} is the sum of the surface diffusion terms K and D and its value depends on the strength of these two independent terms [46]. In order to account for the larger wavelength of the ripples induced by Cs sputtering, one needs to increase D_{tot} by including an additional term. Sarkar *et al.* [60] have referred to this term as D^{IS} and call it the ion-specific surface diffusion constant. The continuum equation would, therefore, be changed to

$$\partial h/\partial t = -K\nabla^4 h - D\nabla^4 h - D^{IS}\nabla^4 h \tag{3.7}$$

The difference in D^{IS} between Xe and Cs should explain the results of Fig. 3.16(b).

Essentially being a diffusion term, an empirical expression for D^{IS} would be of the form [68]

$$D^{IS} = \frac{DC\gamma\Omega^2}{k_B T} \tag{3.8}$$

$$DC = A\exp[-E_d/k_B T] \tag{3.9}$$

Here, D is the surface diffusivity, C is the concentration (number/area) of mobile primary species that participates in surface diffusion, γ is the surface energy per area, Ω is the atomic volume, A is a constant, E_d is the activation energy for diffusion, k_B is the Boltzman's constant, and T is the temperature. For the present scenario, the Cs surface concentrations $C^{surf}_{Cs,n}$ (n refers to a particular bombardment energy) would be different at different bombarding energies, which were calculated using the method as described in ref. [67]. The calculation takes care of desorption effects on the bombarded surface and thereby arrives at a saturation concentration of cesium depending on the bombarding energy. The concentration of mobile primary species that effectively take part in diffusion is assumed to be $C_n = C^{surf}_{Cs,n} \times const.$ The value of the *const.* $\leq$ 1 since C_n cannot exceed $C^{surf}_{Cs,n}$. Figure 3.18 shows that at lower energies, a higher concentration of effective mobile species gives a much higher wavelength. Since the surface Cs concentration varies with the impact energy, an effective atomic volume is considered for each of these energies. This is assumed to vary with $C^{surf}_{Cs,n}$ in such a way that at the highest impact energy (i.e., lowest surface Cs concentration) the value for zero Cs concentration [46, 59] is obtained. Earlier studies on fractional coverages of Cs adsorbed on surfaces demonstrate that E_d = 0.47 eV for Si(001) and 0.7 eV for GaAs [69, 70]. A higher value for the GaAs surface is primarily ascribed to its polar nature, which generates microscopic in-plane electric fields on the surface and thus hinders the adatom diffusion. It would, therefore, be meaningful in our case to estimate the value of E_d slightly higher than that in the case of Si(001). Figure 3.18 shows that lower the value of E_d, higher is the wavelength. These values of E_d are much smaller when compared to cases in the absence of Cs [46, 59]. It is, therefore, understood that a considerable amount of surface Cs concentration can alter the various diffusion parameters thereby changing the ripple wavelength by an appreciable amount.

Therefore, the mobility of primary ions upon their arrival on the surface can play a vital role in deciding the fate of the ripples that evolves during erosion of the sample surface. The ripple wavelength is greater for a higher Cs/Xe ratio. This effect is enhanced for lower energies and can be explained by considering the differences in Cs retention for different bombardment energies and the concentration dependent Cs diffusivity. These studies will prove to be extremely useful for fine tuning dimensions of corrugated nanostructures by low energy ion beam methods.

3.5 Conclusion and Outlook

In summary, work concerning formation of ripples on SiGe surfaces carried out for the last few years covers different aspects of the self-organization process. Although the basic phenomenon can be described by the KPZ model, several experimental findings that result from nonlinear effects have to be explained by incorporating additional terms in the existing growth equation.

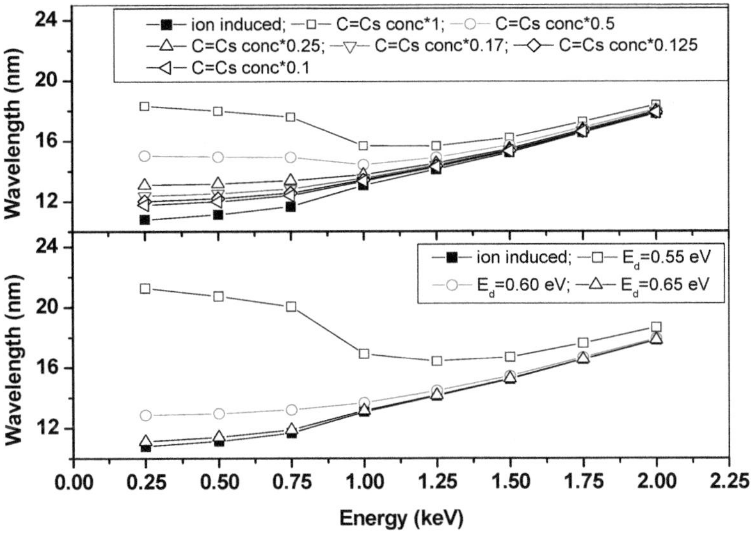

Figure 3.18 Top: Variation of wavelength with C_n, $E_d = 0.56$ eV. Bottom: Wavelength vs. energy for different E_d.

However, this topic requires a special attention since phenomenon like preferential sputtering plays a substantial role in the formation of the nanostructures. The enrichment of one species over the other needs to be considered in detail in the resulting growth equation. A number of recent studies, both theoretical and experimental, have clearly indicated this phenomenon. It is also interesting to note that ripple characteristics on such alloy surfaces also depend on the starting surface morphology and the mobility of the incident primary ions on the bombarding surface. Furthermore, even the current experimental findings till date needs to be substantiated by a good amount of theoretical work by researchers in this area.

Acknowledgements

It is a pleasure to acknowledge W. Vandervorst of IMEC for many thought-provoking discussions during our collaborations and encounters over the years. This work owes a major debt to my colleagues at IMEC, especially A. Franquet, A. Moussa, and B. V. Daele for fruitful discussions on this topic.

References

1. V. Rotello, (2004) *Nanoparticles: Building Blocks for Nanotechnology*, Springer, USA.
2. M. Navez, C. Sella, and D. Chaperot (1962) *C. R. Acad. Sci. Paris*, **254**, 240.
3. T. M. Mayer, E. Chason, and A. J. Howard (1994) *J. Appl. Phys.*, **76**, 1633.
4. M. V. R. Murty, T. Curcic, A. Judy, B. H. Cooper, A. R. Woll, J. D. Brock, S. Kycia, and R. L. Headrick (1988) *Phys. Rev. Lett.*, **80**, 4713.
5. J. L. Whitton, G. Kiriakidis, G. Carter, G. W. Lewis, and M. J. Nobes (1984) *Nucl. Instrum. Meth. Phys. Res. B*, **2**, 640.
6. S. Facsko, T. Dekorsky, C. Koerdt, C. Trappe, H. Kurz, A. Vogt, and H. L. Hartnagel (1999) *Science*, **285**, 1551.
7. O. Plantevin, R. Gago, L. Vazquez, A. Biermanns, and T. H. Metzger (2007) *Appl. Phys. Lett.*, **91**, 113105.
8. B. Ziberi, F. Frost, and B. Rauschenbach (2006) *Appl. Phys. Lett.*, **88**, 173115.
9. C. C. Umbach, R. L. Headrick, and K. C. Chang (2001) *Phys. Rev. Lett.*, **87**, 246104.
10. B. Ziberi, F. Frost, and B. Rauschenbach (2006) *J. Vac. Sci. Technol. A*, **24**, 1344.
11. F. Frost, B. Ziberi, T. Hoche, and B. Rauschenbach (2004) *Nucl. Instrum. Methods B*, **216**, 9.
12. W. L. Chan and E. Chason (2007) *J. Appl. Phys.*, **101**, 121301.
13. U. Valbusa, C. Boragno, and F. B. de Mongeot (2002) *J. Phys.: Condens. Matter*, **14**, 8153.
14. R. M. Bradley and J. M. E. Harper (1988) *J. Vac. Sci. Technol. A*, **6** (4), 2390.
15. P. Sigmund (1969) *Phys. Rev.*, **184**, 383.
16. P. Sigmund (1973) *J. Mater. Sci.*, **8**, 1545.
17. M. C. Cross and P. C. Hohenberg (1993) *Rev. Mod. Phys.*, **65**, 851.
18. M. A. Makeev, R. Cuerno, and A. L. Barabasi (2002) *Nucl. Instrum. Methods B*, **197**, 185.
19. I. Koponen, M. Hautala, and O. P. Sievanen (1997) *Phys. Rev. Lett.*, **78**, 2612.

20. K. Wittmaack (1990) *J. Vac. Sci. Technol. A*, **8**, 2246.
21. J. Erlebacher, M. J. Aziz, E. Chason, M. B. Sinclair, and J. A. Floros (2000) *J. Vac. Sci. Technol. A*, **18**, 115.
22. J. J. Vajo, R. E. Doty, and E.-H. Cirlin (1988) *J. Vac. Sci. Technol. A*, **6**, 76.
23. S. Rusponi, G. Costanini, C. Boragno, and U. Valbusa (1998) *Phys. Rev. Lett.*, **81**, 2735.
24. E. A. Eklund, R. Bruinsma, J. Rudnick, and R. S. Williams (1991) *Phys. Rev. Lett.*, **67**, 1759.
25. H.-N. Yang, G.-C. Wang, and T.-M. Lu (1994) *Phys. Rev. B*, **50**, 7635.
26. S. Park, B. Kahng, H. Jeong, and A. L. Barabasi (1999) *Phys. Rev. Lett.*, **83**, 3486.
27. W. Bolse (1994) *Mater. Sci. Eng. R*, **12**, 53.
28. S. Habenicht, W. Bolse, H. Feldermann, U. Geyer, H. Hofsass, K. P. Lieb, and F. Roccaforte (2000) *Europhys. Lett.*, **50**, 209.
29. J. F. Ziegler, M. D. Ziegler, and J. P. Biersack Srim ver. 2006. URL http://www.srim.org.
30. M. A. Makeev and A. L. Barabasi (1997) *Appl. Phys. Lett.*, **71**, 2800.
31. H. H. Andersen (1979) *Appl. Phys.*, **18**, 131.
32. G. Carter and V. Vishnyakov (1996) *Phys. Rev. B*, **54**, 17647.
33. E. Chason, T. M. Mayer, B. K. Kellerman, D. T. McIlroy, and A. J. Howard (1994) *Phys. Rev. Lett.*, **72**, 3040.
34. J. M. Garcia, M. Castro, and R. Cuerno (2006) *Phys. Rev. Lett.*, **96**, 086101.
35. V. B. Shenoy, W. L. Chan, and E. Chason (2007) *Phys. Rev. Lett.*, **98**, 256101.
36. R. M. Bradley and P. D. Shipman (2010) *Phys. Rev. Lett.*, **105** (14), 145501. doi: 10.1103/PhysRevLett.105.145501.
37. V. Tuboltsev, P. Jalkanen, M. Kolodyazhnaya, and J. Räisänen (2005) *Phys. Rev. B*, **72** (20), 205434. doi: 10.1103/PhysRevB.72.205434.
38. M. Z. Hossain, J. B. Freund, and H. T. Johnson (2008) *J. Appl. Phys.*, **103**, 073508.
39. T. Bobek, S. Facsko, T. Dekorsky, and H. Kurz (2001) *Nucl. Instrum. Methods B*, **178**, 101.
40. T. Allmers, M. Donath, and G. Rangelov (2006) *J. Vac. Sci. Technol. B*, **24**, 582.
41. G. S. Lau, E. S. Tok, R. Liu, A. T. S. Wee, and J. Zhang (2004) *Nucl. Instrum. Methods B*, **215**, 76.
42. S. W. MacLaren, J. E. Baker, N. L. Finnegan, and C. M. Loxton (1992) *J. Vac. Sci. Technol. A*, **10**, 468.

43. A. Kanjilal, S. Prucnal, M. Minniti, W. Skorupa, M. Helm, and S. Facsko (2010) *J. Appl. Phys.*, **107** (7), 073513. doi: 10.1063/1.3369391.
44. K. Takahiro, K. Zhangf, F. Rotter, D. Schwen, C. Ronning, H. Hofsass, and J. Krauser (2007) *Nucl. Instrum. Methods B*, **256**, 378.
45. A. Cuenat, H. B. George, K. C. Chang, J. M. Blakely, and M. J. Aziz (2005) *Adv. Mater.*, **17**, 2845.
46. S. Sarkar, B. V. Daele, and W. Vandervorst (2008) *New. J. Phys.*, **10**, 083012.
47. H. Chen, Y. K. Li, C. S. Peng, H. F. Liu, Y. L. Liu, Q. Huang, J. M. Zhou, and Q.-K. Xue (2002) *Phys. Rev. B*, **65** (23), 233303. doi: 10.1103/PhysRevB.65.233303.
48. D. C. Houghton (1991) *J. Appl. Phys.*, **70**, 2136.
49. G. Carter (1999) *J. Appl. Phys.*, **85**, 455.
50. Z. Jiang, K. Kim, J. Lerma, A. Corbett, D. Sieloff, M. Kottke, R. Gregory, and S. Schauer (2006) *Appl. Surf. Sci.*, **252**, 7262.
51. T. K. Chini, F. Okuyama, M. Tanemura, and K. Nordlund (2003) *Phys. Rev. B*, **67**, 205403.
52. K. B. Winterbon, P. Sigmund, and J. B. Sanders (1970) *Mat. Fys. Medd.*, **37**, 5.
53. M. Moore, N. Kalyanasundaram, J. Freund, and H. Johnson (2004) *Nucl. Instrum. Methods B*, **225**, 241.
54. M. Kardar, G. Parisi, and Y. C. Zhang (1986) *Phys. Rev. Lett.*, **56**, 889.
55. A. L. Barabasi and H. E. Stanley (1995) *Fractal Concepts in Surface Growth*, Cambridge University Press, Cambridge.
56. F. Family and T. Vicsek (1991) *Dynamics of Fractal Surfaces*, World Scientific, Singapore.
57. M. A. Makeev and A. L. Barabasi (2002) *Nucl. Instrum. Methods B*, **222**, 316–335.
58. D. P. Datta and T. K. Chini (2007) *Phys. Rev. B*, **76**, 075323.
59. J. Erlebacher, M. J. Aziz, E. Chason, M. B. Sinclair, and J. A. Floro (2000) *J. Vac. Sci. Technol. A*, **18**, 115.
60. S. Sarkar, A. Franquet, A. Moussa, and W. Vandervorst (2011) *Appl. Surf. Sci.*, **(In print)**.
61. F. Schulz and K. Wittmaack (1976) *Rad. Eff. Solids*, **29** (1), 31–40. doi: 10.1080/00337577608233481.
62. D. P. Datta and T. K. Chini (2004) *Phys. Rev. B*, **69**, 235313.
63. P. Akhter and J. A. Venables (1981) *Surf. Sci.*, **103**, 301–314.

64. U. Bangert, P. J. Goodhew, C. Jeynes, and I. H. Wilson (1986) *J. Phys. D: Appl. Phys.*, **19**, 589–603.
65. W. M. Lau, I. Bello, L. J. Huang, X. Feng, M. Vos, and I. V. Mitchell (1993) *J. Appl. Phys.*, **74** (12), 7101.
66. B. Kierren and D. Paget (1997) *J. Vac. Sci. Technol. A*, **15** (4), 2074.
67. B. Berghmans, B. V. Daele, L. Geenen, A. Franquet, and W. Vandervorst (2008) *Appl. Surf. Sci.*, **255**, 1316–1319.
68. W. W. Mullins (1959) *J. Appl. Phys.*, **30**, 1.
69. R. H. Milne, M. Azim, R. Persaud, and J. A. Venables (1995) *Surf. Sci.*, **336** (1–2), 63–75. ISSN 0039-6028.
70. V. L. Alperovich and D. Paget (1997) *Phys. Rev. B*, **56** (24), R15565–R15568.

Chapter 4

Patterning of Ionic Insulator Surfaces with Low-Energy Ion Beams

Franciszek Krok,[a] Salah R. Saeed,[a,b] Marek Kolmer,[a] and Marek Szymonski[a]

[a] *Research Center for Nanometer-Scale Science and Advanced Materials (NANOSAM), Faculty of Physics, Astronomy and Applied Computer Science, Jagiellonian University, ul. Reymonta 4, 30-059 Krakow, Poland*

[b] *Salah R. Saeed, Sulaimani University College of Basic Education (Science), Chamchamal-Kurdistan Region, Iraq*

Franciszek.krok@uj.edu.pl

Surface modification of single-crystal alkali halides [KBr(001), RbI(001)] induced by low-energy ion beam irradiation has been investigated in ultrahigh vacuum (UHV) by means of high resolution atomic force microscopy (AFM) in contact and noncontact modes of operation. Surface morphology development has been studied as a function of ion beam parameters (ion mass: Ar and He; angle of incidence: 0–80° off-normal; ion energy: 1–5 keV; ion fluence: 10^{13}–10^{18} ions cm^{-2}) and the target temperature (300 to 600 K). Several new features of the process have been found, directly related to the ionic nature of halide surfaces, such as formation of two-dimensional (2D) pits and rectangular 2D epitaxial adislands on the initially atomically flat terraces. At low bombardment fluence, the evolution of 2D pits proceeds along main surface

Nanofabrication by Ion-Beam Sputtering: Fundamentals and Applications
Edited by Tapobrata Som and Dinakar Kanjilal

ISBN 978-981-4303-75-0 (Hardcover), 978-981-4303-76-7 (eBook)
www.panstanford.com

crystallographic directions. Such behavior is typical for the electron stimulated desorption (ESD) process, well known from electron and photon irradiation experiments. No epitaxial adisland formation by ESD, or by ion impact, has been reported so far. For prolonged ion bombardment, surface topography transforms into a regular network of grooves and rims (a ripple structure) oriented parallel to the incident beam. Such a structure has been observed for many other materials (metals, semiconductors), with a major distinction, however, that for KBr, the ripples are composed of small nano-size crystallites with a persistent long range order on the (001) surface. It is demonstrated that the fluence threshold for transition from a random network of 2D pits and adislands into a well-oriented crystalline nanoripple structure is directly related to the balance between the electronic and the ballistic stopping of the impinging ions. The interpretation of these new experimental observations does require a new atomistic approach to the ion-solid interactions at irradiated surfaces.

4.1 Introduction

Surface patterning on the nanometer scale due to ion bombardment has attracted wide interest as a potentially simple technique for large area template and interface fabrication for various important applications in contemporary electronics, materials science, molecular biology, or medicine [1]. Moreover, low-energy ion bombardment is routinely used in cleaning and surface processing of samples for research investigations and industrial processes. Therefore, a detailed knowledge of the dynamic response of the crystal surface to stimulation by the ion beam is of utmost importance. Regarding self-organization effects and spontaneous pattern formation during ion sputtering, the phenomenon of ripple formation on the irradiated surfaces has been intensely studied for a few recent decades [2]. Periodic patterns on ion-irradiated surface were reported for the first time by Navez and coworkers in 1962 [3]. Since then, nanoscale patterns have been found on a variety of materials, such as metals [4–6], semiconductors [7–9], and insulators [10–12]. The main processes that determine the surface morphology of amorphous solids subjected to ion bombardment were outlined for the first time by Bradley and Harper [13].

According to the Bradley–Harper theoretical model, the initial surface modulation is given by the microscopic roughness of the surface and the stochastic nature of the sputtering process as proposed by Sigmund [14]. The probability that an atom is removed from the irradiated surface is dependent on the amount of deposited energy by all penetrating projectile ions. It implies that the trough regions are eroded faster than the crests, which finally leads to destabilization of the surface topography and increase of the surface roughness. Such an instability process competes with a smoothing mechanism, e.g., surface diffusion or viscous flow, both being of thermal [15] or ion bombardment enhanced origin [16]. The interplay between those two processes, i.e., the ion-induced surface instability and the surface relaxation, determines a characteristic wavelength of the created surface patterns [13]. Although the Bradley–Harper theory was proposed to explain the pattern formation on surfaces of amorphous solids, it was also applied with some success to predict morphological structures developing on single-crystal surfaces, which become amorphized during ion irradiation (e.g., semiconductor crystals). One of the most valuable achievements of the Bradley–Harper model is the ability to predict the dependence of the ripple alignment on the projectile angle of incidence. However, apart from undisputed achievements, the model has also several deficiencies. For example, it can explain neither experimentally observed occurrence of a saturation of the surface roughness for large ion fluencies, nor a linear dependence of the ripple wavelength on the projectile's kinetic energy. In contrast to those experimental observations, the Bradley–Harper model predicts an unlimited exponential increase of the ripple amplitude and a decrease in their wavelength with the increasing primary ion energy. Thus, several modifications of the Bradley–Harper model have been proposed over the years. For example, the pattern stabilization for long sputtering times [17], nanodot formation for the isotropic case [18], and, more recently, the observation of ordering [19] and coarsening of the pattern features [20] were explained by adding additional components to the basic Bradley–Harper model.

Very recently, Aziz and coworkers showed that the erosion-based paradigm of the Bradley–Harper approach could be essentially irrelevant for the determination of the surface instability, which finally leads to pattern formation on the ion-irradiated surfaces [21]. Instead, they proposed that in the process of pattern formation, ion-impact-induced atom redistribution on the surface is the decisive effect that overwhelms the erosion due to sputtering.

4.2 Remarks on Experimental Details

The essential part of the results presented in this chapter was obtained in a multichamber UHV system developed at the Department of Physics of Nanostructures and Nanotechnology of the Jagiellonian University allowing for sample preparation, its analysis with surface averaging techniques, and high resolution scanning probe microscopy (SPM) imaging under the same UHV conditions, i.e., the base pressure in the vacuum chamber was of 5×10^{-11} mbar in all stages of the experiment, including SPM characterization. Maintaining very good vacuum conditions is known to be of crucial importance for studying alkali halide surfaces [22]. Alkali halide crystals purchased from TBL-Kelpin (Neuhausen, Germany) were cleaved in air parallel to the (001) cleavage plane and have been clamped with tungsten springs on a molybdenum plate mounted on a copper sample holder. After mounting, the samples were immediately transferred into the vacuum system for proper cleaning by annealing in UHV at 450 K for several hours. The annealing process is essential in order to remove the adsorbed impurities and to reduce the stress and charge accumulation after sample cleaving [23]. The samples were exposed to focused beams of 1–5 keV Ar^+ and He^+ ions having the spot size of 1 mm^2. The beam has been rastered over the entire surface area. The average ion current density was 1 μA cm^{-2} and the ion fluence, Φ_{ion}, was changed between 10^{13} and 10^{18} ions cm^{-2}. The angle of incidence, α, of the ion beam was varied in the range 0–80° with respect to the surface normal. The resulting surface topographies were analyzed by AFM (VP2 Park Scientific Instruments device) in a contact mode (cAFM) and a noncontact mode (nc-AFM — in literature often called dynamic force microscopy: DFM) under UHV conditions using the piezoresistive silicon cantilevers. For the nc-AFM measurements, cantilevers of the resonance frequency of 358,4 kHz were used. The amplitude of oscillation and a nominal tip radius were 10 nm and 20 nm, respectively. The frequency shift (detuning) relative to the cantilever resonance frequency was set in the range of −5 to −20 Hz. For better statistics, the nc-AFM measurements have been done at several different areas of the irradiated sample surface.

4.3 Surface Morphology Changes at the Initial Stages of Ion Bombardment

The materials most frequently studied with respect to surface structuring with ion beam are metals and semiconductors. In such systems, surface atom ejection occurs due to ballistic collisions. It has been reported that nanoscale dots were created at close to normal beam incidence [7], whereas ripple-like structures were formed at more oblique angles of incidence [2]. Although there is a general consensus about evolution of the ripple structures as being due to the competition between surface roughening introduced by ion sputtering and surface smoothing created by various diffusion related processes, details of such mechanisms are still under discussion. Depending on particular experimental conditions, pure thermally assisted diffusion [13, 15], viscous flow [8, 24], ion-enhanced or ion-inhibited diffusion [25, 26], and preferential sputtering without net mass transfer [16, 27] are considered important contributors to the effective surface smoothing. There is, however, a wide range of materials like alkali halides, oxides, and frozen noble gases in which erosion can be dominated by electronic processes [28]. Alkali halide crystals are particularly interesting class of materials due to their simple crystallographic and electronic structures. It is also relatively easy to obtain well-characterized single-crystal surfaces. Finally, the relative contribution of electronic and ballistic processes can be easily tailored in those crystals by changing the mass of the projectile [29], as the relative contribution of electronic processes increases for lighter primary ions. Alkali halides can be used, therefore, as a model system to study development of surface morphology at single-crystal surface under conditions of coexisting electronic and ballistic erosion mechanisms. Within the electronic desorption channel, the excitation energy deposited by the impinging ions into the crystal electronic system is used primarily for the creation of Frenkel centers [30]. Subsequent diffusion and recombination of those centers at the surface result in emission of its atomic constituents. Such an electronic desorption process occurs in a manner that predominantly reflects symmetry of the lattice structure and, therefore, it is insensitive to the direction of the primary ion beam. On the other hand, for oblique incidence, the ballistic processes initiated by a direct momentum transfer are increasingly important for retaining the memory of the initial impact direction. As a result, under those conditions, ballistic processes can induce anisotropy of the resulting surface morphology correlated with the ion beam direction.

Surface morphology evolution of RbI(001) single crystals for initial stages of ion bombardment with 5 keV He^+ has been investigated by Krok *et al.* [31] with DFM at UHV conditions. DFM images taken for RbI(001) surface after consecutive ion irradiations at the crystal temperature of about 370 K are reproduced in Fig. 4.1.

The erosion starts by creation of randomly spread rectangular pits of monatomic depth on the surface (Fig. 4.1b). As the ion fluence increases, the erosion proceeds by growth and linking of the pits in the first, upper-most layer similarly to the ESD process described above, but the second layer is eroded before the first layer is completely removed (Fig. 4.1c). The multilayer mode of desorption is even more distinctly visible in Fig. 4.1d where the pits in the third layer, together with the remained part of the first one, are observed.

One of the very first observations that the grazing ion incidence leads to ripple-like pattern formation on ionic crystal surfaces was the experiment by Batzill *et al.* [32]. They reported the fabrication of periodic wire arrays by 4.5 keV Ar^+ bombardment of a CaF_2(111) crystal at grazing incidence (4–6° with respect to the surface plane) at room temperature (Fig. 4.2). Their measurements clearly showed that the wires are oriented along azimuthal direction of ion beam incidence; therefore, they concluded that the ion beam is primarily responsible for orienting the wires. Apart from morphological evo-

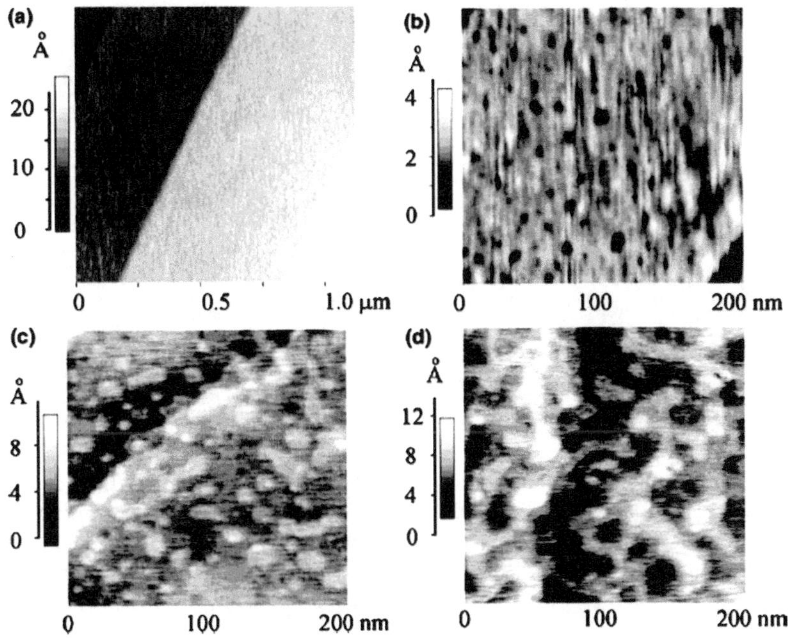

Figure 4.1 DFM images of cleaved RbI(001) surface: (a) as cleaved, (b) irradiated with an ion fluence of 0.43×10^{13} ions cm^{-2}, (c) irradiated with an ion fluence of 1.27×10^{13} ions cm^{-2}, (d) irradiated with an ion fluence of 2.62×10^{13} ions cm^{-2}. The irradiation was performed with 5 keV He^{+} ion beam with an angle of incidence 60° off-normal. The sample temperature during the irradiation was 370 K. Reproduced from [31].

lution, ion bombardment can also induce changes in the elemental composition of irradiated multicomponent surfaces. According to theoretical predictions [33], partial sputtering yields depend on atomic masses and surface binding energies of the constituents. The results obtained by Batzill also indicate that the calcium fluoride surface is sputtered with preferential emission of fluorine, thus excess of calcium at the surface assembles into periodic wire-like stress domains.

The postulated self-organization mechanism invokes minimization of the system free energy and the elongation of the wires along the beam direction due to angle-dependent anisotropic sputtering of the wires. However, it should be noted that the morphology of

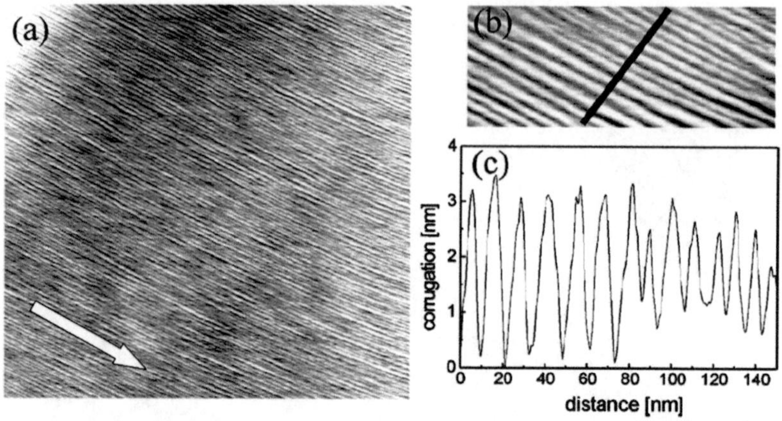

Figure 4.2 AFM image of irradiated $CaF_2(111)$ surface with 4.5 keV Ar^+ ion beam at a glancing angle of 10° and the total ion fluence of 2.63×10^{15} ions cm^{-2}. The white arrow in (a) is the projection of the ion beam direction on the surface. (a) Contact-mode AFM topography image of $1 \times 1\ \mu m^2$, (b) enlarged portion of the same surface, $260 \times 100\ nm^2$, (c) cross section through the line indicated in (b). Reproduced from [32].

the irradiated $CaF_2(111)$ surfaces was investigated *ex situ* with an ambient AFM operating in contact mode. Thus one should be careful in the interpretation of these data as it is known that the surface enriched in Ca is highly reactive when exposed to air.

4.4 Ripple Formation for Prolonged Irradiation with Off-Normal Incidence Ion Beam

Mussi *et al.* [34] have investigated nanostructuring and optical activation of the randomly oriented lithium fluoride polycrystalline samples by 0.8 keV Ar ion sputtering at different ion fluences. The angle of incidence was fixed at 35° off-normal and the sample temperature was 80°C (Fig. 4.3).

Although the AFM images were taken only for three different ion fluences, the authors concluded that the trend in ripple height and the mean periodicity seem to follow the predictions of a so-called "continuum theory" and the scaling low predicted by Makeev, Cuerno, and Barabasi [35]. The linear continuum theory (LCT),

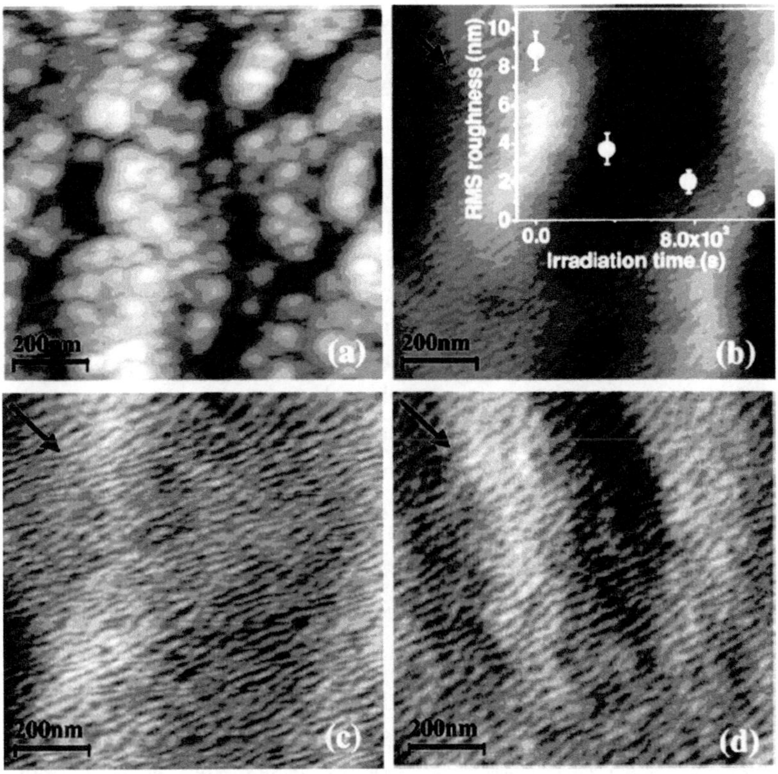

Figure 4.3 1 × 1 μm cAFM images of the surface of 4 LiF initially identical samples: not sputtered (a) (z dynamic range: 51 nm), irradiated for 1 h (b) (z dynamic range: 27 nm), 2 h (c) (z dynamic range: 12 nm), and 3 h (d) (z dynamic range: 7 nm) with 800 eV Ar^+ ions at an angle of about 35°. The black arrow indicates the direction of the projected ion beam on the crystal surface. The graph reported in the inset shows the mean *rms* roughness as a function of the irradiation time. Reproduced from [34].

initially outlined by Bradley and Harper [13], has been commonly used for explaining the process of ion-induced ripple formation in amorphous samples. Basic assumptions of the continuum theory imply that the local erosion yield is continuously and smoothly varying with the local surface morphology and, as shown by Sigmund [14], the erosion is faster in depressions than at elevations due to a spatial extension of the ion-induced collision cascades. This ballistic sputtering process causing surface coarsening is balanced with diffusion-type surface-smoothing processes of thermal nature [13]. Since the linear theory could account for a limited number of experimental findings only, several attempts have been made to introduce additional effects and corresponding terms into the continuum theory description. In particular, Makeev, Cuerno, and Barabassi [35] generalized Bradley and Harper's second-order linear equation by adding nonlinear and fourth-order terms to address the inadequacies of LCT. The generalized theory introduced an ion-induced smoothing mechanism via preferential sputtering without mass movement on the surface. The ion-induced smoothing becomes important when the thermal diffusion is negligible and, therefore, makes the ripple wavelength at low temperatures independent of the ion flux. In addition, this theory has many persuasive trends, including a linear increase in the ripple wavelength with ion energy and a stabilization of the ripple amplitude for a prolonged ion bombardment. Some other experiments, however, indicate lack of the wavelength flux dependence over a certain range of fluxes [36, 37]. In addition, it is reported that for certain ion substrate combinations, a certain minimum angle of incidence (the angle between the ion beam and the surface normal) is required for formation of the ripple structure. Such findings question the generality of the Makeev–Cuerno–Barabass theory even for amorphous materials (or the materials amorphized due to ion bombardment).

4.5 Temperature-Dependent Patterning of Ionic Single-Crystal Surfaces

The temperature dependence of the morphology of irradiated surfaces can provide a direct insight into the mechanisms of surface modification processes. In Fig. 4.4, the set of AFM images obtained for Ar ion irradiated KBr(001) surface at various temperatures is shown [38]. In the range from room temperature to 473 K, the ripples are quite regular and their wavelengths and amplitudes are not changing with temperature too much. In contrast to this athermal regime of ripple formation, a sudden increase in the surface root means square (*rms*) roughness by a factor of 8 occurs

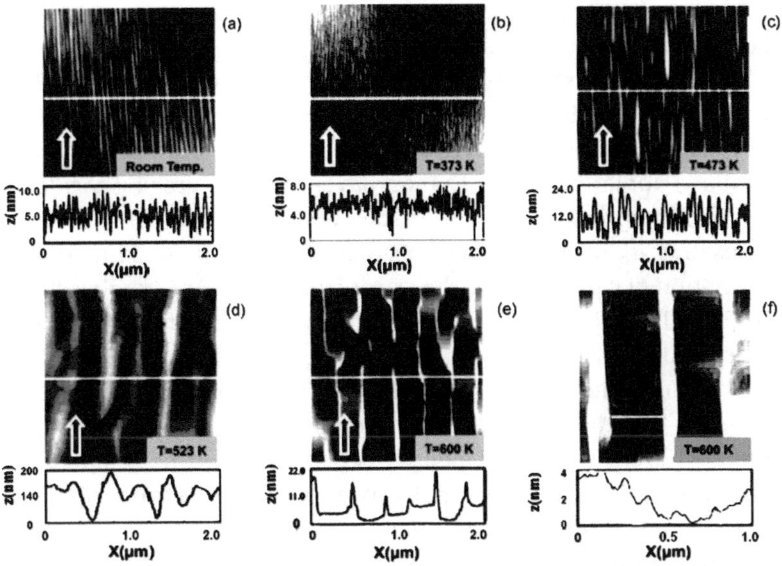

Figure 4.4 KBr(001) surface modified with 5×10^{17} 5 keV Ar^+ ions cm^{-2} striking at 75° off-normal. A set of 2×2 μm^2 cAFM images obtained for samples exposed to the ion beam at (a) room temperature, (b) 373 K, (c) 473 K, (d) 523 K, and (e) 600 K; (f) represents the zoomed 1×1 μm^2 area of the image (e). The cross-cut profiles along the white lines on the images are shown below. Reproduced from [38].

in a relatively small temperature interval of 50 K. The wavelength of the ripples is dramatically increasing with temperature too, i.e., from 55 nm to 460 nm. This sudden transition from athermal to thermal behavior of the ripple formation mechanism is accompanied by a pronounced change in the ripple appearance from V-like to U-like shape of the valleys (see the horizontal cross-cut profiles below the AFM images in Fig. 4.4). Further increase in the temperature results in equally fast drop of the surface roughness, whereas the structure wavelength is only slightly affected. However, the overall U-shape of the structure is preserved. A regular inversed-pyramid shape of the pits located at the flat areas between the ridges is clearly visible (Fig. 4.4f). We must stress that in the highest temperatures investigated in our experiment, a very substantial desorption of the surface material takes place, which prevents us from performing such measurements for a long time.

For single crystals, there are at least two noticeable reports that could not be explained by the continuum theory [37, 39]. The work of van Dijken *et al.* [39] offers an atomic-scale description of the very shallow ripple formation (nanogrooves) upon a grazing incidence ion bombardment of a single crystal Cu (001) surface. More recently, Hansen *et al.* [37] also have considered an atomic-scale model for formation of ripples under the grazing incidence Ar^+ bombardment of Pt(111) surface. One of the main experimental findings of that work indicates a striking difference between an athermal (200–470 K) and thermal (470–700 K) pattern formation regime. It is stressed that the two temperature regimes are distinct not only by their different ripple wavelength dependence on the temperature but also by the different pattern appearance —from ripple V-shaped valleys to ripple ridge one (U-shaped valleys). Furthermore, at the bottom of U-shaped valleys (high temperatures), some compact vacancy islands are seen bonded by the steep walls oriented along the ion beam direction, whereas only narrow vacancy grooves are seen at the bottom of V-shaped valleys (low temperatures).

Hansen *et al.* [37] postulated that large differences in step edge versus terrace damage are crucial for the ripple formation under the grazing incidence. In the low temperature (athermal) regime, initially the vacancy grooves are formed along the beam direction due to rapid motion of the irradiated step edges. Subsequent interaction of the neighboring grooves through repulsion of adatoms and the depletion effect prevent formation of the vacancy grooves on narrow adatom pads separating the neighboring grooves. At elevated temperatures (the thermal regime) alignment of vacancy structures is achieved by the preferential coalescence of compact vacancy clusters along the beam direction as a consequence of damage effects caused by the planar subsurface channeling. A large step edge barrier for vacancies to heal at ascending steps, as postulated by Hansen *et al.*, causes preferential production of new vacancy structures at the bottom of the vacancy grooves, transforming the initial pattern into well-developed U-shaped ripple structure with an increased roughness.

Surprisingly, it has been found that the temperature dependence of the ion-induced KBr(001) surface morphology is quite similar to the Pt(111) case, despite large differences in properties of the two crystals. The two temperature regimes found in ref. [37] are clearly seen for KBr too, including the U-valley formation at elevated temperatures, each one covered at the bottom with rectangular vacancy islands bound in the step edges along the low index direction (Fig. 4.4).

In KBr, the change between athermal and thermal regimes of ripple formation seems to occur at the temperature corresponding to the onset of the ground state F-centre mobility (above 500 K) [40]. At this temperature, the halogen vacancies (F-centers) could migrate thermally to the surface and coalesce into larger vacancy agglomerates at the flat areas on the surface. The vacancy island equilibrium shape is rectangular with edges along the low index direction [100] and [010]. In our previous work, we reported on the step edge barrier for alkali halide molecule migration over the descending steps [41]. This barrier is responsible for pyramid-like shape of the alkali halide islands grown epitaxially by molecular beam epitaxy. Our vacancy islands covering the flat bottom areas of the U-shaped valleys have the inverted-pyramid shape, confirming the existence of such a step edge barrier.

Furthermore, it was shown that the equilibrium surface morphology (for high fluences) critically depends on the angle of incidence as shown in the work by Saeed *et al.* [38]. Only for oblique angles (larger than 60° off-normal), the ripples are very regular and nearly parallel to the projection of the incident beam. Therefore, the apparent similarity in ripple formation mechanism between platinum and potassium bromide single crystals would not be surprising.

However, within the investigated energy range, the ion-sputtering processes of alkali halides are dominated by the electronic processes rather than the ballistic collisions [31, 42] despite a substantial nuclear stopping of keV heavy ions. This is due to a high probability of the valence excitations in those materials and a very high efficiency of radiation-less decay of those excitations with a subsequent transfer of the released energy into the lattice constituents (Frenkel pairs, direct halogen emission, etc.).

4.6 Electronic versus Ballistic Processes on Ion-Irradiated Ionic Surfaces

To address the mechanism of ripple formation on the surface of single crystal that is simultaneously eroded by both electronic and ballistic processes, the KBr(001) surface morphology development was monitored step by step as a function of the ion beam fluence and the mass of ion projectile as it is shown in Fig. 4.5. Figure 4.5 depicts the development of the surface morphology on KBr(001) single crystal subjected to 4 keV Ar^+ [Fig. 4.5a–c] and 4 keV He^+ [Fig. 4.5d–f] bombardment with the off-normal angle of incidence $\alpha = 75^\circ$ at room temperature. Figure 4.5a presents KBr surface exposed to a low fluence of Ar^+ ions ($\Phi_{ion} = 3.0 \times 10^{13}$ ions cm^{-2}). The white arrow in Fig. 4.5a indicates the projection of the ion beam direction on the irradiated surfaces. This direction was chosen not to coincide with the main [100] and [010] crystallographic surface directions, e.g., there is an azimuthal angle of about 45° between the projection of the ion beam and the [100] direction. Such geometry was chosen in order to distinguish between the influence of the electronic and ballistic processes on the surface morphology development, as it is known that electronic desorption leads to removal of surface atoms along main surface crystallographic directions [43]. As shown in the image, the surface is composed of extended atomic terraces, which are separated by monoatomic cleavage steps running across the image. The step contours exhibit a fiord-like shape with elongated clusters of pits oriented in the direction normal to the steps. On the terraces, pits of a monolayer depth coexist with monoatomically high rectangular KBr adislands. Both these features have a rugged shape with local step edges aligned along the main crystallographic directions of the [100] and [010] directions. The atomically resolved nc-AFM imaging reveals that the KBr adislands are of perfect crystallographic order and stoichiometry (inset in Fig. 4.5a). After additional ion irradiation, we observe progressing elongation of individual pits followed by their agglomeration in a form of grooves aligned parallel to the projection of the ion beam direction (Fig. 4.5b, $\Phi_{ion} = 3.4 \times 10^{15}$ ions cm^{-2}). As in the case of adislands, the walls of the grooves have a

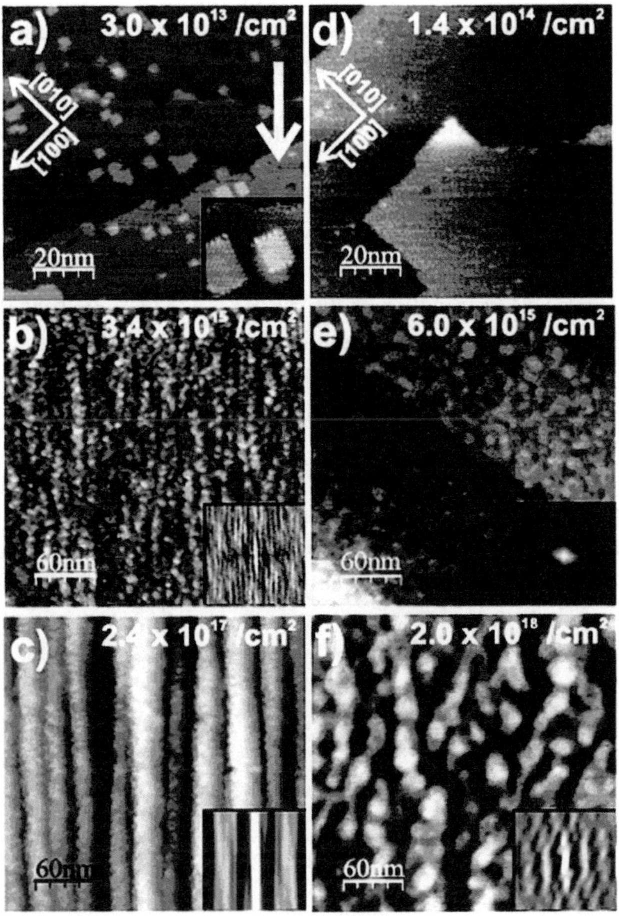

Figure 4.5 Nc-AFM topographies of KBr(001) surface under 4 keV Ar^+ (a–c) and He^+ (d–f) irradiation as a function of ion fluence [(a) -3.0×10^{13} ions cm^{-2}, (b) -3.4×10^{15} ions cm^{-2}, (c) 2.4×10^{17} ions cm^{-2}, (d) 1.4×10^{14} ions cm^{-2}, (e) 6.0×10^{15} ions cm^{-2}, and (f) -2.0×10^{18} ions cm^{-2}]. The irradiation was performed at room temperature and ion beam has an angle of incidence of $\alpha = 75^\circ$ with respect to the surface normal. The white arrow indicates the projection of the ion beam on the irradiated surfaces. The image sizes are 100×100 nm^2 in (a) and (d) and 300×300 nm^2 in (b), (c), (e), and (f). Insets: in (a) atomically resolved image of the KBr adislands (image size 11×11 nm^2), in (b)–(c) and (e)–(f) 2D self-correlation function maps of the topographies shown. Reproduced from [44].

zigzag shape with the local edges bounded to the main, e.g., [100] and [010] crystallographic surface directions. The grooves are separated by the remains of the original topmost layer of the crystal with the adislands located on their top. A closer inspection of Fig. 4.5b shows that at the bottom of the extended pits, new vacancy sites are already forming. The extended irradiation leads, finally, to a creation of the ripple structure of the wavelength of 25 nm and the valley-to-peak amplitude of 4 nm, oriented along the ion beam direction (Fig. 4.5c, $\Phi_{\text{ion}} = 2.4 \times 10^{17}$ ions cm^{-2}). The high resolution nc-AFM imaging of the ripple ridges shows that they are formed from multilayer atomically high terraces. The terraces are of well-ordered crystalline structure with local edges oriented along the [100] and [010] crystallographic surface directions.

In order to investigate how the relative ratio of the electronic and ballistic processes influences the dynamics of the morphology development, the KBr(001) surface has been irradiated with lighter, He^+ ions beam of 4 keV at $\alpha = 75°$ (Fig 4.5d–f). The evident consequence of a smaller projectile mass is much smaller damage of the irradiated surface. Only for the He^+ ion fluence of 1.4×10^{14} ions cm^{-2} (Fig. 4.5d), the appearance of small pits, few nm in lateral size, is observed. Further irradiation leads to expansion of the pits, which ultimately results in a removal of the topmost layer. This process is accompanied by simultaneous erosion of the next atomic layer due to creation of vacancies at the bottom of pits, as it was observed during Ar^+ irradiation. Also as in case of Ar^+ irradiation, the local edges of both the remaining parts of the topmost layer and of the created pits are bound to the main crystallographic directions of the surface. However, in contrary to Ar^+ irradiation, the developed morphology does not exhibit any overall anisotropy (Fig. 4.5e, $\Phi_{\text{ion}} = 6.0 \times 10^{15}$ ions cm^{-2}). Only a very prolonged irradiation leads to creation of shallow groove- and ridge-like structures elongated parallel to the ion beam direction (Fig. 4.5f, $\Phi_{\text{ion}} = 2.0 \times 10^{18}$ ions cm^{-2}). In this case, however, the ripple-like structure is much weaker than the one observed after Ar^+ bombardment with comparable fluence. It has a wavelength of about 50 nm and the valley-to-peak amplitude of 3 nm.

Within the investigated energy range, the ion-sputtering mechanism of alkali halides is usually dominated by electronic processes rather that by ballistic collisions [31] and leads to the surface erosion in a layer-by-layer mode. Therefore, such a process is not able to produce adparticles with equal amount of K and Br atoms. Thus, the adislands observed for the oblique Ar^+ irradiation (Fig. 4.5a) are formed due to the ballistic component of sputtering. These particles can easily diffuse and finally agglomerate into KBr adislands. As the agglomeration takes place on the KBr surface, it is natural to expect for the formed islands to retain KBr surface symmetry. As shown in Fig. 4.5a, for 4 keV Ar^+, the adislands cover about 8% of the imaged area, which proves that there is an efficient production of admolecules on the irradiated surface. The contribution of adislands is negligible for 4 keV He^+ ion irradiation. Such behavior is, however, expected for much lighter projectile as the ballistic energy transfer decreases with an increase in the mass difference between colliding partners.

We recognize twofold effect of the ballistic component of sputtering on erosion of KBr surface. First, impinging projectile can produce separate vacancies on the perfect terraces. This effect can be achieved by either ejection or relocation of the surface atoms from their original positions. Most of these vacancies are one layer deep. However, deeper structures can be also created if the bottom of the pits formed by agglomerated vacancies is hit by another projectile and another atom is removed. The created vacancy serves as a trapping site for diffusing-excited F-centers, which leads to ejection of alkali atoms and, consequently, to a development of additional pits. This sputtering channel creates numerous precursor sites for more efficient electronic desorption. Such a mechanism implies that the surface erosion proceeds in a multilayer mode (simultaneous erosion of a few atomic layers) and retains surface lattice symmetry. Since the memory of the initial ion beam direction is lost during F-center diffusion, this type of surface roughening is insensitive to the direction of the primary ion beam.

However, there is also a second process initiated by a ballistic energy transfer that will lead to a development of surface morphological structures that are dependent on the direction of the ion beam. Namely, as shown by MD simulations [44], the sputtering efficiency of the ascending and descending step edges illuminated by the oblique ion beam is different by about factor 4. This is similar to what observed by Hansen *et al.* for Ar^+ bombardment of a single crystal of Pt [37]. The phenomenon is attributed to a planar subsurface sputtering that forces the projectile to move in the subsurface volume. During this movement, the projectile stimulates ejection or relocation of numerous surface atoms along its path. Some of these events can occur quite far away from the step edge. This mechanism takes place only for irradiation of the ascending edge. Projectiles irradiating descending edge or side walls of the pits are not channeled. The vacancy sites created during subsurface channeling serve as a trapping site for diffusing excited F-centers. This process will additionally enhance total sputtering yield during bombardment of ascending edge, leading to the formation of numerous pits along the path of channeled projectile. These pits will grow fast and merge together due to electronic desorption channel. Both more efficient ejection of atoms from the ascending edge and formation of expanding pits along trajectory of the channeled projectile will lead to a development of surface anisotropy related to the ion beam direction, as found in our experiments. It should be emphasized that the development of KBr surface anisotropy will be always related to the ion beam direction independently of the ion beam azimuthal direction. As we have already pointed out, the ripple ridges are formed from terraces with local edges aligned along the main [100] and [010] directions. Because the angle of incidence is "close" to [110] direction, both local edge directions are equally populated. Then the change in the azimuthal angle will result in depopulation of one local direction of the ridge in favor of the other main direction. However, the overall direction of the ripples will always follow the ion beam direction due to large difference in the erosion rate for the ascending and descending surface step edges.

In order to quantitatively characterize the observed ion-induced morphologies, we have calculated a so-called anisotropy factor of the surface morphology and the *rms* roughness of the rippled surface for various ion fluences. The results are shown in Fig. 4.6. The surface anisotropy is defined as the ratio of the parallel and perpendicular, with respect to the ion beam direction, width of the central peak in a 2D self-correlation function map calculated from the nc-AFM images (insets in Fig. 4.5). Thus the value of the anisotropy higher than 1 corresponds to surface structures, which are elongated along the ion

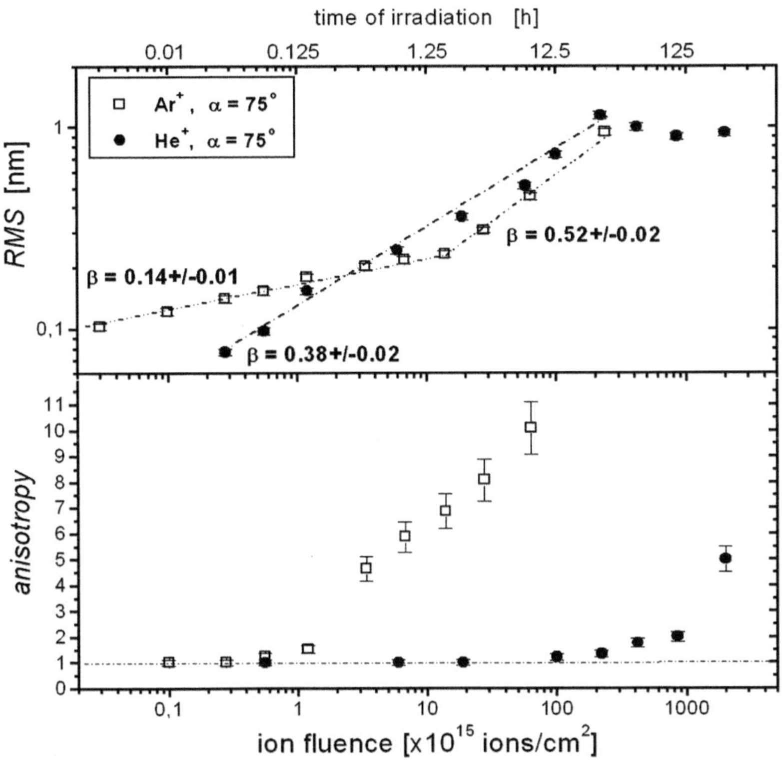

Figure 4.6 (a) *rms* and (b) anisotropy vs the total ion fluence (equivalent to the total time of irradiation) of the ion-induced morphologies of KBr(001) surface under bombardment with 4 keV and $\alpha = 75°$ Ar ions (□) and He ions (•) beams. The lines in (a) represent the fits according to the scaling power law $rms \sim t^{\beta}$. The corresponding values of growth exponent β are shown in the graph. The anisotropy factor is defined as the ratio of the parallel and perpendicular, with respect to the ion beam direction, width of the central peak in a 2D self-correlation function map. Reproduced from [44].

beam direction, whereas the anisotropy lower than 1 corresponds to the elongation in the direction perpendicular to the ion beam. It is clear from Fig. 4.6b that the overall behavior of the surface bombarded by 4 keV Ar^+ and He^+ ion projectiles is somewhat similar, as the development of the surface anisotropy parallel to the primary ion beam direction is visible in both cases. However, there is also a significant difference. The threshold for the development of anisotropy is shifted to much larger fluence for He^+ projectile. This difference can be attributed to a lower contribution of the ballistic sputtering, a lower probability of the surface defect creation by ballistic atom relocation due to a planar channeling in the subsurface volume, and a smaller difference of sputtering yields for ascending and descending edges.

Even larger differences in behavior of the irradiated surface are visible in Fig. 4.6a in which the *rms* versus sputtering time (t) is presented. The data indicate that, in general, the *rms* scales as $rms \sim t^{\beta}$, where β is the growth exponent [45]. The KBr surface evolution under the bombardment with He^+ ions can be described with a single $\beta = 0.38 \pm 0.02$ (before saturation of the *rms*). Interestingly for Ar^+ irradiation, this dependence cannot be described by a single $\tilde{\beta}$. Instead, there is a crossover in the scaling law from $\beta = 0.14 \pm 0.01$ at low fluence regime (up to 2×10^{16} ions cm^{-2}) to $\beta = 0.52 \pm 0.02$ at high fluence irradiation.

How can we understand the observed scaling behavior of the ion-irradiated KBr surfaces? The data presented in Fig. 4.6b show that the irradiation with He^+ ions leads to isotropic surface erosion for a considerably long irradiation time. At this stage of irradiation, the ion-induced morphology develops in a manner reflecting the initial surface symmetry. The anisotropic effects related to the ion beam direction are small due to a low contribution of the ballistic process to the total sputtering of the crystal. Creation of new vacancy defects at the bottom of the pits forms new trapping centers, which can further expand by alkali atom removal due to trapping of diffusing excited F-centers, seems to be the main contribution of the ballistic process to the surface erosion. These defects are growing fast as the process is controlled by efficient electronic desorption. This mechanism, in turn, implies that with the increasing time of irradiation, there is an increase in the *rms* value due to the increasing number of simultaneously eroded atomic layers. Finally, at the high ion fluence, a rough morphology with a considerable concentration of the pit/terrace edges develops allowing for the effect of anisotropic projectile-induced etching of the step edges to be noticeable. As a result, anisotropic structures are created on irradiated KBr surfaces. The presented above scenario of KBr(001) surface erosion under the He ion impact is supported by the results of Monte Carlo simulations concerning the erosion of carbon surface with 4 keV Ar ions performed by Koponen *et al.* [46]. They have found that *rms* of the surface eroded only by atom removal without any relaxation processes scales with the growth exponent of $\beta = 0.33$–0.34. These values are very close to the ones we got for He^+ irradiation where the KBr surface erosion proceeds predominately via electronic processes leading to step atom removal. The simulations performed by Koponen *et al.* show also that the presence of efficient relaxation processes, inducing smoothing of the surface during the ion irradiation, affects the growth exponent shifting it to a lower value $\beta = 0.14$–0.17. As shown in Fig. 4.6a, our results for low fluence irradiation by Ar^+ projectiles can be also described by the same growth exponent. As shown by our MD simulations in this stage of irradiation, efficient formation of adatoms and admolecules by ballistic collisions is important. It is postulated, therefore, that surface diffusion of these particles is responsible for efficient surface smoothing. The restructuring of the ion beam illuminated steps as well as filling of the ion-induced terrace damages by the diffusing admolecules can be a plausible explanation for the slower roughening of the irradiated surfaces.

For a prolonged irradiation with Ar ions at $\alpha = 75°$, we observe the change of a growth characteristics. After a crossover (total fluence about 2×10^{16} ions cm^{-2}), the growth exponent changes its value to $\beta = 0.52 \pm 0.02$. In Fig. 4.7, the KBr surface topographies acquired just before and after crossover point are shown. The image collected before the crossover (Fig. 4.7a, $\Phi_{ion} = 1.4 \times 10^{16}$ ions cm^{-2}) shows the relatively smooth KBr surface, as seen in the cross section profile of the image. At this fluence, the pits are already elongated; however, they are still separated. The effect of a preferential erosion of illuminated (ascending) steps resulting in the elongation of the pits in the direction of ion beam dominates the morphology evolution. It seems that the key difference between surfaces irradiated with low and high ion fluence is the agglomeration of separate pits into three-dimensional vacancy agglomerates, i.e., groove-like structures. We would like to

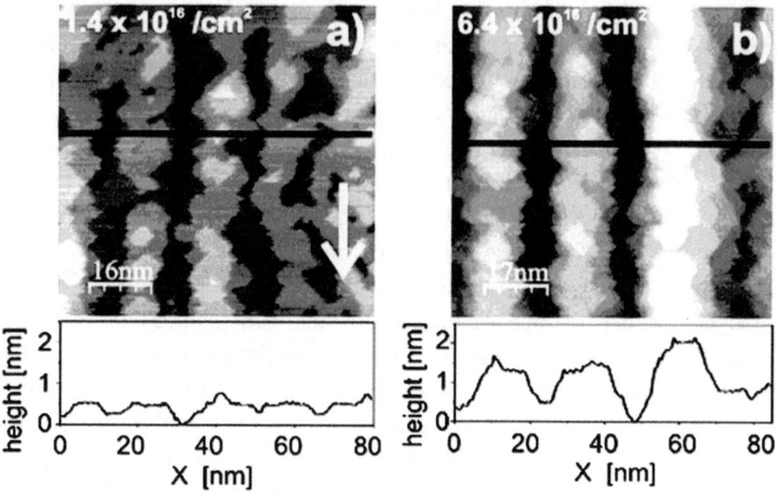

Figure 4.7 Nc-AFM images of the ion-induced topographies of KBr(001) surface (4 keV Ar^+ ions with $\alpha = 75°$) at the two irradiation time periods: (a) initial stage of irradiation (total ion fluence 1.4×10^{16} ions cm^{-2} and $\beta = 0.14 \pm 0.01$) and (b) the later stage after crossover time (total ion fluence 6.4×10^{16} ions cm^{-2} and $\beta = 0.52 \pm 0.02$). The cross-section profiles taken along the lines indicated in the topography images are shown below. Reproduced from [44].

emphasize that desorption mechanism operating in the low fluence regime by its nature leads to amplification of the surface roughness due to the electronic roughening; the deeper are the grooves, the larger is the number of step edge precursor sites, the higher is the desorption efficiency leading to new vacancy production. At a certain stage of the ion irradiation (the crossover time), separated grooves begin to overlap leading to creation of deep, groove-like structures of sub-micrometers length. The appearance of the long grooves is associated with a substantial decrease in the total length of ascending, step edges (Fig. 4.7b, $\Phi_{ion} = 6.4 \times 10^{16}$ ions cm^{-2}) and, consequently, with a decrease of the efficiency of KBr admolecules creation. As a consequence, the smoothing of the surface by diffusing adparticles is reduced and the electronic roughening mechanism starts to dominate the surface evolution causing change in the dynamics of the *rms* growth.

4.7 Concluding Remarks

In this overview, we report on the low-energy ion-beam-induced nanostructuring process of single-crystal ionic surfaces. We have found that for the angle of incidence close to normal, isotropic surface morphology develops with symmetry, which predominantly reflects the symmetry of the lattice structure. On the other hand, for oblique incidence, the ballistic processes initiated by a direct momentum transfer retain the memory of the initial impact direction. As a result, on the irradiated surfaces, anisotropic morphology is developed in the form of periodic ripples. We have shown that ion-irradiated alkali halides remain their crystallinity on the surface even for a prolonged bombardment at room temperature. This is due to high efficiency of the electronic processes, including defect diffusion and recombination of vacancies with the structural defects such as step edges, kinks and corners, preserving a high degree of the surface crystallographic order even upon large fluence irradiation.

It is worth to note that the above explanation of the mechanism of the ripple development on alkali halide surfaces, in some way, implies analogy between the curvature-dependent sputtering by isotropic collision cascades and the electronic roughening observed in case of ion-induced erosion of ionic crystal surfaces. Similar to the curvature-dependent sputtering, the electronic roughening acts as a destabilizing mechanism for the irradiated surface. The deeper are the pits, the larger is the number of the step edge precursor sites resulting in the higher desorption efficiency. This, in turn, will lead to the creation of new pits. But this is where the analogy ends. Due to the dominance of the electronic erosion channel, the irradiated surfaces remain well ordered, even for very large ion fluences. Consequently, due to preserved surface crystallinity, there is a lack of smooth and continuous variation of the local erosion yield with the surface morphology. These findings together with the lack of the exponential growth of the surface roughness (instead, we observe the scaling behavior of the *rms*) imply that none of the basic Bradley–Harper assumptions is fulfilled, emphasizing the inapplicability of the continuum theory for the description of the ion-induced ripple formation on alkali halide crystal surfaces.

Acknowledgments

The work was supported by the Polish National Science Centre Program No. DEC-2011/03/B/ST5/01567 and partially by the Polish Ministry of Science and Higher Education (268/N-ESF/2008/0).

References

1. Chan, W. L., and Chason, E. (2007). *J. Appl. Phys.*, **101**, 121301.
2. Valbusa, U., Boragno C., and Buatier de Mongeot, F. (2002). *J. Phys.: Condens. Matter.*, **14**, 8153.
3. Navez, M., Sella, C., and Chaperot, D. (1962). *Compt. Rend.*, **254**, 240.
4. Rusponi, S., Boragno, C., and Valbusa, U. (1997). *Phys. Rev. Lett.*, **78**, 2795.
5. Ramana Murty, M. V., Curcic, T., Judy, A., Cooper, B. H., Woll, A. R., Brock, J. D., Kycia, S., and Headrick, R. L. (1999). *Phys. Rev. B*, **60**, 16956.
6. Redinger, A., Hansen, H., Linke, U., Rosandi, Y., Urbassek, H. M., and Michely, T. (2006). *Phys. Rev. Lett.*, **96**, 106103.
7. Facsko, S., Dekorsy, T., Koerdt, C., Trappe, C., Kurz, H., Vogt, A., and Hartnagel, H. L. (1999). *Science*, **285**, 1551.
8. Chason, E., Mayer, T. M., Kellerman, B. K., McIlroy, D. T., and Howard, A. J. (1994). *Phys. Rev. Lett.*, **72**, 3040.
9. Carter, G., Vishnyakov, V., and Nobes, M. J. (1996). *Nucl. Instrum. Methods B*, **115**, 440.
10. Mayer, T. M., Chason, E., and Howard, A. J. (1994). *J. Appl. Phys.*, **76**, 1633.
11. Datta, A., Yuh-Renn, Wu., and Wang, Y. L. (2001). *Phys. Rev. B*, **63**, 125407.
12. Heller, R., Facsko, S., Wilhelm, R. A., and Möller, W., (2008). *Phys. Rev. Lett.*, **101**, 096102.
13. Bradley, R. M., and Harper, J. M. E. (1988). *J. Vac. Sci. Technol. A*, **6**, 2390.
14. Sigmund, P. (1973). *J. Mater. Sci.*, **8**, 1545.
15. Mullins, W. W. (1957). *J. Appl. Phys.*, **28**, 33.
16. Makeev, M. A., and Barasi, A. L. (1997). *Appl. Phys. Lett.*, **71**, 2800.
17. Cuerno, R., and Barabasi, A.-L. (1995). *Phys. Rev. Lett.*, **74,** 4746.
18. Kahng, B., Jeong, H., and Barabási, A.-L. (2001). *Appl. Phys. Lett.*, **78**, 805.
19. Facsko, S., Bobek, T., Stahl, T. A., Kurz, H., and Dekorsy, T. (2004). *Phys. Rev. B*, **69**, 153412.

20. Castro, M., Cuerno, R., Vázquez, L., and Gago, R. (2005). *Phys. Rev. Lett.*, **94**, 016102.

21. Madi, C. S., Anzenberg, E., Ludwig Jr., K. F., and Aziz, M. J. (2011). *Phys. Rev. Lett.*, **106**, 066101.

22. Wilson, R. M., and Williams, R. T. (1995). *Nucl. Instrum. Methods B*, **101**, 122.

23. Barth, C., and Henry, C. R. (2004). *Nanotechnology*, **15**, 1264.

24. Erlebacher, J., Aziz, M. J., Chason, E., Sinclair, M. B., and Floro, J. A. (1999). *Phys. Rev. Lett.*, **82**, 2330.

25. MacLaren, S. W., Baker, J. E., Finnegan, N. L., and Loxton, C. M. (1992). *J. Vac. Sci. Technol. A*, **10**, 468.

26. Rossnagel, S. M., Robinson, R. S., and Kaufman, H. R. (1982). *Surf. Sci.*, **123**, 89.

27. Ditchfield, R., and Seebauer, E. G. (1999). *Phys. Rev. Lett.*, **82**, 1185.

28. Postawa, Z., Maboudian, R., El-Maazawi, M., Erwin, M. H., Wood, M. C., and Winograd, N. (1992). *J. Chem. Phys.*, **96**, 3298.

29. Postawa, Z. (1994). *Radiat. Eff. Defects Solids*, **128**, 107.

30. Szymonski, M. (1990). *Nucl. Instrum. Methods B*, **46**, 427.

31. Krok, F., Kolodziej, J. J., Such, B., Czuba, P., Piatkowski, P., Struski, P., and Szymonski, M. (2004). *Nucl. Instrum. Methods B*, **226**, 601.

32. Batzill, M., Bardou, F., and Snowdon, K. J. (2001). *J. Vac. Sci. Technol. A.*, **19**, 1829.

33. Sigmund, P. (1969). *Phys. Rev.*, **184**, 383.

34. Mussi, V., Granone, F., Boragno, C., Buratier de Mongeot, F., Valbusa, U., Marolo, T., and Montereali, R. M. (2006). *Appl. Phys. Lett.*, **88**, 103116.

35. Makeev, M. A., Cuerno, R., and Barabasi, A. L. (2002). *Nucl. Instrum. Methods B*, **197**, 185.

36. Frost, F., Schindler, A., and Bigl, F. (2000). *Phys. Rev. Lett.*, **85**, 4116.

37. Hansen, H., Redinger, A., Messlinger, S., Rosandi, Y., Urbassek, H. M., Linke, U., and Michely, T. (2006). *Phys. Rev. B*, **73**, 235414.

38. Saeed, S. R., Sinha, O. P., Krok, F., and Szymonski, M. (2008). *Appl. Surf. Sci.*, **255**, 1766.

39. van Dijken, S., Jorritsma, L. C., and Poelsema, B. (1999). *Phys. Rev. Lett.*, **82**, 4038.

40. Salminen, O., Riihola, P., Ozols, A., and Viitala, T. (1996). *Phys. Rev. B*, **53**, 6129.

41. Goryl, M., de Mongeot, F. B., Krok, F., Vevecka-Priftaj, A., and Szymonski, M. (2007). *Phys. Rev. B*, **76**, 075423.

42. Szymonski, M. (1982). *Nucl. Instrum. Methods* B, **194**, 523.
43. Bennewitz, R., Schar, S., Barwich, V., Pfeiffer, O., Meyer, E., Krok, F., Such, B., Kolodziej, J., and Szymonski, M. (2001). *Surf. Sci.*, **474**, L197.
44. Krok, F., Saeed, S. R., Postawa, Z., and Szymonski, M. (2009). *Phys. Rev. B*, **79**, 235432.
45. Barabasi, A.-L., and Stanley, H. E. (1995) *Fractal Concepts in Surface Growth*, Cambridge University Press, Cambridge.
46. Koponen, I., Hautala, M., and Sievanen, O.-P. (1996). *Phys. Rev. B*, **54**, 13502.

Chapter 5

Nanostructures on Thin Films by keV Ion Beams

Prasanta Karmakar
Variable Energy Cyclotron Centre, 1/AF, Bidhannagar, Kolkata 700 064, India
prasantak@vecc.gov.in

The low-energy ion beam is a promising candidate for fabrication and characterization of nanostructures. The formation of ripples, dots, and kinetic roughening on a number of thin films, e.g., Au, Pt, Ag, Cu, Co, ZnO, and bulk surfaces of Si(100) and TiO_2(110) under keV energy ion bombardment are discussed. The influence of initial surface perturbations and the importance of local ion impact angle on the development of ion induced surface morphology are studied and the beginning of nanostructure formation by ion bombardment is explored. Regularly spaced electric and magnetic stripe formation and the potential sputtering by multicharged Ar ions of oxidized part of prefabricated Si nanoripples are also revealed.

5.1 Introduction

To investigate the mechanism of ion-induced nanostructure formation, a large number of works on bulk surfaces have been performed

Nanofabrication by Ion-Beam Sputtering: Fundamentals and Applications
Edited by Tapobrata Som and Dinakar Kanjilal

ISBN 978-981-4303-75-0 (Hardcover), 978-981-4303-76-7 (eBook)
www.panstanford.com

[1–3]. The nanostructure formation on thin films deposited on substrate is more interesting because of industrial applications of such nanostructured thin films. Moreover, ion-induced isolated quantum nanostructure formation is possible by sputtering of thin films. Metal nanostructures produced by ion bombardment on metal thin films are also attractive for plasmonic applications.

Thin films of metals are deposited on polished Si(100) wafers. The film thicknesses were in the range of 30–200 nm. These samples were then sputtered with mass analyzed keV ion beams to produce nanostructures. To explore the basic mechanism of ion-induced nanopatterning, bulk surface of Si(100) and TiO2(110) were also studied under ion bombardment.

5.2 Variation of Surface Patterns

A simple way of changing the pattern of ion-induced nanostructures is the variation of ion incidence angle. It has been found (Fig. 5.1) that at near normal Ar^+ incidence, dot-like structures are developed, while at larger angles of incidence (20–60°), checkerboard and mound structures are developed. At 70° a remarkable transition of the topography occurs, showing the development of arrays of

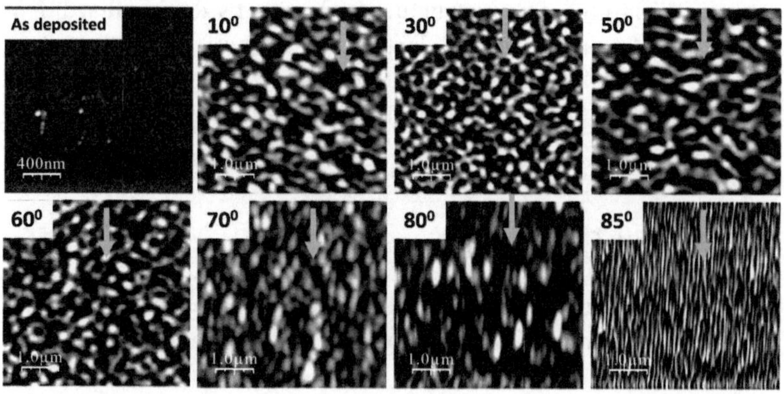

Figure 5.1 AFM images showing a sequence of evolution of the surface topography of the Pt film with increasing bombarding angle. The projectile is 16.5 keV Ar^+ with a fluence of 5×10^{16} ions cm^{-2}. The ion beam direction is from the bottom to the top.

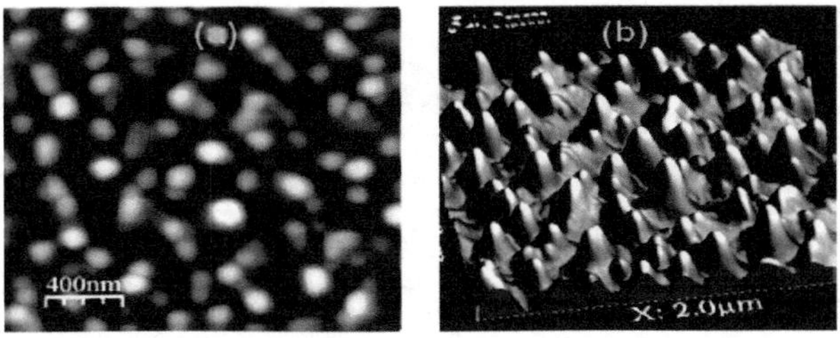

Figure 5.2 Surface morphology (a) 2D view and (b) 3D view showing dot-like structures after 16.5 keV Ar^+ ion bombardment at an incident angle of 10° on Co thin film. The ion fluence is 5×10^{16} ions cm^{-2}.

mounts, inclined along the ion beam direction. Further increase in the incidence angle shows the beginning of distinct changes of the surface morphology, which ultimately ends up with regular ripple structures at grazing ion incidence (80–85°). The results are not only typical to Pt films but also valid for other metallic films, namely, Co, Cu, Ag, and Au.

5.2.1 *Metal Dot Formation*

We experimentally found similar dot-like structures of different polycrystalline metal thin films under normal or near-normal Ar^+ ion bombardment at room temperature. These structures are found to be stable for a long time at room temperature. A thin film of Co deposited on Si(100) was bombarded with 16.5 keV Ar^+ ion at an angle of 10°. As shown in Fig. 5.2, dot-like structures are formed at an ion dose 5×10^{16} ions cm^{-2}. Similar structures are also observed for 16.5 keV Ar^+ ion-bombarded Ag thin film at an angle 10° and for Pt thin film at 0° ion incidence [4].

5.2.2 *Ripple Formation on Thin Films*

Figure 5.3 shows some selected atomic force microscopy (AFM) images of the sputtered Co, Cu, Ag, Pt, and Au films, respectively, at the angles of 60°, 70°, and 80°. At 60°, weakly pronounced ripples with wave vector parallel to the ion beam direction, especially in Co and Cu films, appear at projection of the ion beam direction. Finally

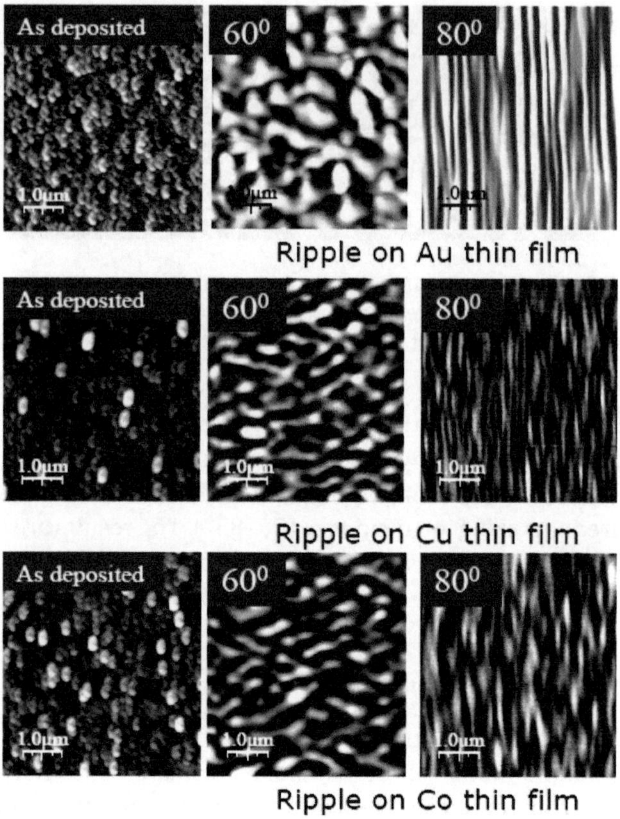

Figure 5.3 AFM images of the unbombarded and 16.7 keV Ar^+ sputtered Au, Cu, and Co surfaces at different angles of incidence θ as indicated. The bombarding fluence is 1×10^{17} ions cm^{-2}. See also Color Insert.

at 80°, regular ripple-like surface instability, with the wave vector perpendicular to the ion beam direction is developed.

5.3 Origin of Ion-Induced Nanostructure Formation

The Bradley–Harper model [5], based on Sigmund's sputtering theory [6], assumes an ellipsoidal shape of the collision cascade. If the surface has local curvature, more energy from the collision cascades reaches to the valleys than the hills and therefore, the preferential sputtering of the valleys generates instability. In addition to the sputtering-induced instability, thermal diffusion was also considered, which is responsible for the smoothening of the surface during bombardment.

It is obvious that to initiate the curvature-dependent instability, the presence of local surface curvature is essential. Therefore, for an initial flat surface and fixed angle of ion incidence, no modulated structures could be expected. But experiments with initial flat surface produces ripple structures. Bradley and Harper assumed an initial perturbation present on the surface, and sputtering increases the amplitude of the perturbation, but the origin of perturbation was not clearly mentioned. The possible origins of perturbations are (i) stochastic nature and in-homogeneity of the incident ion beam; (ii) initial surface roughness; (iii) preferential sputtering [7, 8]; and (iv) chemical reactivity of projectile and presence of surface impurities [9]. These are described in the following subsections.

5.3.1 *Initial Roughness*

Single ion impact on clean single crystal surface showed [10] that at the point of impact, both vacancies and adatoms are created. Because of stochastic nature of incident ion beam, these defects are expected to be randomly distributed on the surface resulting in a rough surface at the early stage of ion bombardment or prior to the regular structure formation. Thus, rough surface, which is supposed to be the origin of initial perturbations, is necessary for ion-induced curvature-dependent structure formation. Therefore, if the initial roughness is incorporated by any simple way, the ion-induced structure formation is expected to be influenced.

Here we have studied the effect of initial roughness on well-known and well-studied ripple formation on Si(100) surfaces. We have incorporated random roughness on Si(100) by simple chemical process and studied the 16.7 keV molecular oxygen ion-beam-induced ripple pattern on various initially rough surfaces and compared the results with initial flat surface. We observe that the ripple structures are formed at relatively lower ion fluence in presence of initial roughness on the surface. For higher values of initial roughness, additionally we observe the process of kinetic roughening and

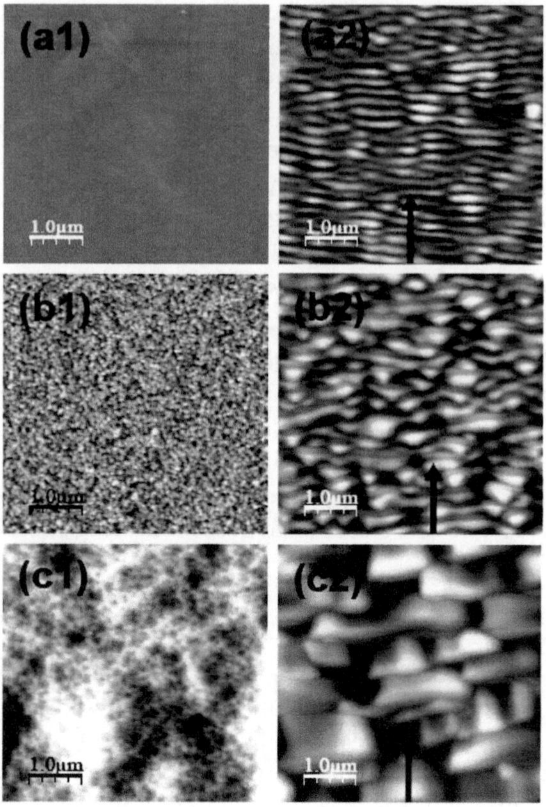

Figure 5.4 (a1) Flat Si(100) of *rms* roughness 0.1 nm. Chemically generated initial rough Si surfaces of *rms* roughness (b1) 3.7 nm, (c1) 31 nm. (a2), (b2), and (c2) represent the corresponding AFM topographies after 16.7 keV O_2^+ bombardment at fluence 6×10^{17} atoms cm^{-2}, respectively. The arrows indicate the beam direction.

rotation of surface pattern at moderate ion fluence, which are only observed at very high fluence in case of an initial flat surface.

Si(100) samples were cleaned with trichloroethelene followed by methanol in an ultrasonic bath. The cleaned and dried samples were then immersed in a prepared solution of 5 mol L^{-1} HNO_3 and 12 mol L^{-1} HF at room temperature. The samples were bombarded by 16.7 keV O_2^+ at an angle of 60° with respect to the surface normal. The ion fluence was varied from 2×10^{16} to 1.1×10^{19} atoms

cm^{-2}. For comparison, the morphologies of the unetched Si(100) surface before and after sputtering at the same fluence are shown in Fig. 5.4 (a1) & (a2).

To investigate the fluence-dependent structure formation on initially rough surfaces, we bombarded the flat Si (roughness 0.1 nm) and the etched Si of roughness 7 ± 2 nm, 9 ± 2 nm, and 28 ± 2 nm, respectively, at various ion fluences. Figures 5.5a–f show AFM images obtained from the bombarded Si samples with initial roughness 7 ± 2 nm. The ripple structures start to form at fluence

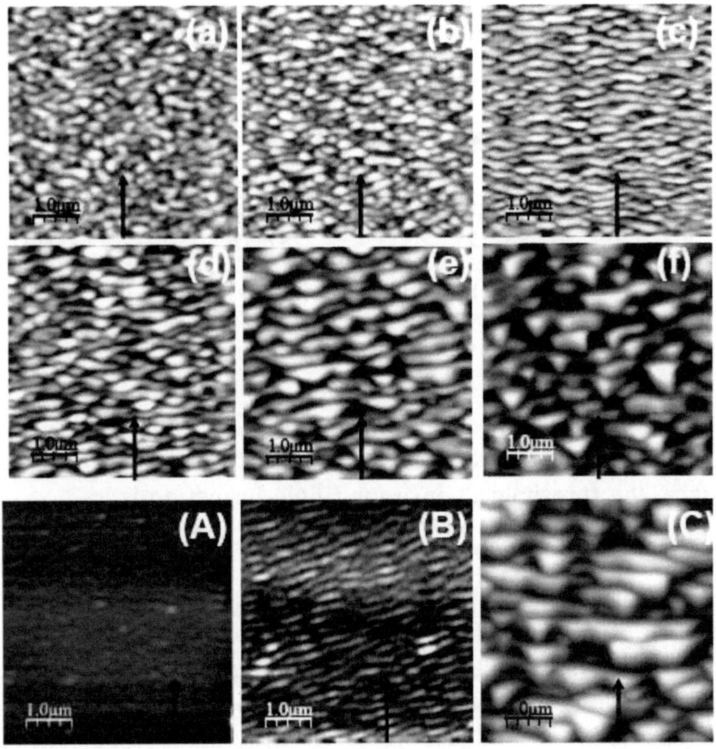

Figure 5.5 AFM topographies of 16.7 keV O_2^+ ion-bombarded Si of initial roughness 7 ± 2 nm at fluences (a) 3×10^{16} atoms cm^{-2}, (b) 5×10^{16} atoms cm^{-2}, (c) 1×10^{17} atoms cm^{-2}, (d) 2×10^{17} atoms cm^{-2}, (e) 4×10^{17} atoms cm^{-2}, and (f) 6×10^{17} atoms cm^{-2}. Morphology of the flat Si(100) surface after bombardment at (A) 2×10^{17} atoms cm^{-2}, (B) 3×10^{17} atoms cm^{-2}, and (C) 1.4×10^{18} atoms cm^{-2}. The arrows indicate the beam direction.

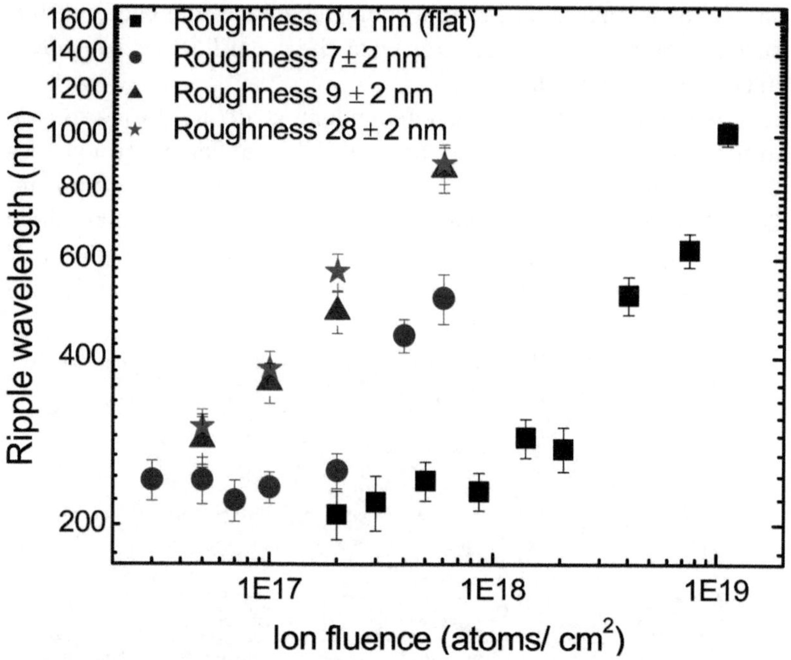

Figure 5.6 Ripple wavelength vs. ion fluence for flat and different initially rough Si surfaces.

of 3×10^{16} atoms cm^{-2} only; while for the initial flat Si (100) surface, the threshold fluence for ripple formation is $\approx 3 \times 10^{17}$ atoms cm^{-2} (Fig. 5.5B). If ion sputtering is continued on both the initially flat and 7 ± 2 nm rough surfaces, the ripples grow with fluence, but at high fluence the ripples starts to disintegrate leading to kinetic roughening of the surface [11]. Similar formation of ripple structures on flat Si(100) surface and their dynamics were reported before for the same ion beam parameters [12]. It is interesting to note that the kinetic roughening starts around the fluence 1.4×10^{18} atoms cm^{-2} for the flat surface (Fig. 5.5C), while it is 4×10^{17} atoms cm^{-2} for the 7 ± 2 nm rough surface (Fig. 5.5f). Figure 5.6 shows the variation of ripple wavelength with ion fluence for both flat and initially rough surfaces. For the flat surface, the wavelength does not change with the fluence at lower ion fluence regime following the Bradley–Harper theory. For the 7 ± 2 nm rough surface, the identical initial ripple wavelength invariance is also observed, but the ripple starts to develop at lower ion fluence than that for the flat surface. However, for higher initial rough surfaces, e.g., 9 ± 2 nm and 28 ± 2 nm, invariance of ripple wavelength is not observed; instead, the wavelength shows an increase from the beginning of ion bombardment (Fig. 5.6).

For long-time ion bombardment, kinetic roughening and saturation of ripple amplitude were predicted by the nonlinear extension of the linear Bradley–Harper theory [11, 13]. Experimentally, the nonlinear effect has been reported in case of Ar^+-bombarded Si(100) [14] and graphite (HOPG) [15] surfaces. We also reported earlier [12] the kinetic roughening in case of O_2^+-bombarded flat Si(100) (for long ion bombardment time). For the case of 7 ± 2 nm initial rough surface, the observation of kinetic roughening (Fig. 5.2f) and the increase in ripple wavelength with ion fluence (Fig. 5.6) indicate that the nonlinear effects start at moderate ion fluence compared to that in the flat surface. The initial invariance of ripple wavelength with ion fluence for both flat and 7 ± 2 nm rough surfaces (Fig. 5.6) is in qualitative agreement with the Bradley–Harper linear theory, which predicts the ripple wavelength to be independent of ion fluence. In case of higher initial rough surfaces (9 ± 2 nm and 28 ± 2 nm), the increase in ripple wavelength from the beginning of ion bombardment (Fig. 5.6) indicates that the nonlinear effects start at the early stages of ion bombardment.

5.3.2 *Preferential Sputtering*

When ions bombard binary compound materials, one component is sputtered faster than the other because of difference in mass, binding energy, and/or vapor pressure [16]. Such preferential sputtering of one component generates faster initial perturbation in terms of vacancy generation on the surface.

Titanium dioxide (TiO_2) is one of the most studied metal oxides due to its use as an efficient photocatalytic material and as a support for catalytically active metal nanoclusters. In this section, nanometer-size vacancy clusters are created by Ar^+ sputtering TiO_2(110) and shown how preferential sputtering plays a role in nanopatterning [7]. The experiments were performed in an ultrahigh vacuum chamber with a base pressure of 1×10^{-10} Torr. A $TiO_2(110)$ single crystal was attached to a Ta sample holder and heated by electron beam heating from behind. The crystal was cleaned by cycles of ion bombardment and annealing (500 eV Ar^+ for 10 min, 975 K for 10 min) until a sharp (1×1) low-energy electron diffraction pattern was obtained and XPS showed no contamination. Clean TiO_2 was then sputtered at room temperature by normally incident 0.5 keV Ar^+ (current density = 1 μA cm^{-2}) at various ion fluences. XPS spectra and scanning tunneling microscope (STM) images were collected after each sputtering and post-annealing step.

The morphology of clean and ion-sputtered TiO_2 at various Ar^+ fluences is shown in the STM images of Fig. 5.7. The terraces and steps of virgin $TiO_2(110)$ are visible in Fig. 5.7a, which is similar to that seen by Fischer *et al.* [17]. Figure 5.7b was collected after

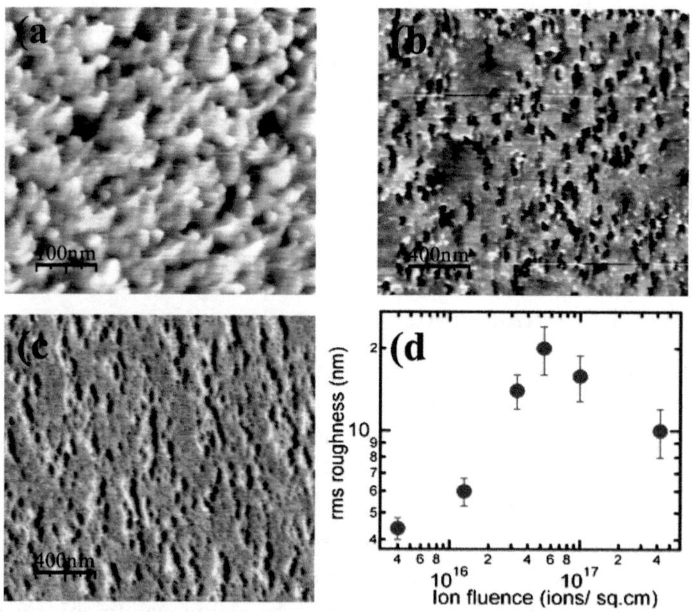

Figure 5.7 STM topography of $TiO_2(110)$ (a) virgin surface, (b) after sputtering with 0.5 keV Ar^+ at 3.3×10^{16} ions cm^{-2}, (c) 4.2×10^{17} ions cm^{-2}, and (d) the *rms* roughness as a function of ion fluence.

bombardment by 3.3×10^{16} ions cm^{-2} of Ar^+ and shows defects and nanometer-scale cavities. Following prolonged bombardment, the vacancies are elongated along the $[1\bar{1}0]$ direction and extend into the subsurface layer, as seen in Fig. 5.7c. Defects and cavities increase the surface roughness. The average roughness, calculated from a collection of STM images, is plotted as a function of sputtering fluence in Fig. 5.7d. The roughness initially increases up to a fluence of 6×10^{16} ions cm^{-2} and then decreases following additional bombardment.

Figure 5.8 shows XPS data for $TiO_2(110)$ collected after each sputtering cycle and then after post-annealing. The Ti $2p$ spectra are shown in Fig. 5.8, in which the $2p_{3/2}$ binding energies for different oxidation states are depicted by vertical lines [18, 19]. Only the Ti^{4+} component is present on the virgin surface, while Ti^{3+}

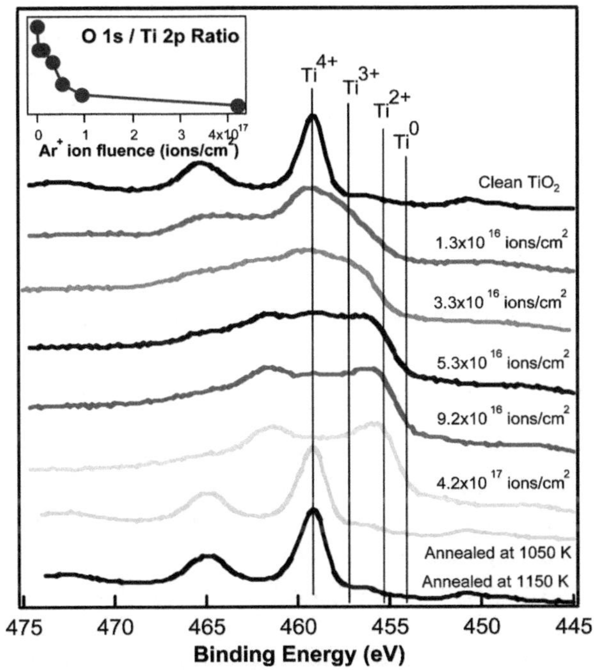

Figure 5.8 Ti $2p$ XPS spectra collected after sputtering at different Ar^+ fluences and post-annealing. The inset shows the O $1s$ to Ti $2p$ ratio as a function of ion fluence.

and Ti^{2+} emerge with sputtering. At the highest Ar^+ fluence, the Ti^{4+}component has vanished. The appearance of Ti^{3+} and Ti^{2+}, in addition to Ti^{4+}, was previously reported for TiO_2(100) bombarded by 2 keV Ar^+ [20].

The O 1*s* level decreases in intensity and shifts slightly to higher binding energy with ion bombardment. The inset in Fig. 5.8 shows how the intensity ratio of oxygen to titanium decreases with sputtering. Oxygen is preferentially sputtered due to its lower mass, and the initial fast decrease suggests that the oxygen atoms from the outermost layer are the first to be removed. This was also seen for TiO_2 by Pan *et al.* when the stoichiometry of the sputtered material was measured with XPS and LEIS [21].

The initiation of ion-induced structure formation on TiO_2 by Ar^+ ion impact can be described by random and discrete arrival of ions and preferential sputtering of oxygen. The flux in the present experiment is 6.2×10^{12} ions cm^{-2} sec^{-1}, so that, on an average, one ion is incident in a 16 nm^2 surface area per second. The lateral extension of the collision cascade for a 0.5 keV Ar^+ ion in TiO_2 is ~1 nm [22], and the lifetime of a collision cascade is on the order of 10^{-13}–10^{-12} s [23]. Thus, the overlap of collision cascades initiated by sequentially implanted ions is extremely rare and ion impact can be considered a discrete and random process. In addition, preferential sputtering of oxygen increases the initial instability, as random atomic vacancies, primarily of oxygen, are created at the onset of sputtering. This initial defect formation disrupts the flatness of the surface.

5.3.3 *Chemical Reactivity/Surface Contamination*

The development of the nanostructure also depends on the reactivity of the projectile ions with the sample. When Si(100) is bombarded with projectiles like N^{1+} and O^{1+}, the initial bombardment develops different phases on the sample surface due to the reactive property of the projectile ions. Further bombardment leads to the development of nanostructure due to the nonuniform sputtering of the same surface because of their compositional change [24]. In case of bombardment of oxygen and nitrogen ions on Si(100), the ripple formation is well defined compared to argon and helium ions. Due to higher reactivity of the N^{1+} and O^{1+} ions, Si_xN_y and SiO_z are

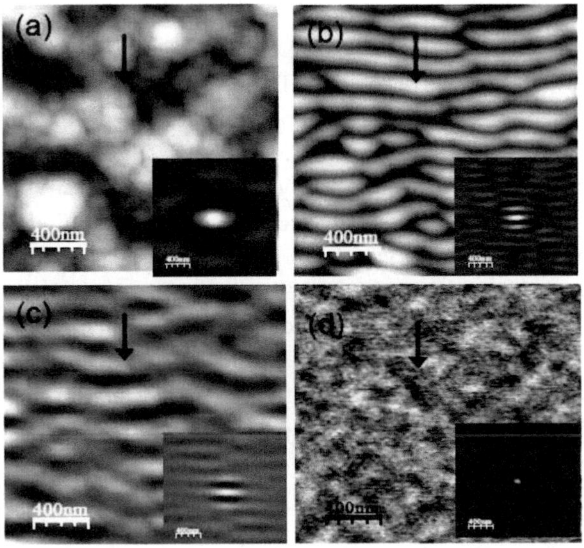

Figure 5.9 AFM images of the samples bombarded with 8 keV (a) He^+, (b) N^+, (c) O^+, and (d) Ar^+ ions at an angle of 60° for ion fluence of 1×10^{18} ions cm^{-2}.

formed, respectively. The change in chemical composition leads to nonuniform sputtering yield of the sample surface, which results in perturbation and consequent quick ripple structure formation [24]. The Fourier transform infrared spectroscopy (FTIRS) study of superficial Si layers, formed by 9 keV N_2^+ ion bombardment demonstrated the existence of Si_3N_4 absorption centers [25]. The formation of SiO_x for 10 keV O_2^+ bombardment was also shown in ref [26].

AFM images of the samples bombarded at an incidence angle of 60° with surface normal by 8 keV He^{1+}, N^{1+}, O^{1+}, and Ar^{1+} ions for the ion fluence of 1×10^{18} ions cm^{-2} are shown in Fig. 5.9a–d. The figures show that for two ion fluences. Moreover, for the same fluence and energy, the developed structures differ for different projectiles.

Ar^+ ions form a damaged layer with the original composition of Si(100). Argon being an inert gas, the retention of Ar on Si is much lower as compared to oxygen [24] and nitrogen [27]. Elst *et al.* [24] reported that no surface topography is formed under 8 keV

Ar^+ bombardment, while a rough topography is observed under O_2^+ bombardment. Carter *et al.* [28] observed no waves or ripple trains on Si by 5 or 10 keV Ar^+ irradiation at room temperature. In the present case, we observed no well-defined ripple topography for 8 keV Ar ion bombardment, while reactive projectile forms well-defined ripple in this energy regime. The possible reason is that argon being an inert gas could not form sufficient surface perturbation in this energy regime, which is necessary for the ripple formation, whereas reactive gases can do by generating different phases on the surface, which aids the ripple formation. As well, having the closer mass of oxygen and nitrogen, Neon could not produce any major topographic features even at fluences greater than 10^{20} ions cm^{-2} [29]. However, with further increase in either energy or fluence, it is possible to develop ripple structure with the Ar ion on Si. We observed ripple formation with 32 keV Ar^{2+} ion bombardment [30]. Chini *et al.* [31] reported ripple formation by 30 and 60 keV Ar ion bombardment on Si at ion incidence angles of 58° and 60°, respectively. Also Carter *et al.* [29] reported ripple-like structures on Si at room temperature by 20 keV Ar^+ at a very high fluence of 1.7×10^{19} ions cm^{-2}.

Helium being an inert gas, the nanostructure formation is observed because of its very low mass. With an advantage of blister formation, He^+ ion could develop nanostructure on Si(100) [32]. Earlier, Yamauchi *et al.* [33] showed the formation of bubble structure on Si by He ion bombardment. At room temperature, bubbles of 200 nm radius were formed and also the retention of He with silicon increases with the ion fluences at room temperature.

The present section showed that the generation of initial perturbation by pre-roughening, impurities, reactive sputtering, and preferential sputtering helps the ion beams to skip the process of surface roughening for local curvature development. In case of inert ion bombardment on a flat, contamination-free, pure single elemental sample surface, the ion beam itself generates the initial random rough surface. If the local curvature on the surface is readily present or generated by other means, the Bradley–Harper curvature-dependent sputtering takes place from the very beginning of ion bombardment, and ripples are observed at relatively lower ion fluences.

5.4 Dependence on Local Ion Impact Angle

The angle between projectile ion and the target surface is a crucial parameter that determines the shape of collision cascades. The amount of energy to be distributed among the surface atoms for removal or diffusion depends on the shape and position of the collision cascades that leads to the nanostructure formation on the surface. The effective ion incidence angles may differ from the global ion incidence angle if the surface has a local curvature. Here the nanostructures are developed on rough surface and showed how the local ion impact angles control the nanostructure formation.

ZnO nanostructures have been put to a wide range of applications ranging from field emission sources to optical devices such as solar cells, efficient blue light emitters, etc. [34], as well as it is an environment-friendly material.

We have formed ripple structures from rough ZnO thin films deposited on Si(100) substrate by 10 keV O^+ bombardments at impact angles of 80° and 60°. The importance of local ion impact angle toward the perpendicular and parallel ripples are discussed.

Thin films of about 115 nm thick ZnO were deposited by dc magnetron sputtering on clean Si(100) substrates. Prior to the deposition, the Si(100) substrates were cleaned and degreased in ultrasonic bath. The samples were bombarded with mass analyzed 10 keV O^+ ions at incidence angles of 80° and 60° with respect to the surface normal. To investigate the evolution of the surface morphology, the fluence was varied over a wide range from 5×10^{16} ions cm^{-2} to 1×10^{18} ions cm^{-2}.

The surface morphology as a function of ion fluence for the 60° and 80° incidence angles are shown in Fig. 5.10. The AFM image of the unbombarded ZnO film is shown in Fig. 5.10a, whereas Fig. 5.10b shows the morphology of the samples after the bombardment with ion fluence of 5×10^{17} ions cm^{-2} at oblique incidence where the ripple structures are formed in the direction perpendicular to the

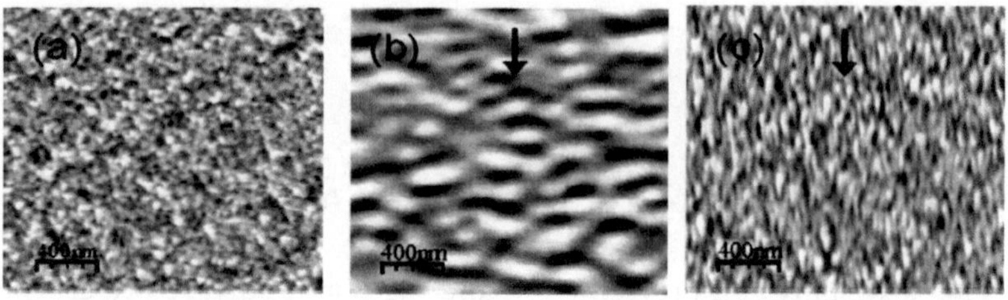

Figure 5.10 AFM images of (a) unbombarded sample, (b) bombarded at 60° with fluence 5×10^{17} ions cm^{-2}, (c) bombarded at 80° with the fluence of 1×10^{18} ions cm^{-2}.

ion beam direction. Similarly, Fig. 5.10c shows the AFM images of the morphology with ion fluence of 1×10^{18} ions cm^{-2} for the grazing angle (80°).

Here the observed perpendicular ripple formation (Fig. 5.10b) on ZnO film at 60° ion bombardment is consistent with the explanations of previous section. The parallel ripple formation (Fig. 5.10c) at grazing ion incidence requires more insight into the ion impact effects on the surface having initial surface features. Although, the Bradley–Harper model predicts parallel ripple formation at grazing angle bombardment, the Sigmund sputtering theory, which is the basis of the Bradley–Harper model, does not describe well the grazing incidence ion bombardment [5, 6]. Therefore, to describe the ripple formation at grazing ion incidence, the distinctiveness of grazing ion surface interaction is to be taken into account. For incident ion beam of energy E, the energy component perpendicular to the surface is $E_{\perp} = E \cdot \cos^2\theta$, where θ is the ion incidence angle with respect to surface normal. Although, $E_{\perp}$ decreases with θ, the sputtering yield first increases with θ and then decreases for θ close to 90°. This is because the shape of the collision cascades changes with ion incidence angles. For 10 keV O^+ ion impact on ZnO surface, the projected range along the depth and radial direction is calculated by stopping and range of ions in matter (SRIM) [22] and shown as a function of incidence angle in Fig. 5.11. It is seen that with increasing θ, the longitudinal range decreases and radial range increases. The ratio (lateral/longitudinal), which could be approximated as the aspect ratio of the collision cascade zones, is about 1.5 at ion incidence angle 80°. Additionally, in grazing incidence, the cascade

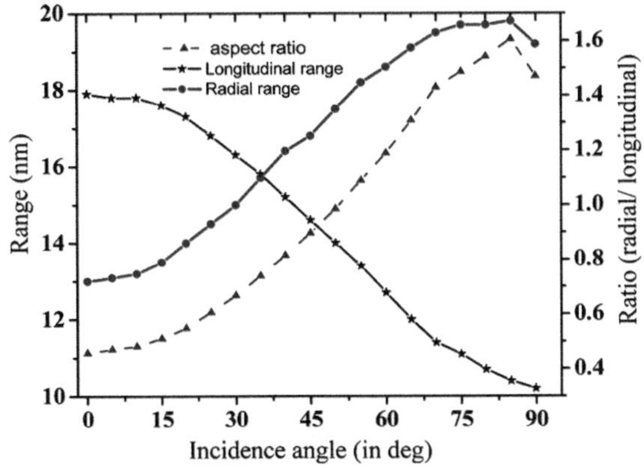

Figure 5.11 Variation of the longitudinal and radial range as a function of incidence angle.

remains closer to the surface resulting the maximum adatom and vacancy creation very near to the surface. Further increase in θ reduces the value of $E_{\perp}$. If the angle of incidence θ exceeds a critical angle, $E_{\perp}$ becomes so small that energy transfer to the sample atom is no longer sufficient to initiate a collision cascade, not to sputter atoms from the surface; therefore, the ions are reflected out from the surface [13]. In case of 10 keV O^{1+} ions bombarding on an atomically flat surface at an angle of 80°, the perpendicular energy component $E_{\perp}$ is 301 eV only; therefore, most of the incident ions will either create adatoms and vacancies on the surface along the beam direction or reflected back instead of notable sputter etching. It is also reported that elastic losses are low compared to inelastic losses at grazing angle [35]. Therefore, sputtering-induced nanostructure formation at grazing angle ion bombardment on ZnO is forbidden for a sample with ideally flat surface. In general, ripple formation at oblique angle is observed in a number of materials [12, 31, 36], while observation of parallel ripple at grazing ion bombardment with the same sample and experimental condition is very rare because of small perpendicular energy component and flatness of the sample.

In the present case, oxygen ions are bombarded on ZnO thin film at oblique (60°) and grazing angle (80°) but as grown ZnO

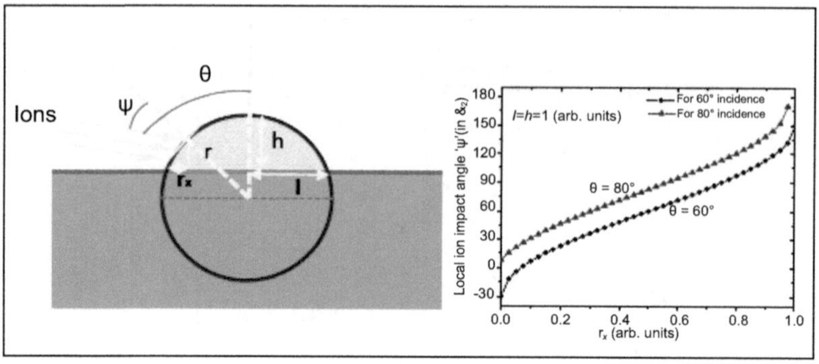

Figure 5.12 Variation of local ion impact angle (ψ) with ion impact point of a mount structure of radius "l" and height "h" for ion incidence angle (θ) 60° and 80°.

surfaces are rough enough with grain structures on the surface. Therefore, the effective angle of impact changes with the position of impact point as shown in Fig. 5.12. It varies from normal to grazing incidence depending on the curvature and shape of the grains. For a hemispherical grain with height h and lateral width $2l$, the local ion impact angle $\psi = \cos^{-1}\{1 - [(2hr_x)/(l^2 + h^2)]\} - (\pi/2) + \theta$, where r_x is the lateral length at the point of ion impact and varies from r_l to $r + l$, r being the radius of curvature of the mount and can be calculated following: $r = (l^2 + h^2)/2h$.

It is seen from Fig. 5.12 that the effective ion incidence angle for 60° case, a major part of the mount is bombarded at oblique angle ($|\psi| \leq 64°$); therefore, perpendicular ripples are expected. In case of 80°, the top section of the mount is bombarded at grazing angle, whereas small bottom portion is either near normal or obliquely bombarded. Due to grazing ion incidence on the top portion, the adatoms and vacancies generated by ion impact will have high probability of displacement in the ion beam direction. When the generated adatoms and vacancies move on the surface in the direction of ion beam, the curvature-dependent transfer of energy leads to more sputtering at the dents than the hills present on the surface [5, 6]. Therefore, the combination of the displacement of adatoms and vacancy on the top surface of the mount along the ion beam direction and curvature-dependent sputter erosion leads to the parallel ripple structures at grazing ion bombardment on

ZnO thin film (Fig. 5.10b). Similar ripple structures along the ion beam direction were reported earlier on initially rough Si and metal surfaces [37–39] where the present explanation may be applicable. Also Chan *et al.* reported parallel ripple on Cu(001) for 70° incidence [40] and Erlebacher *et al.* have demonstrated parallel ripple on Si samples heated at temperatures between 400°C and 700°C [14, 41]. Very recently, Redinger *et al.* [42] reported the alignment of surface vacancies and adatoms along the ion trajectory for keV noble gas ions bombardment at glancing angle on Pt(111).

5.5 Spatially Resolved Magnetic and Electric Zone Formation

Semiconducting–insulating and magnetic–nonmagnetic segments are produced using prefabricated ion-induced nanoripples. Ripple structures of Si are formed by oblique angle keV energy oxygen or argon ion bombardment. Further oxygen ion implantation leads to insulating–semiconducting zone formation, whereas Fe ion implantation helps to produce nonmagnetic–magnetic stripes. Conducting AFM, magnetic force microscopy, and energy dispersive X-ray spectroscopy (EDS) shows the coexistence of semiconducting–insulating and magnetic–nonmagnetic zones. Grazing angle keV ion bombardment fabricates parallel metal ripples from polycrystalline thin films. C-AFM measurements show that the crests of the ripples are regularly spaced one-dimensional conducting wire.

Here two techniques are explored to produce regularly spaced isolated conducting and magnetic structures without masking. In the first technique, the projectile ions are implanted in the selective regions of prefabricated Si ripples by exploiting local ripple geometry and angle-dependent implant concentration. The ripple structures are prefabricated on Si by oblique angle keV energy Ar^+ ion bombardment. Insulating zones are formed by oxygen ion implantation, whereas magnetic zones are formed by Fe ion implantation. In another technique, isolated conducting Pt ripples are formed by bombarding thin film of Pt deposited on Si substrate. The crests are made of metallic Pt, while the valleys are exposed Si.

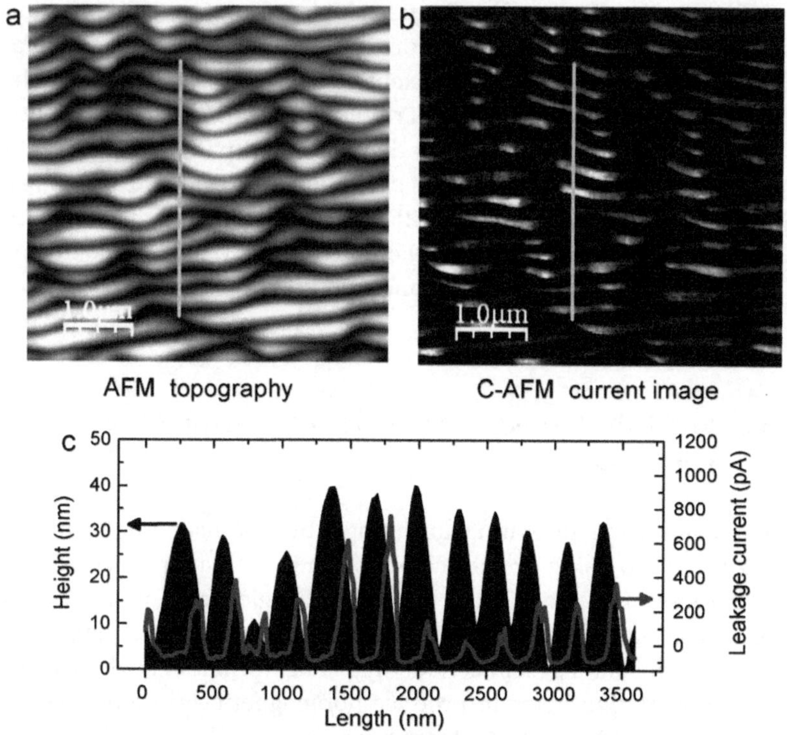

Figure 5.13 (a) AFM images; (b) c-AFM images of selectively oxidized Si ripples produced by high fluence oxygen ion bombardment at oblique angle; (c) superimposed line profile of AFM and c-AFM cross section along the marked line. The projection of ion beam direction on the surface is from top to bottom. See also Color Insert.

Si samples were bombarded with mass analyzed 10 keV O^+ or 32 keV Ar^+ ions at incidence angle 60° to produce ripple structure on the surface. To create selective oxide zones, the ripple structures were bombarded at higher oxygen ion fluence keeping the same ion target geometry. To form spatially selective magnetic structures, Fe ions were implanted at the same ion incident angle subsequent to ripple formation by Ar^+ or O^+. Self-organized conducting nanowires of Pt were generated by 16 keV Ar^+ ion bombardment on Pt/Si thin films at grazing angles (80°) with respect to the target surface normal. The Pt films of thickness

200 nm were deposited on clean Si(100) surface by magnetron sputter deposition. AFM, c-AFM, and magnetic force microscopy (MFM) were used for topography, surface conductivity, and magnetic structure measurements, respectively.

The ripple structures generated by oblique angle bombardment are shown in Fig. 5.13(a). Periodic ripple structures are formed. For further oxygen ion bombardments, the front sides of the ripples are oxidized whereas other sides of the ripples are less oxidized or remain semiconducting. C-AFM measurements show (Fig. 5.13(b) & (c)) the nonconducting front sides of the ripples. Figure 5.14 shows the magnetic structure formation after Fe implantation and annealing. In both cases, the change of local ion impact angle plays the crucial role for spatially selective structure formation. The ion beam is incident at oblique angle with respect to the global target surface normal, but the local angle changes depending on the prefabricated structure on the surface. The front side of the ripples faces the incident ions at normal or near normal, whereas the ions

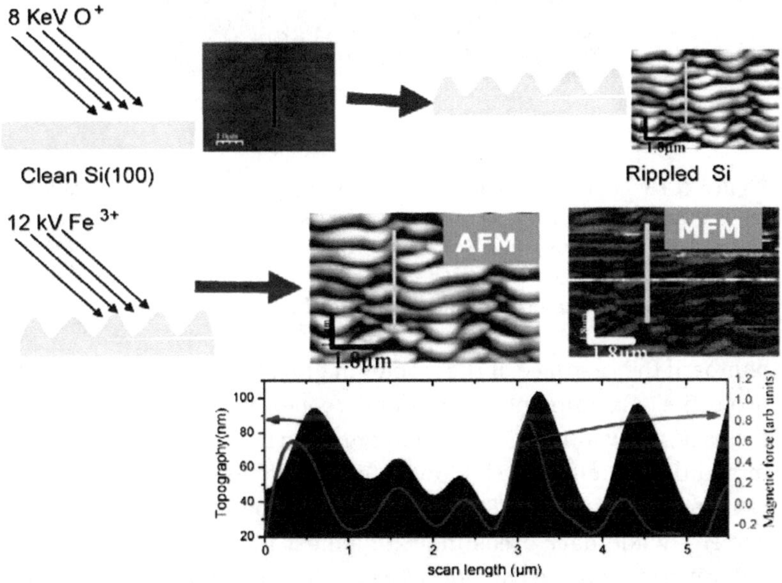

Figure 5.14 Formation of Si ripples on flat Si(100) surface. Fe ions are implanted on the ripple structures at oblique angle to form Fe-rich front surface of the ripples. See also Color Insert.

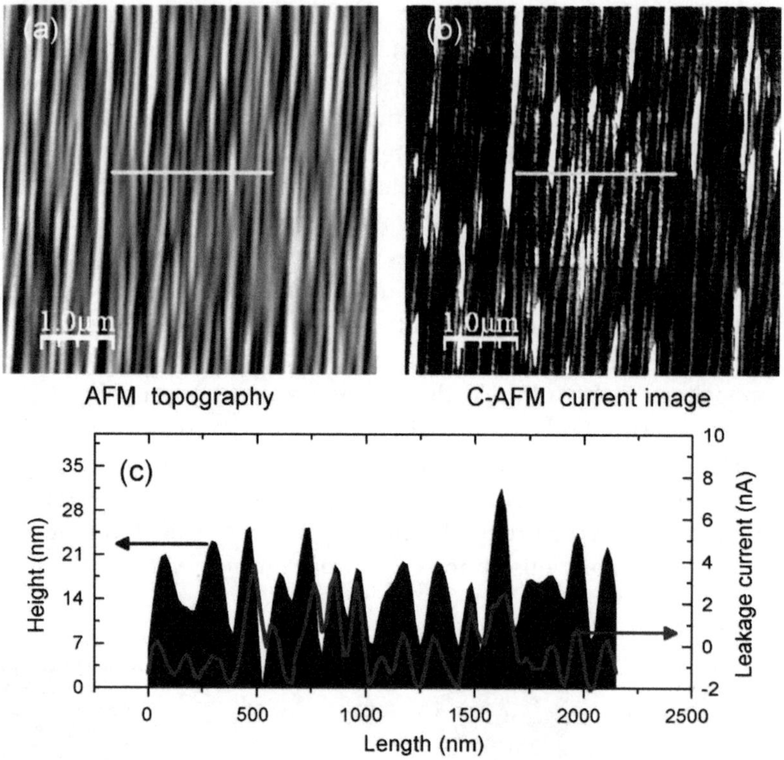

Figure 5.15 (a) AFM images of parallel Pt ripple formed by 16 keV Ar^+ ion at grazing angle; (b) c-AFM images of the same shows the unidirectional isolated Pt wire formation; (c) superposition of cross-sectional AFM and c-AFM along the marked line. The projection of ion beam direction on the surface is from top to bottom. See also Color Insert.

bombard the rear face of the ripples at grazing angle. As a result, the projectiles are implanted more and deeper at the front surface. At the back, a few ions are either implanted on the surfaces or reflected. In addition, due to the shadowing effect, the lower part of the rear remains unirradiated. The bright areas represent higher leakage current, while dark areas indicate almost no leakage current. The difference of leakage current between the faces of ripple represents preferential oxygen implantation and consequent oxidation at the beam facing ripple surface (Fig. 5.13). Similar effect was also observed earlier for silicon and aluminum ripples [36, 43, 44].

Exploiting the same mechanism of preferential ion implantation, equally spaced periodic magnetic structure formation is possible. Figure 5.14 shows the AFM and MFM images of Fe implanted and annealed Si ripples. A week magnetic structure is formed at the front side of the ripples where Fe ions are preferentially implanted. The EDS measurements show favored Fe ion implantation at front face of the ripples and SQUID measurements shows weak ferromagnetism after annealing, which is described elsewhere [30].

Figure 5.15 shows the self-organized conducting Pt ripple formed from Pt thin film deposited on Si(100) by grazing angle Ar^+ ion bombardment. C-AFM measurements show isolated conducting track along the ripples. No leakage current in between the ripples identifies the exposed Si substrate.

Uniformly spaced conducting and magnetic structure in nanometer scale can thus be formed directly by low-energy ion beams. In case of preferential implantation, the variation of local ion impact angle plays the major role. The relative concentration of implanted oxygen is very much dependent on the ion impact angle. The implant concentration is maximum up to 30° following a sharp fall with further increase of incident angle [36, 43–45]. Similar scenario is also expected for Fe implantation into Si ripples.

The parallel Pt ripple formation (Fig. 5.15a) at grazing ion incidence is discussed earlier (Section 5.2.2). Here the conductivity along the hills implies the isolated one-dimensional conducting wire-like structures. Similarly, bombardment of magnetic thin films at grazing angle leads to formation of nanorod or nanowires exhibiting uniaxial magnetic anisotropy [46].

5.6 Coulomb Explosion Sputtering of Spatially Oxidized Nanostructures

In recent years, the interactions of multicharged ions with solid surfaces have been the subject of active research, both for fundamental interest to understand various emission processes as well as future technological applications. In contrast to the well-known kinetically induced ion beam sputtering process, the erosion of surfaces exploiting the potential energy of the projectile is called

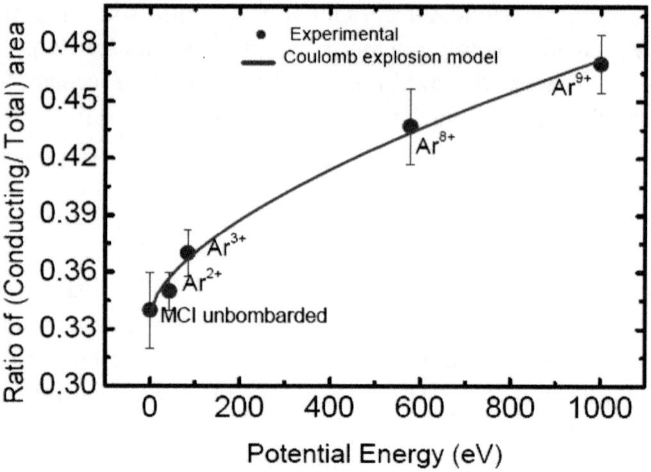

Figure 5.16 Ratio of conducting (Si) to total area as a function of projectile (Ar^{q+}) potential energy. See also Color Insert.

Coulomb explosion or potential sputtering. As per the Coulomb explosion model, the potential sputtering depends on the surface conductivity of the target. Here we have used the partially oxidized Si ripple (section 5.5) having different conductive zone in nanometer range and showed the enhanced sputtering of oxidized zone due to multicharged Ar ion impact.

The dependence of potential sputtering on surface conductivity is shown by comparing the sputtering erosion of coexisting semiconducting and poor conducting surfaces by topographic and conductivity imaging as a function of MCI charge state (Fig. 5.16) [45]. This experiment establishes clearly one of the fundamentals of MCI — surface interaction, i.e., the dependence of potential energy erosion on the conductivity of the ion impact site by comparing the sputtering of coexisting oxide and nonoxide surfaces under identical conditions.

5.7 Summary

Low-energy ion beams thus generate a variety of nanostructures on various technologically important thin films and material surfaces.

By changing the ion beam parameters, shape and size of the nanostructures can be controlled. It is also observed here that introduction of perturbation by chemical or physical means nanostructure formation is possible on the materials where normally ion-induced nanopatterning is forbidden. Perturbation generated by such chemical or physical methods also speeds up the initiation and growth of such structures.

Local ion incidence angle plays a crucial role when the surface is rough enough or structures are formed on the surface. Exploiting the local angle effect, variable conducting or magnetic zone formation in nanometer scale is possible. Further, highly charged ion-beam-induced tailoring of nanostructures or gentle selective sputtering of surface is promising.

Acknowledgments

The author is highly grateful to Prof. D. Ghose and Prof. A. Chakrabarti for their constant support and encouragement. The author would also like to acknowledge Sampa Bhattacharjee for her contribution.

References

1. M. A. Makeev, R. Cuerno, and A. L. Barbasi (2002) *Nucl. Instrum. Methods B*, **197**, 185.
2. U. Valbusa, C. Boragno, and F. B. d. Mongeot (2002) *J. Phys.: Condens. Matter*, **14**, 8153.
3. W. L. Chan and E. Chason (2007) *J. Appl. Phys.*, **101**, 121301.
4. P. Karmakar (2005) Thesis, Saha Institute of Nuclear Physics, Jadavpur University, Kolkata.
5. R. M. Bradley and J. M. E. Harper (1988) *J. Vac. Sci. Technol. A*, **6**, 2390.
6. P. Sigmund (1969) *Phys. Rev.*, **184**, 383.
7. P. Karmakar, G. F. Liu, and J. A. Yarmoff (2007) *Phys. Rev. B*, **76**, 193410.
8. S. Facsko, T. Dekorsy, C. Koerdt, C. Trappe, H. Kurz, A. Vogt, and H. L. Hartnagel (1999), *Science*, **285**.

9. G. Ozaydin, A. S. Özcan, Y. Wang, K. F. Ludwig, H. Zhao, R. L Headrick, and D. P. Siddons (2005) *Appl. Phys. Lett.*, **87**, 163104.
10. U. Valbusa, C. Boragno, and F. B. d. Mongeot (2002) *J. Phys.: Condens. Matter*, **14**, 8153.
11. S. Park, B. Kahng, and A. L. Barabasi (1999) *Phys. Rev. Lett.*, **83**, 3486.
12. P. Karmakar and D. Ghose (2005) *Nucl. Instrum. Methods B*, **230**, 539.
13. R. Cuerno and A.-L. Barabási (1995) *Phys. Rev. Lett.*, **74**, 4746.
14. J. Erlebacher, M. J. Aziz, E. Chason, M. B. Sinclair, and J. A. Floro (2000) *J. Vac. Sci. Technol. A*, **18**, 115.
15. S. Habenicht (2001) *Phys. Rev. B*, **63**, 125419.
16. B. J. Garrison (1982) *Surf. Sci.*, **114**, 23.
17. S. Fischer, A. W. Munz, K. D. Schierbaum, and W. Gopel (1996) *J. Vac. Sci. Technol. B*, **14**, 961.
18. J. F. Moulder and W. F. Stickle (1992) *Handbook of X-ray Photoelectron Spectroscopy*, Perkin-Elmer, Physical Electronics Division, Eden Prairie.
19. J. Mayer, E. Garfunkel, T. E. Madey, and U. Diebold (1995) *J. Electron Spectrosc. Relat. Phenom.*, **73**, 1.
20. S. Hashimoto, A. Tanaka, A. Murata, and T. Sakurada (2004) *Surf. Sci.*, **556**, 22.
21. J.-M. Pan, B. L. Maschhoff, U. Diebold, and T. E. Madey (1992) *J. Vac. Sci. Technol. A*, **10**, 2470.
22. J. Ziegler, J. Biersack, and U. Littmark (1985) *The Stoping and Range of Ions in Solids*, Pergamon, New York.
23. W. Möller, *Fundamentals of Ion-Surface Interaction*, lecture notes found at www.fz-rossendorf.de/FWI.
24. K. Elst and W. Vandervorst (1994) *J. Vac. Sci. Technol. A*, **12**, 3205.
25. V. I. Bachurin, A. B. Churilov, E. V. Potapov, V. K. Smirnov, V. V. Makarov, and A. B. Danilin (1999) *Nucl. Instrum. Methods B*, **147**, 316.
26. W. Reuter (1986) *Nucl. Instrum. Methos B*, **15**, 173.
27. V. I. Bachurin, P. A. Lepshin, and V. K. Smirnov (2000) *Vacuum*, **56**, 241.
28. G. Carter, V. Vishnyakov, and M. J. Nobes (1996) *Nucl. Instrum. Methods. B*, **115**, 440.
29. G. Carter, V. Vishnyakov, Y. V. Martynenko, and M. J. Nobes (1995) *J. Appl. Phys.* **78**, 3559.
30. S. Bhattacharjee, P. Karmakar, A. K. Sinha, and A. Chakrabarti (2011) *AIP Conf. Proc.*, **1349**, 611.

31. T. K. Chini, D. P. Datta, and S. R. Bhattacharyya (2009) *J. Phys.: Condens. Matter*, **21**, 224004.
32. S. Yamamoto, S. Nagata, A. Takayama, and M. Yoshikawa (2006) *Nucl. Instrum. Methods. B*, **249**, 374.
33. Y. Yamauchia, Y. Hirohataa, T. Hinoa, and M. Nishikawab (2001) *Appl. Surf. Sci.*, **169–170**, 626.
34. V. A. Coleman, *et al.* (2006) *Zinc Oxide Bulk, Thin Films and Nanostructures*, edited by C. Jagadish and S. J. Pearton, Elsevier, Oxford.
35. M. Kato, T. Kurose, N. Inoue, T. Hasegawa, and K. Matsuzawa (1992) *Nucl. Instrum. Methods B*, **67**, 316.
36. P. Mishra, P. Karmakar, and D. Ghose (2006) *Nucl. Instrum. Methods B*, **243**, 16.
37. P. Karmakar and D. Ghose (2004) *Surf. Sci.*, **554**, L101.
38. S. A. Mollick, P. Karmakar, and D. Ghose (2008) *Proc. DAE Solid State Phys. Symposium, India*, **53**, 705.
39. S. A. Mollick and D. Ghose (2009) *J. Appl. Phys.*, **106**, 044309.
40. W. L. Chan, N. Pavenayotin, and E. Chason (2004) *Phys. Rev. B*, **69**, 245413.
41. J. Erlebacher, M. J. Aziz, E. Chason, M. B. Sinclair, and J. A. Floro (1999) *Phys. Rev. Lett.*, **82**, 2330.
42. A. Redinger (2010) *Phys. Rev. Lett.*, **104**, 075501.
43. Y. Homma, A. Takano, and Y. Higashi (2003) *Appl. Surf. Sci.*, **203–204**, 35.
44. B. Gautier, B. Fares, G. Prudon, and J.-C. Dupuy (2004) *Appl. Surf. Sci.*, **231–232**, 136.
45. P. Karmakar, S. Bhattacharjee, V. Naik, A. K. Sinha, and A. Chakrabarti (2010) *J. Phys. Cond. Matt.*, **22**, 175005.
46. K. Zhang, M. Uhrmacher, H. Hofsäss, and J. Krauser (2008) *J. Appl. Phys.*, **103**, 083507.

Chapter 6

Surface Nanopatterns on Si(100) by Normal-Incidence Ion Sputtering with Simultaneous Metal Incorporation

Raúl Gago,[a,*] José A. Sánchez-García,[a]
Andrés Redondo-Cubero,[a,b] and Luis Vázquez[a]

[a]*Instituto de Ciencia de Materiales de Madrid, Consejo Superior de Investigaciones Científicas, c/Sor Juana Inés de la Cruz 3,*
Campus de Cantoblanco Madrid, E-28049, Spain
[b]*Instituto Tecnológico e Nuclear, Instituto Superior Técnico,*
Universidade Técnica de Lisboa, 2686-953 Sacavém, Portugal
*r.gago@icmm.csic.es

Normally, self-organized surface nanopatterns produced by ion beam sputtering (IBS) result in nanodot or nanoripple morphologies depending on geometrical considerations. However, recent studies show that additional contributions should be considered when multielemental targets are involved due to the eventual development of a modulation between the surface composition and its morphology. This fact can be extended to targets irradiated with the presence of intentional or accidental impurities (mainly metals) on the surface that, for example, may be responsible for triggering the pattern formation at (near) normal ion incidence on monoelemental semiconductors such as silicon. In this chapter, we deepen in the

Nanofabrication by Ion-Beam Sputtering: Fundamentals and Applications
Edited by Tapobrata Som and Dinakar Kanjilal

ISBN 978-981-4303-75-0 (Hardcover), 978-981-4303-76-7 (eBook)
www.panstanford.com

role of simultaneous metal incorporation in pattern formation on Si(100) surfaces by normal-incidence IBS with 1 keV Ar^+. We show that the level of metal incorporation can additionally be used to tune the pattern morphology and, in particular, selectively produce unconventional nanohole patterns for metal levels higher than those present in the standard nanodot pattern counterpart. We examine the parameter window for the production of different morphologies to verify the morphological and compositional correlation. Finally, we highlight relevant mechanisms that could drive to such pattern selectivity.

6.1 Introduction

IBS has become an attractive route to spontaneously induce self-organized nanostructures on many different surfaces [10]. In particular, this technique has been successfully applied to produce ordered patterns, mainly consisting of either nanoripple or nanodot periodic nanostructures, on single-crystal, polycrystalline, or amorphous surfaces [6, 30, 40]. The potential of IBS as a competitive nanofabrication method relies on the single-step process that induces surface patterns in very short times (just a few minutes of irradiation are required) and over relatively large areas.

The studies on IBS pattern formation by different groups during the last few years have contributed to deepen the knowledge of its underlying physical mechanisms. However, at the same time, they clearly show that there are still open issues concerning IBS surface patterning from the theoretical and experimental points of view. In the first case, although the models have been progressively generalized [4, 5, 13, 29] starting from the seminal work of Bradley and Harper [1], it is still not clear whether the underlying mechanisms should be revised or some additional physical mechanisms must be still incorporated to account for the wide range of experimental observations. Among plausible mechanisms still to be fully implemented in the models, we can highlight the possible role played by surface stress [32, 43] and the existence of a viscous surface layer [5]. From the experimental point of view, the open issues refer mainly to the reproducibility of IBS nanopatterning, the possibility of tuning the pattern morphologies, and the eventual enhancement (and control) of the pattern ordering. Besides, a major disparity relies on the predicted universal character of IBS nanopatterning from a theoretical point of view and, in contrast, the diversity (or even lack) of patterns obtained experimentally on different materials. This controversy could be partially attributed to the specific characteristics of each experimental system since, for example, small changes in the beam configuration can lead to very different surface morphologies [44].

A critical issue related to the mentioned lack of universality found experimentally refers to the problems usually found to induce nanodot patterns by (near) normal ion incidence on single-element semiconductor surfaces, particularly, silicon [9, 25]. This fact contrasts with the relatively easy patterning of compound semiconductors, such as GaSb, where preferential sputtering effects of one species could be relevant [12]. Although the successful production of nanodot patterns by IBS has been reported on silicon surfaces [15, 45], the lack of reproducibility has motivated some additional considerations, such as the necessity of surface impurities (typically metals), to effectively trigger the pattern formation. This fact was first reported by Ozaydin *et al.* [31], where Mo seeding on the surface was found to be a necessary condition to induce nanodot patterns on Si(100) under normal ion irradiation with a 1 keV Ar^+ ion beam. It was suggested that Mo induces a certain surface roughness due to preferential sputtering effects, which may be required to initiate and sustain the patterning process of monoelemental semiconductors. Recent investigations taking care of keeping the Si surface clean from external impurities have shown that pattern formation on Si surfaces only takes places for incidence angles above 45–50° [27, 28]. The lack of pattern formation at near-normal incidence conflicts with the predictions of the Bradley–Harper model [1]. Therefore, additional relaxing mechanisms need to be invoked. In particular, in analogy with studies for high-energy irradiation (keV) of Si [3], the presence of ion-induced downhill surface currents could be responsible for surface smoothening without the presence of such impurities. This prediction has remained relatively unconnected from IBS nanopatterning although further credit has been recently gained based on the data reported in the work by Madi *et al.* [27, 28].

The relevance of impurities and composition on pattern formation and evolution has then attracted much theoretical interest. For example, Shenoy *et al.* [36] correlated the role played by preferential sputtering in modulating the surface composition with respect to the topography under off-normal incidence. The same argument has been recently used under normal incidence in conjunction with ion-induced downhill currents to describe the spontaneous nanodot pattern formation in binary compounds [2]. This model can also be extended to single-element targets in the presence of impurities. In fact, recent observations [23] have also shown that pattern formation by IBS in GaSb is triggered by compositional segregation, in a similar way that surface impurities would act in Si surfaces.

From the above, the relevance of simultaneous metal incorporation on the production of surface patterns by IBS is clear, especially, on silicon surfaces. This problem has remained unaddressed partially due to the lack of systematic chemical characterization of the irradiated surfaces. For example, there are still open issues concerning the main role played by metal incorporation and the threshold content necessary for pattern formation. All these results indicate the importance of the chemical characterization of the IBS-patterned surfaces to evaluate the (un)intentional metal incorporation. The main aim of this chapter is to review our recent investigations [34, 35] on the effect of simultaneous metal incorporation on IBS nanopatterning of Si(100) surfaces and to deepen the understanding of the mechanisms involved in such processes. This study is not focused on the need of metal incorporation for pattern formation but, instead, we explore the impact of the level of impurities on the pattern morphology. For this purpose, we have done an extensive morphological and compositional analysis of Si(100) surfaces subjected to ion irradiation in a wide range of process parameters. Interestingly, we find the production of nanohole or nanodot patterns with similar features (roughness, wavelength, and ordering), despite the different morphologies. Moreover, the pattern selectivity is driven by the metal incorporation since nanohole (nanodot) patterns are promoted at higher (lower) levels of residual metals on the surface. Finally, we have generalized our findings by using two different types of ion sources.

6.2 Experiment

6.2.1 *Sample Preparation: Irradiation of Si(100) Targets by IBS*

Single-crystal *n*-type Si(001) targets (Sb doped, 8–20 $\times$ 10^{-3} Ωcm) were bombarded with 1 keV Ar^+ ions at normal incidence in a high-vacuum chamber with a base pressure below 5 $\times$ 10^{-7} mbar. For most irradiations, the ion beam was extracted from an alternating cold cathode ion source (ACC-IS) (JENION GmbH, http://www.jenion.de) with a beam diameter of 4 cm and located ~20 cm away from the target. In order to search for general trends, additional IBS experiments were performed with a commercial 3 cm beam-diameter CSC Kaufman-type ion gun (Veeco©) under similar irradiation conditions as those performed with the ACC-IS. In both cases, the targets were bombarded with current densities (J) between 40 and 340 μA cm^{-2} and ion fluencies up to ~4 $\times$ 10^{18} ions cm^{-2}. Prior to the IBS process, the ion current density on the target was adjusted with a Faraday Cup located on a movable shutter, which prevents unwanted irradiation on the target. Several minutes were required for beam stabilization after warming up of the source.

In the gas discharge for beam generation with the ACC-IS, the Fe cathodes are progressively eroded by the plasma. This fact provides to the ion beam an unavoidable component of Fe atoms toward the target. This apparent drawback is used, instead, as a way to study the role of metal incorporation on the nanopatterning process. In this case, the Fe flux can be tuned by the source operation since it is proportional to the discharge power or, equivalently, beam currents. For further control on the metal incorporation, a secondary source of metal impurities was added by surrounding the targets with metallic parts. For this purpose, a mask consisting of a 1 mm thick metal (Fe, Mo) foil with a hole at the center was placed on top of the targets. This mask is additionally used to fix the target to the sample holder and to create a step between exposed and masked parts. This step is later used to compute the erosion rate, as measured with a Dektak 150 (Veeco©) mechanical profilometer, and to further verify the nominal ion current density. The size of the hole in the mask is used to tune the metal incorporation rate from this secondary source. In the irradiations with the Kaufman source, the metal supply coming directly from the ion gun operation is negligible and metals are only to be incorporated through the metallic elements surrounding the target. Consequently, the metal supply during IBS can be simplified to one single source. In order to reach similar metal contents with both type of ion guns, we used foils with 18 mm and 10 mm holes for the ACCS-IS and Kaufman setups, respectively.

6.2.2 *Morphological and Compositional Analysis of Irradiated Targets*

The resulting surface morphology was imaged *ex situ* by tapping mode atomic force microscopy (AFM) with a Nanoscope IIIa equipment (Veeco©). Silicon cantilevers with a nominal radius of 10 nm were employed. In order to characterize the pattern wavelength (λ), we derived the power spectral density (PSD) from the AFM images. This function reads as

$$\mathrm{PSD}(k, t) = \langle H(\mathbf{k}, t) H(-\mathbf{k}, t) \rangle \tag{6.1}$$

where $\boldsymbol{k}$ is the spatial frequency in reciprocal space, H($\boldsymbol{k}$, t) is the Fourier transform of the surface height function $h(r, t)$, and $\langle \ldots \rangle$ denotes averaging in reciprocal space. In the case of a pattern, the existence of a morphological instability associated with a characteristic length will appear in the PSD curve as a clear local maximum at $k = 1/\lambda$. In addition, the full width at half maximum (FWHM) of the PSD peak is inversely proportional to the correlation length (ξ) of the pattern (i.e., mean size of the ordered domains). Another useful parameter to characterize the morphologies is the surface skewness S, which is a measure of the vertical asymmetry of the surface [39] and, hence, it is sensitive to the presence of protruding or hollow features. If $\hbar = \langle h(\mathbf{r}) \rangle$ is the average height along the surface, S is defined as

$$S = \langle (h(r) - \hbar)^3 \rangle / w^3 \tag{6.2}$$

where w is the root mean square (rms) surface roughness and, here, the angular brackets denote spatial averaging. In our case, PSD functions, w, S, and $\hbar$ values are obtained directly by the software of the AFM equipment.

Since metal incorporation can play a relevant role on pattern formation and morphology, the residual composition of the irradiated surfaces was determined routinely by means of Rutherford backscattering spectrometry (RBS) with a 1 MeV He^+ probing beam. The spectra were acquired with a Si solid-state detector located at a scattering angle of 170° and with an energy resolution of 15 keV. The measurements were conducted in random and channeling (RBS/C) geometries, the latter by aligning the probing beam along the main crystallographic direction of the Si(100) target. In this way, complementary compositional and structural data are obtained. The latter is mainly related to the formation of a top most amorphous surface layer induced by the ion bombardment.

6.3 Nanopattern Formation with a Cold-Cathode Ion Source

6.3.1 *Description of the Different Morphologies*

Figure 6.1 shows the different morphologies that have been found during IBS with an ACC-IS. In particular, we have obtained flat and smooth surfaces (Fig. 6.1a) with a roughness similar to that of the bare Si(100) targets (w ~0.2 nm), patterns with either nanohole (Fig. 6.1b) or nanodot (Fig. 6.1c) structures and, eventually, a mixed pattern (Fig. 6.1d) consisting of both nanohole and nanodot morphologies. The insets of Fig. 6.1b,c show the autocorrelation functions of 400 nm wide AFM images revealing the short-range hexagonal order of both nanohole and nanodot patterns. As shown in Fig. 6.1e, the height distribution of each AFM image leads to different signs of the S values. This parameter will be very useful to classify (quantitatively) the different patterns. In particular, nanodot and nanohole patterns are clearly catalogued for $S > 0.2$ and $S < -0.2$, respectively.

The nanodot patterns displayed in Fig. 6.1c are rather similar to those reported previously with a Kaufman-type ion gun [15–18]. In this case, typical nanodot heights are 2–3 nm with a diameter close to 25 nm, leading to w ~0.7 nm. On the contrary, nanohole morphologies are quite unconventional in terms that, so

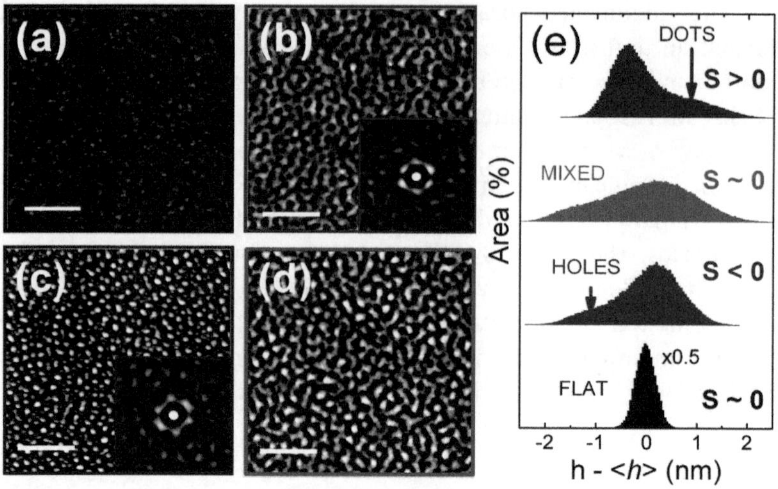

Figure 6.1 *Left:* AFM 1 × 1 μm^2 images showing the different morphologies obtained after IBS of Si(100) with an ACC-IS at normal incidence: (a) flat, (b) nanoholes, (c) nanodots, and (d) mixed nanohole/nanodot structures. The horizontal bar corresponds to 200 nm and the vertical scale is 6.5 nm for all the images. The insets in (b) and (d) display the autocorrelation functions of 400 nm wide areas, showing a short-range hexagonal order. *Right:* (e) Height distributions of the AFM images showing the height distribution asymmetry that results in positive (dots) and negative (holes) values of the skewness (S). See also Color Insert.

far, they have not been reported extensively. In this case, nanoholes are typically 1.5–3 nm deep and 20–40 nm wide, depending on the pattern density (note that the nanohole dimensions could be underestimated because of tip convolution effects). Interestingly, the nanohole and nanodot patterns share common features, as derived from the PSD functions displayed in Fig. 6.2. Both PSD curves display a maximum at a similar position, indicating that λ is very similar for both patterns (~50 nm). Moreover, the FWHM of both PSD curves indicates the analogous degree of (short-range hexagonal) ordering in terms of ξ. Finally, the PSD regions for k values below that corresponding to the pattern wavelength are quite alike, which indicates that a similar kinetic roughening is taking place at large length scales in both types of morphologies. Then, both

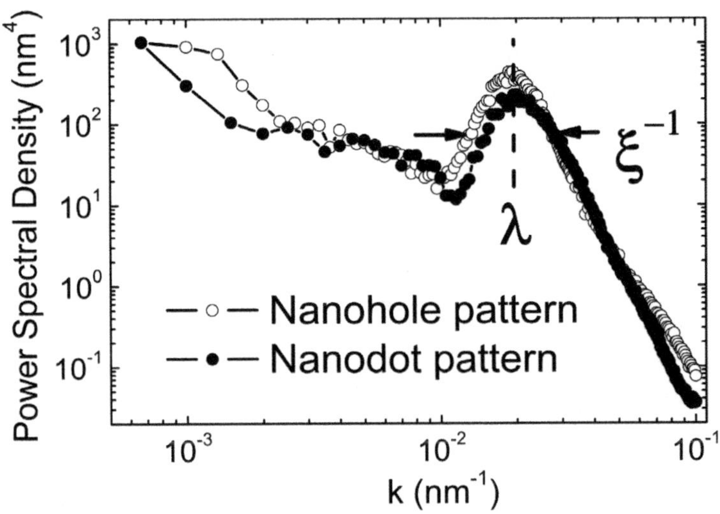

Figure 6.2 PSD curves versus k obtained from the AFM images of nanoholes (open circles) and nanodot (solid circles) dense patterns. The comparison yields similar morphological characteristics between both patterns in terms of roughness, wavelength, and degree of short-range order.

dense nanohole and nanodot patterns can be quite alike in terms of homogeneity, surface roughness, wavelength, and short-range order.

6.3.2 *Parameter Window for the Production of Different Morphologies*

In order to gain deeper understanding on the nanohole/nanodot pattern formation, we have done an extensive survey of the resulting surface morphologies as a function of two process parameters, namely, J and sputtering time (fluency). For this purpose, the irradiated surfaces were characterized by AFM and their topography ascribed to either flat (□), nanohole (○), mixed (■), or nanodot (•) morphologies following the criterion given by the skewness derived from the AFM images. The results from this analysis are shown in Fig. 6.3. Visual inspection of the symbols' distribution clearly leads to the following conclusions:

(i) Flat surfaces are obtained for low current densities (<50 μA/cm^2) and low irradiation times indicating that there is an incubation time required for surface roughening and pattern formation onset.
(ii) Nanohole structures are only formed for low J values at moderate fluency values (10^{17}–10^{18} ions cm^{-2}) and they are never observed for current densities J >100 μA cm^{-2}.
(iii) In contrast, nanodot patterns are more favorable at higher J conditions and they are always produced, for a given J, above a certain fluency onset.
(iv) Finally, there is transition between the nanohole and nanodot morphologies by increasing J and/or the sputtering time that takes place through an intermediate mixed pattern.

In any case, Fig. 6.3 indicates that the most likely pattern to be found in IBS experiments with simultaneous metal incorporation would be the nanodot morphology, whereas nanohole patterns are only observed within a narrow parameter window, mostly for low ion fluxes. This result is consistent with the scarce observation of nanohole patterns by other groups. In the following sections, we analyze the morphological evolution for a set of samples irradiated at different J values with fixed fluence (dashed lines in Fig. 6.3) and for constant J for different sputtering times (vertical arrows in Fig. 6.3).

6.3.3 *Influence of the Ion Flux on Nanodot/Nanohole Pattern Features*

Figure 6.4 shows the surface topography of Si(100) targets after irradiation with an ACC-IS for increasing J values at a constant ion fluency of ~4×10^{17} ions cm^{-2} (bottom dashed line in Fig. 6.3). Qualitatively, five different surface topographies are distinguished for increasing J values. First, as indicated in the previous section, flat surfaces (not shown) are obtained for J < 50 μA cm^{-2}. Second, scattered nanohole structures (Fig. 6.4a), 15–35 nm wide, over flat hole-free plateau regions appear for J between 50–70 μA cm^{-2}. This nanohole pattern evolves to a denser array (Fig. 6.4b,c) by increasing J up to 100 μA cm^{-2}. Above this value, a mixed pattern (Fig. 6.4d,e)

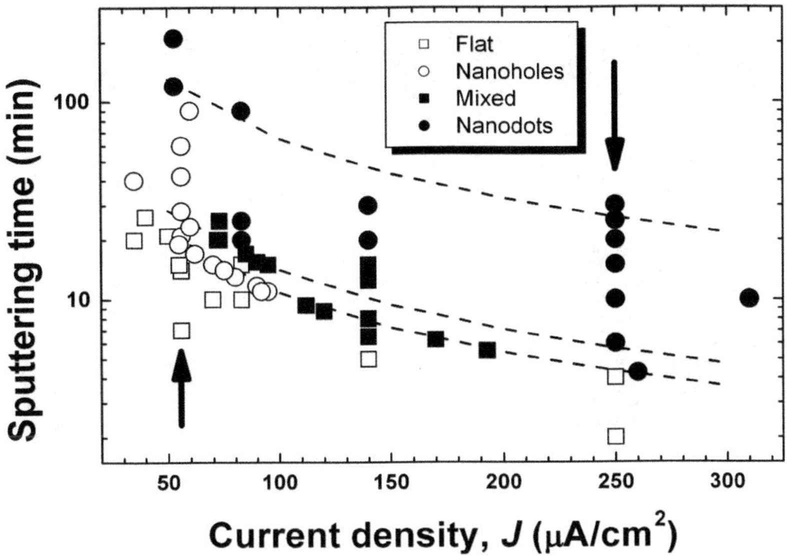

Figure 6.3 Mapping of the surface morphology of Si(100) targets after IBS for different J and sputtering times. The sample topography is displayed as: flat (□), nanohole (○), mixed (■), or nanodot (•). The dashed lines and vertical arrows indicate the pattern evolution at constant ion fluency and flux conditions, respectively.

is produced since, as mentioned above, both nanohole and nanodot structures coexist. Finally, dense nanodot patterns (Fig. 6.4f) are produced for $J > 200$ μA cm^{-2}.

The pattern evolution described above can be followed quantitatively in Fig. 6.5 where S, w, and λ are plotted as a function of J for the three different ion fluencies that are marked by the dashed lines depicted in Fig. 6.3. Again, the S value of the resulting images is used to classify the different morphologies into flat, nanohole, mixed, and nanodots. Figure 6.5a shows the dependence of S on J. In case of low fluency (4×10^{17} ions cm^{-2}), S initially drops sharply from zero to negative values for J values higher than 50 μA cm^{-2} when, as illustrated by Fig. 6.4, some nanohole structures appear scattered on the surface. This situation corresponds to the formation of deep and wide isolated nanoholes over the flat surface that leads to the abrupt decrease in S. As J increases, the nanohole patterns become denser

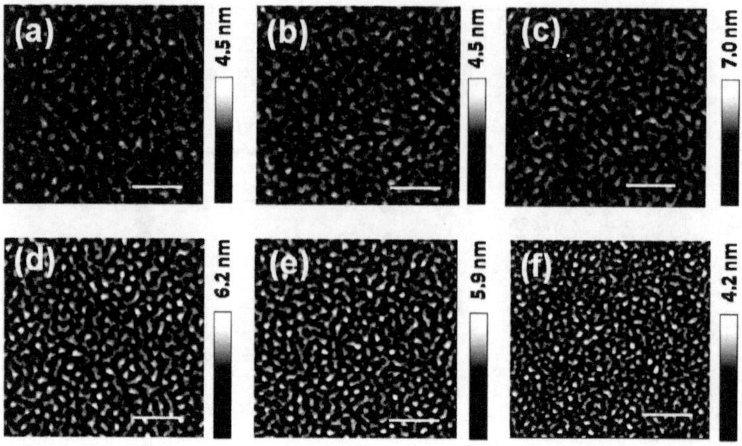

Figure 6.4 AFM 1 × 1 μm^2 images for Si(100) surfaces after IBS with an ACC-IS at 1 keV Ar^+ for a total fluency of 4 × 10^{17} ions cm^{-2} and current densities J of: 60 μA cm^{-2} (a), 75 μA cm^{-2} (b), 90 μA cm^{-2} (c), 120 μA cm^{-2} (d), 170 μA cm^{-2} (e), and 240 μA cm^{-2} (f). The horizontal scale bar corresponds to 200 nm. See also Color Insert.

with smaller and shallower holes (although tip convolution effects can also contribute to the smaller nanohole dimensions derived from AFM) leading to a negative S but with a higher value than in case of more isolated nanohole structures. The larger density of nanoholes is also reflected in a higher w (it increases up to ~1 nm) and a decrease in λ from 70 to ~50 nm ($\lambda \sim J^{1/2}$), as shown in Fig. 6.5b,c, respectively. Eventually, for a further increase in J, the nanoholes start to coexist with nanodots driving to a mixed pattern characterized by S values close to zero. This situation is accompanied by the saturation in w and λ. Finally, for even higher J values, the nanodot patterns develop, yielding clear positive S values (>0.2). In this case, w decreases with respect to the mixed pattern and λ remains constant at ~50 nm. Thus, as J increases, the irradiated surfaces evolve from an initial flat morphology to a nanohole-mixed-nanodot pattern sequence.

As shown in Fig. 6.5, increasing the ion fluency shifts the pattern formation and transition toward lower J values and, eventually, for the highest fluency (2.5 × 10^{18} ions cm^{-2}), nanohole or mixed

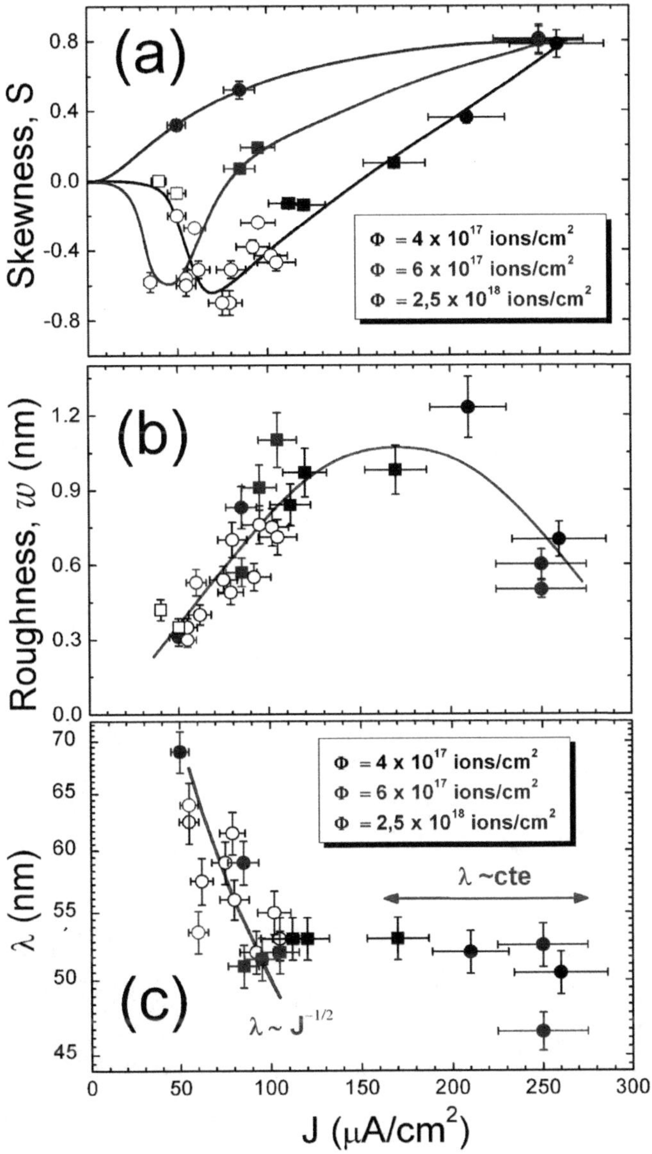

Figure 6.5 AFM analysis of Si(100) surfaces after IBS with an ACC-IS at 1 keV Ar^+ for a total fluency of 4×10^{17} ions cm^{-2} (black symbols), 6×10^{17} ions cm^{-2} (red symbols), and 25×10^{17} ions cm^{-2} (blue symbols). Flat (□), nanohole (◯), mixed (■), and nanodot (•) morphologies are represented by the usual symbols. Lines are only guides to the eye.

patterns are not observed. It is interesting to note that *w* and λ behaviors with *J* are independent of the final ion fluency.

6.3.4 *Pattern Formation Dynamics*

As shown schematically in Fig. 6.3, at a constant ion current density, there is a pattern transition for increasing irradiation times. To follow this change quantitatively, we have studied the evolution of the morphological features for low and high ion fluxes (marked by vertical arrows in Fig. 6.3), namely 55 and 250 μA cm^{-2}.

The AFM images for selected samples irradiated under these conditions are shown in Fig. 6.6. For low *J* (55 μA cm^{-2}), the target surface remains flat for irradiation times below 15 min. After ~20 min, a dense nanohole pattern is observed, as shown in Fig. 6.6a. This type of structure is still produced for a relatively long temporal window (up to 90 min, as shown in Fig. 6.6b) but, as the irradiation time is increased, the nanohole structures become scarcer and more scattered over the surface. After 120 min of irradiation, the surface smoothens and the nanohole pattern disappears (Fig. 6.6c). Finally, for the longest irradiation time

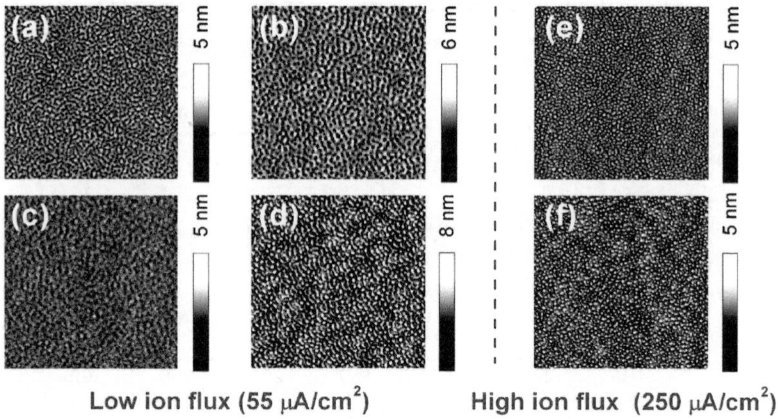

Figure 6.6 AFM 2 × 2 μm^2 images of Si(100) surfaces irradiated with 1 keV Ar^+ under normal incidence from an ACC-IS for low (55 μA cm^{-2}) and high (250 μA cm^{-2}) ion current densities for different times: *Low J*: (a) 28 min, (b) 90 min, (c) 120 min, and (d) 210 min. *High J*: (e) 10 min and (f) 30 min. See also Color Insert.

(210 min), a clear nanodot structure develops. On the contrary, for high J (250 μA cm^{-2}), the pattern formation onset takes place earlier ($t > 6$ min) but only nanodot structures (Fig. 6.6e) are observed for irradiation times above this threshold. It should be noted that whereas the nanohole pattern seems to be unstable after prolonged irradiation, the nanodot counterpart is quite stable for high fluencies, where only a long wavelength corrugation appears for long enough irradiation times. This larger lateral corrugation displays kinetic roughening properties and has also been observed in nanodot patterns produced with a Kaufman ion source [15], as well as predicted theoretically by a continuum two-field hydrodynamic model [4].

Figure 6.7 shows the PSD curves (left panel) and a detailed quantitative analysis (right panel) of the AFM images for Si(100) targets sputtered at different times. The emergence of a pattern on the surface leads to a clear peak in the PSD curves (Figure 6.7a,b), which, as anticipated, occurs earlier for the higher J condition. In case of low flux, the S values reveal the formation of nanohole patterns ($S < -0.2$) up to an ion fluency of $\sim 10^{18}$ ions cm^{-2}. In this situation, the minimum in S is related to the formation of deeper nanoholes, as revealed by the increase in w. Longer irradiation times lead to a mixed pattern that finally evolves into a nanodot one. On the contrary, for high J, the surface roughening (Fig. 6.7d) only derives into a nanodot pattern.

For both fluxes, the peak of the PSD curves shifts with increasing irradiation time toward lower k values (larger distances) indicating that the pattern wavelength coarsens with time (Fig. 6.7e). Note that this coarsening process takes place either for nanohole or nanodot patterns with a similar exponent of $\lambda \sim t^{0.1}$. This coarsening has been previously observed for nanodot patterns produced on Si(100) [17] and GaSb(001) [33]. The coarsening seems to be disrupted in the transition between the nanodot and nanohole patterns at low J.

Finally, it is interesting to note that although nanodot patterns produced at both J values after prolonged irradiation present similar λ and S values, a much higher roughness (nanodot height) is achieved at low J. In this line, one should also note that the dynamics at low J condition yields an ordering increase with fluency. This is derived from the decrease in the FWHM of the PSD peak with

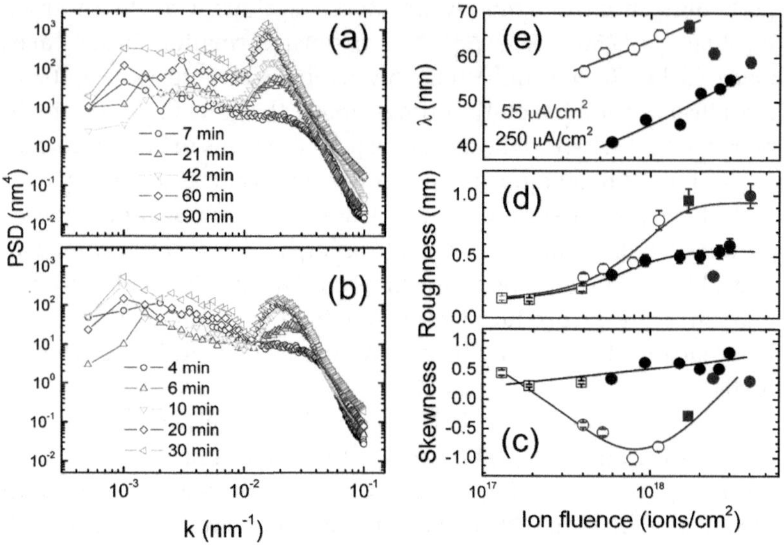

Figure 6.7 Left: PSD functions of the samples irradiated for different times at 55 μA cm^{-2} (a) and 250 μA cm^{-2} (b). Right: Pattern wavelength (c), surface roughness (d), and skewness (e) as a function of ion fluency for samples irradiated at 55 μA cm^{-2} (red symbols) and 250 μA cm^{-2} (blue symbols). See also Color Insert.

sputtering time evidenced in Fig. 6.7a. This ordering enhancement does not take place for the high J conditions since the peaks of the PSD curves in Fig. 6.7b remain rather broad and with similar FWHM for prolonged irradiation times. Interestingly, the order of nanodot patterns produced with a Kaufman source was also enhanced at low ion fluxes due to a reduced erosion rate under this condition [18].

6.4 Role of Metal Incorporation on Pattern Selectivity

As shown in Section 6.3, the formation of nanohole and nanodot patterns can be tuned experimentally by a proper selection of both the ion current density and fluency. Theoretically, the formation of nanohole or nanodot structures by IBS had been predicted to occur [20] on the basis of changes on the shape of the ion collision cascade. However, for the same projectile and target combination, this property can only be tuned with changes in the ion energy. In our experiments, the ion energy has been kept constant at 1 keV and, therefore, this theoretical framework cannot, in principle, explain the pattern selectivity found within the present results. A peculiarity of our configuration relies on the simultaneous incorporation of metals during the irradiation process, which is not taken into account in the existing models. As discussed in the introduction, the incorporation of impurities adds other possible scenarios for IBS nanopatterning. For this reason, we have carried out a systematic compositional analysis on our samples displaying different pattern morphologies.

6.4.1 *Compositional Analysis of Irradiated Surfaces*

The metal content of the different patterns has been assessed by RBS with a 1 MeV He^+ probing beam. Figure 6.8a shows the experimental RBS spectra in channeling mode for silicon surfaces with nanohole and nanodot patterns produced at $J = 70$ μA cm^{-2} and 260 μA cm^{-2}, respectively, for a total fluency of 4×10^{17} ions cm^{-2}. The spectra from an untreated Si(001) target in RBS/C and random geometries are also shown as reference of pure crystalline and amorphous materials, respectively. Note that the intensity in the random spectra is multiplied by a factor of 0.25 and, hence, the yield of backscattered projectiles obtained due to the channeling geometry (RBS/C) for the Si(100) surface is reduced to a few percents. This reduction is a typical indication of good crystalline quality in monocrystalline targets. However, even with such crystals, there is always a surface peak contribution in RBS/C because, obviously, channeling from surface atoms is not possible due to geometrical blocking. The RBS/C geometry with respect to the [100] direction of the target allows us detecting the formation of the topmost amorphous silicon surface layer (a-Si) induced by the ion bombardment. In fact, the comparison of the spectra for irradiated surfaces with that of a non-irradiated pure crystalline one [Si(100)] reveals an increase in the Si signal at the surface region. The thickness of the a-Si layer is within a few nm for both irradiated (nanopatterned) targets, as derived from RBX software fitting results [22]. It should be noted that the topmost layer is

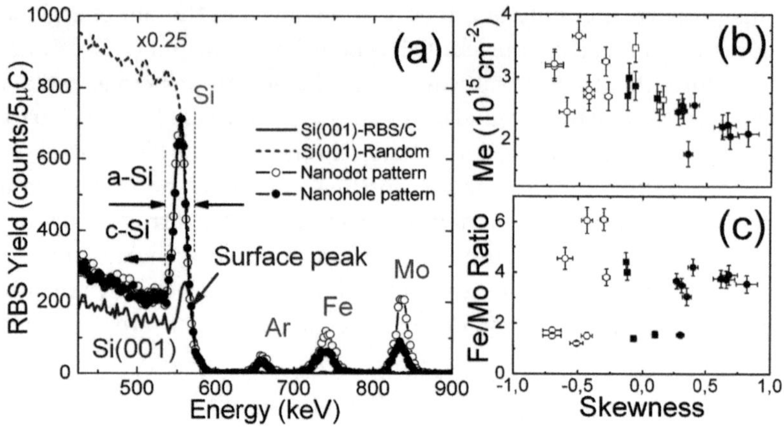

Figure 6.8 (a) Experimental RBS/C spectra along the [100] direction of the irradiated Si(100) targets with nanodot (solid dots) and nanohole (open dots) patterns, together with the random and aligned spectra from an untreated Si(100) surface. (b) Total metal (Me) and (c) relative Fe and Mo contents as a function of the surface skewness.

completely amorphous although the yield does not reach the level of the random spectrum. This fact is due to its reduced thickness, just about the resolution limit of the technique at the surface region (5–10 nm). The higher signal from projectiles scattered from Si below the near surface region (energy channels lower than the surface peak) in the RBS/C spectra of irradiated samples as compared to the bare Si(001) layer provides a further evidence of the ion-induced amorphization. This effect is associated to the dechanneling of the incidence and backscatter projectiles due to trajectory deviations when travelling through the amorphous topmost layer [14].

The RBS spectra displayed in Fig. 6.8a reveal the presence of Fe and Mo metal impurities, as well as implanted Ar ions, at the surface region. Obviously, the detection of Fe and Mo species is related to the erosion of the ACC-IS cathodes and the target mask, respectively. Due to the kinematic factor of the collision with the He^+ projectiles, scattering events from heavier elements located at the surface are detected at higher energies. The increase in the RBS scattering cross section with the square of the atomic number of the scattering element [8] provides higher sensitivity (around 0.1 atom %) to heavy elements such as Fe and Mo, which allows their detection despite the reduced atomic areal density ($<10^{16}$ atom cm^{-2}). The cross section for each element also implies that the detection efficiency for Mo is $\sim$2.6 times higher than that of Fe. This means that despite the higher Mo signal in the spectra, the surface is Fe rich. It is worth noting that the resolution of RBS at the surface is in the range of the amorphous surface layer thickness. Hence, we cannot resolve any compositional profile within the topmost layer. However, the RBS spectra suggest that Fe could present a broader in-depth distribution since the Mo signal is narrower than that of Fe. The different metal incorporation pathways, preferential sputtering effects and ion-beam-induced mixing, diffusion or segregation processes [24] could lead to different Mo and Fe depth distributions. In any case, the depth of interest here is near the resolution limit of RBS and, therefore, further investigation is required to provide solid conclusions in this sense.

Interestingly, Fig. 6.8a shows that the residual metal content on the surface decreases appreciably from nanohole to nanodot patterns. This suggests a possible correlation between the pattern morphology and the metal content that could drive the pattern selectivity found with our sputtering conditions (Fig. 6.3). In order to unveil such hypothesis, we have plotted in Fig. 6.8b the metal content measured by RBS for a large collection of samples as a function of the corresponding surface skewness, as obtained from AFM images. Thus, with such a plot nanohole (nanodot) patterns are to be found at the left (right) part of the graph. Note also that the symbols help to identify different morphologies. From the symbol distribution, a clear correlation between metal content and pattern morphology is observed. In particular, nanohole patterns are produced for higher metal contents (within the 2.5–3.7 $\times$ 10^{15} atoms cm^{-2} range), whereas nanodot patterns are induced with metal levels slightly below such range (1–2.5 $\times$ 10^{15} atoms cm^{-2}). It is important to note that this correlation is general in the sense that it occurs irrespective of the specific J and fluency values used to produce each pattern. In Fig. 6.8c, the ratio of the Fe–Mo content is plotted versus the pattern morphology (i.e., S). This time, there is not any clear trend, which allows us to conclude that the pattern selectivity is mainly tuned by the total metal content on the surface rather than the ratio of the constituents. So far, it is worth noting that most of the experimental studies in IBS nanopatterning of silicon surfaces with either Fe [15–18] or Mo [31, 32] incorporation report nanodot rather than nanohole morphologies. This can be due to the fact that, if an unintentional metal contamination takes place, it usually leads to metal levels within the window for nanodot production. In our case, the additional supply of Fe atoms coming from the ACC-IS operation has facilitated to reach the levels required for nanohole formation. An additional aspect of our setup is the relatively large gun–target distance ($\sim$20 cm), which also limits the operation to low J values. This condition, as shown above, is another relevant prerequisite for the observation of such unconventional morphology.

6.4.2 *Production of Nanohole Patterns with a Kaufman Ion Source*

As commented above, one of the critical experimental issues related IBS nanopatterning lies on the possible influence of the specific experimental setup used on the obtained nanopatterns [44]. Thus in order to assess the generality of our findings about the pattern selectivity through the metal content, it is convenient to reproduce them with other equipment. For this reason, we have extended our work toward the production of both nanodot and nanohole patterns with a Kaufman-type ion source. As mentioned in the experiment section, in this setup the metal supply during IBS can be reduced to a single source by using a mask that partially covers the target. In order to reach similar metal contents than those attained with the ACC-IS (Fig. 6.8b), we have designed mixed Fe/Mo and Fe masks with a reduced diameter of the inner hole (10 mm) (note that the experiments with the ACC-IS were performed with a 18 mm mask). According to the know-how provided by our previous experiments for nanohole production with ACC-IS, we have irradiated Si(001) targets for both low J ($<50\ \mu A\ cm^{-2}$) and fluencies ($3–5 \times 10^{17}$ ions cm^{-2}). Under these conditions, we have been able to successfully produce nanohole patterns with the Kaufman source, as shown in Fig. 6.9a. Note that the pattern also presents short-range hexagonal order as evidenced by the autocorrelation image of the topography (inset of Fig. 6.9a). Figure 6.9b displays the PSDs of typical nanohole

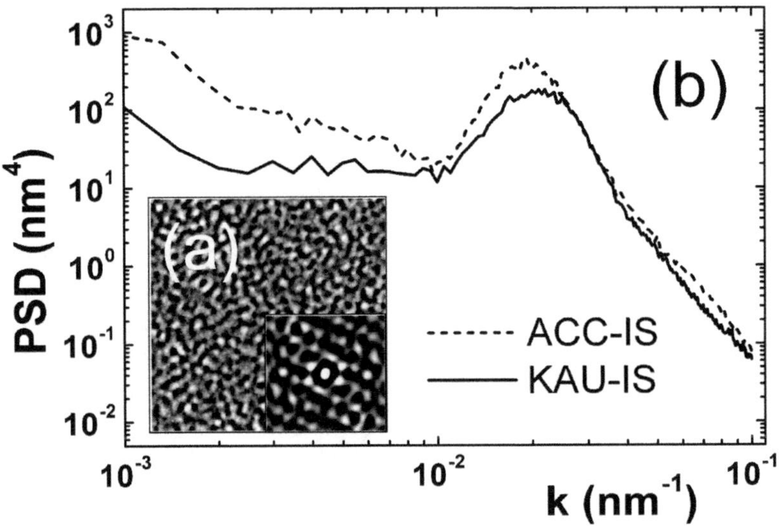

Figure 6.9 (a) 2 × 2 μm² AFM image of a nanohole pattern produced with a Kaufman-type ion source (KAU-IS) with Fe and Mo incorporation during IBS. Note that the nanohole pattern also displays short-range hexagonal order, as derived from the autocorrelation function corresponding to a 400 nm wide AFM image (inset). (b) PSD curves from images of nanohole patterns produced with alternating cold-cathode (dashed line) and Kaufman (solid line) ion sources. Both nanohole patterns display similar features in terms of roughness, wavelength, and short-range ordering.

patterns induced with both ion guns. Clearly, they are rather similar in terms of roughness, short-range ordering, and wavelength.

The next step in this research, according to the scheme followed with the ACC-IS system, consisted in the evaluation of the metal content by RBS for each generated pattern to assess whether the same correlation between metal content and pattern applies. Figure 6.10 shows the values corresponding to the residual metal content in targets irradiated with the Kaufman system together with those obtained previously with the ACC-IS setup (Fig. 6.8b). The agreement between both sets of data is evident since nanohole patterns are obtained for higher metal contents (2.5–5.6 × 10^{15} atoms cm^{-2}) than nanodot morphologies (1.5–2.5 × 10^{15} atoms cm^{-2}) irrespective of the ion gun.

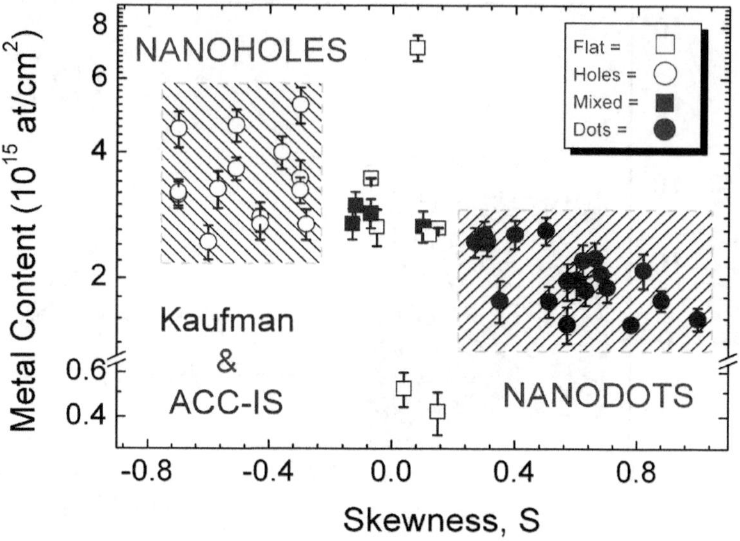

Figure 6.10 General correlation between the residual metal content on the surface of irradiated Si(100) and the resulting pattern as obtained with both ACC-IS (red symbols) and Kaufman (blue symbols) sources. The symbols for the different surfaces morphologies are shown in the label. See also Color Insert.

6.4.3 *Effect of the Metal Nature on Nanohole Pattern Formation*

From the above generalization, it is clear that the total metal content plays a critical role on the pattern selectivity. Further, it seems that the Fe–Mo ratio is not relevant for nanohole pattern formation since such patterns have been successfully produced with different Fe–Mo ratios and even with the sole Fe incorporation in case of a Kaufman ion source. The natural question that follows to this statement is what mechanisms are induced by the metal incorporation or if any impurity, at the proper level, will be able to tune the pattern morphology. The answer to this question is rather complex since, for example, the nature of impurities capable to trigger nanodot pattern formation and the underlying mechanisms are still not properly known. As a first approach, we have done a simple experiment to corroborate whether the metal nature matters

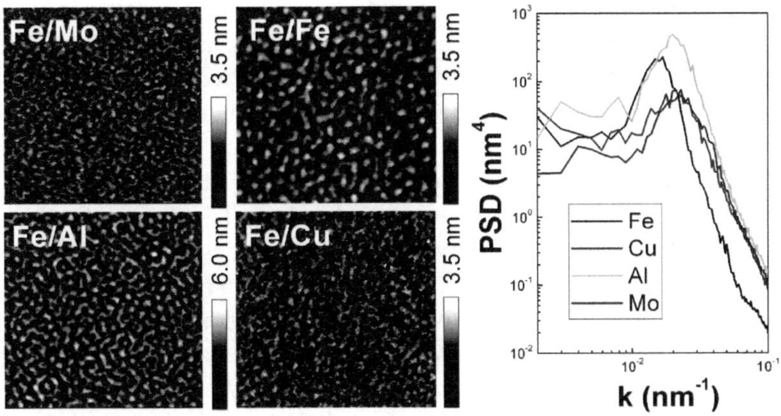

Figure 6.11 AFM 1 × 1 μm^2 of Si(100) targets masked with different metal foils and irradiated with an ACC-IS using 1 keV Ar^+ ions at a current density of 50 μA cm^{-2} and for a fluency of 4 × 10^{17} ions cm^{-2}. In all the cases, the PSD curves (right graph) show the formation of a pattern although its characteristics depend on the metal combination. See also Color Insert.

in the formation of nanohole patterns. For this purpose, we have fabricated masks with an inner hole diameter of 18 mm from metal foils of different materials (Mo, Fe, Al, and Cu). For all these irradiations, we have always the unavoidable source of Fe from ACC-IS. Therefore, we have different metal combinations depending on the mask nature. Obviously, in case of the Fe mask, only Fe impurities are incorporated.

Figure 6.11 shows the resulting surface of Si(100) targets irradiated with different mask materials. The irradiations were performed with 1 keV Ar^+ extracted from the ACC-IS at a current density of 50 μA cm^{-2} and for a fluency of 4 × 10^{17} ions cm^{-2}. Note that these conditions have been selected since they yield clear nanohole patterns with the Mo mask (Fig. 6.3). The total metal content, as derived from RBS, is around 3–4 × 10^{15} atoms cm^{-2} for all the samples, where ∼2 × 10^{15} atoms cm^{-2} of Fe are supplied by the ACC-IS cathodes. This value is consistent with the range expected for nanohole formation in accordance with Fig. 6.10. The AFM images reveal that the surface presents a surface pattern for all the metal combinations, which is evidenced by the peak in the PSD curves shown on the right graph of Fig. 6.11. A nanohole pattern is clear for the case of Fe/Mo- and Fe/Cu-containing surfaces, with a surface skewness well below −0.2. However, the surface rather displays a mixed pattern in case of Fe/Fe and Fe/Al combinations. It is surprising that the sole Fe incorporation does not yield nanohole patterns although maybe the conditions for nanohole formation are slightly shifted (the same could be applied to the Fe/Al combination). Although all these metals have a higher sputtering yield than silicon, their incorporation does not necessarily lead to a nanohole pattern despite the absolute total metal content. Therefore, other mechanism different from preferential sputtering may also take place and influence the eventual pattern formation. Among them, we can mention the different metal surface diffusivities, which have been pointed out recently as a possible mechanism in the morphological and compositional patterning of alloy surfaces [36]. In addition, it must be considered the eventual chemical interaction of metal elements with the silicon target that may induce some changes in the sputtering yield. For example, we have recently shown that in both nanohole and nanodot patterns with Fe/Mo incorporation, metals are mostly in the form of silicides [35]. On the contrary, in case of Cu, which is expected to be immiscible with Si, once could expect the clustering of the metal atoms at the surface. The latter could be a plausible explanation for the different features of nanohole patterns produced with Fe/Mo and Fe/Cu combinations. However, to be able to discriminate between these mechanisms, more simple experiments, in which only one metal element is involved. Such experiments are in progress and we do hope that they will contribute to clarify the scenario of metal-induced IBS nanopatterning.

6.5 On the Mechanisms of Pattern Selectivity and Outlook

As mentioned above, the formation of nanohole or nanodot structures by IBS cannot be explained within the framework of the Bradley–Harper-type models [20], which predict that the pattern selectivity is governed by changes in the shape of the collision cascade. Therefore, in principle, this framework should be ruled out.

Nevertheless, it must be noticed that the mixed metal/silicon nature of the target surface could lead to important distortions, not only of the ion collision cascade, but also of the ion penetration length [34]. Therefore, eventual changes on the surface concentration of the incorporated metal atoms, either vertically or laterally, could lead to differences in the way in which the ion energy is transferred to the target surface.

As our results clearly demonstrate, the pattern selectivity is modulated by the metal content. This proves the important role played by metal incorporation during IBS nanopatterning. In fact, as described in the introduction, the discussion concerning the pattern selectivity observed in our experiments through the role of metal impurities refers to a rather general open issue, both theoretically and experimentally, in IBS nanopatterning. From the seminal paper by Bradley and Harper [1] and further developments, it was accepted that nanodot and rippled patterns induced by IBS should be produced under normal and oblique ion incidence, respectively. Many experimental reports have been consistent with this framework although, also, some unexpected results were also published. As research in the field progressed, it became clear that the experimental data so far reported displayed a large amount of diversity and even a certain lack of the universality predicted by the theoretical models. During these years, some experimental groups [19, 31, 32, 34, 35], worked on the influence of metal incorporation on the IBS nanopatterns as researchers started to be aware that unintentional metal incorporation during IBS, even in short amounts, could alter the final pattern morphology.

The main role that one could expect from metal incorporation on IBS nanopatterning refers to the so-called "seeding" effect, which was reported in the texturization of different surfaces subjected to ion bombardment [41, 37]. In this case, the simultaneous supply of "seed" materials, which are different from the constituent materials of a sputtered target surface, enhances surface texturing. This methodology has been used to grow Mo-seeded Si microcones and nanorods by Ar^+ ion sputtering [37, 38]. It is thought that these seed materials, which should have either a lower sputtering yield or a higher melting point than the target constituents [37], form clusters on the irradiated surface that shield the underlying atoms from being sputtered by the incoming ions, while uncovered areas on the target are eroded faster. Then this process leads to the development of surface protrusions, such as micro(nano)cones, topped with seed atoms. A similar process has been proposed to operate during (un)intentional "seeding" in IBS nanopatterning [31].

As a consequence of the experimental evidences regarding the role of metal impurities on pattern formation, the theoretical interest in understanding IBS nanopatterning of compound targets has increased [2, 36, 43] and different scenarios have been disclosed. From these studies, it seems clear that preferential sputtering of one species over others [42] plays an important role but also their eventual different surface diffusivities. Additionally, surface stress induced by ion bombardment [7, 11, 21] could also be amplified by the presence of impurities. This mechanism has been invoked to be relevant [31, 43], and could even explain the unexpected behavior with irradiation temperature [18], in IBS nanodot patterning with the incorporation of metallic impurities.

According to the theoretical studies of pattern formation on multielemental targets, it becomes evident that IBS can induce not only a morphological pattern but also a compositional pattern that could be in or out of phase depending on the target and irradiation conditions. However, the theoretical models do not necessarily predict if the species with a lower or higher sputtering yield will remain at the top or bottom of the resulting surface (dot, hole, or ripple) morphologies. In this sense, it is clear that future experimental actions should be directed to address the surface composition of irradiated targets with nanoscale resolution, i.e., to be able to discern the surface composition of individual morphologies. This experimental data would be very valuable to improve or validate theoretical models.

Finally, the issue of the influence of metal incorporation on IBS nanopatterning has led the researchers to perform very well-controlled experiments in two senses. First, the compositional characterization of the irradiated surface has become a mandatory experimental routine to assess the quality of the irradiation experiments in terms of eventual contamination during the process. Second, there has been an evident interest in performing "clean" experiments where no contamination takes place. From recent experiments fulfilling such condition [26, 28], a surprising fact has arisen since IBS does not induce any pattern on the surface when the ion beam impinges under angles smaller than 45° with respect to the substrate normal. This unexpected evidence has triggered [5] new interest and efforts to understand theoretically the IBS nanopatterning process for which fully-controlled experiments will be indeed required.

6.6 Conclusion

In conclusion, we have stressed the relevance of metal impurities on nanopattern formation by IBS. In particular, we have shown the successful nanostructuration of Si(001) surfaces with an ACC-IS source, where Fe and Mo impurities are incorporated simultaneously during IBS from the ACC-IS cathode and the target holder, respectively. Under these working conditions, nanohole patterns are produced for low ion fluxes (<100 μA cm^{-2}) and relatively low ion fluencies ($<10^{18}$ ions cm^{-2}), whereas nanodot morphologies are produced above this parameter window. Therefore, nanohole morphologies are obtained in a narrow IBS window and, contrary to the nanodot case, the pattern is not stable under long time irradiation.

RBS compositional analysis of the irradiated surfaces has shown that nanohole morphologies are produced when the metal content on the irradiated surfaces is higher ($2.5–3.5 \times 10^{15}$ atom cm^{-2}) than in case of nanodot patterns ($1–2.5 \times 10^{15}$ atom cm^{-2}). This result suggests the role of metal incorporation as the driving mechanism for the pattern selectivity. Such assumption has been further corroborated by extending the quantitative results obtained with the ACC-IS to a Kaufman-type ion source operating under similar conditions. The pattern selectivity through metal incorporation is rather complex and far from being completely understood although plausible (concurrent) mechanisms that could drive to such pattern selectivity have been discussed. Finally, this work highlights the relevance of chemical analysis of nanopatterned surfaces by IBS to deepen our understanding of this intriguing self-organized surface nanopatterning phenomenon.

Acknowledgments

We are thankful to R. Cuerno, J. Muñoz-García, and M. Castro for helpful discussions and suggestions. The help in sample characterization by Zs. Czigány, R. Caillard, J. A. Martín-Gago, F. J. Palomares, and M. Fernández during the course of this research is also greatly acknowledged. This work has been partially supported by grants FIS2009-12964-C05-04, CSD2007-00010 and CSD2008-00023 from the *Ministerio de Ciencia e Innovación* (Spain), FIS2012-38866-C05-05 from Ministerio de Ciencia y Competitividad (Spain) and S2009/PPQ-1642 (AVANSENS) from the Comunidad Autónoma de Madrid (Spain). Finally, A. R. C. and JASG acknowledge funding from SFRH/BPD/74095/2010 (Portugal) and the I3P program from the CSIC (Spain), respectively.

References

1. Bradley R. M. and Harper J. M. E. (1988). Theory of ripple topography induced by ion bombardment, *J. Vac. Sci. Technol. A*, **6**, 2390–2395.
2. Bradley R. M. and Shipman P. D. (2010). Spontaneous pattern formation induced by ion bombardment of binary compounds, *Phys. Rev. Lett.*, **105**, 145501.
3. Carter G. and Vishnyakow V. (1996). Roughening and ripple instabilities on ion-bombarded Si, *Phys. Rev. B*, **54**, 17647–17653.
4. Castro M., Cuerno R., Vázquez L., and Gago R. (2005). Self-organized ordering of nanostructures produced by ion-beam sputtering, *Phys. Rev. Lett.*, **94**, 016102.
5. Castro M., Cuerno R. (2012). Hydrodynamic approach to surface pattern formation by ion beams, *Appl. Surf. Sci.* **258**, 4171.
6. Chan W. L. and Chason E. (2007). Making waves: Kinetic processes controlling surface evolution during low energy ion sputtering, *J. Appl. Phys.*, **101**, 121301.
7. Chan W. L. and Chason E. (2008). Stress evolution and defect diffusion in Cu during low energy ion irradiation: Experiments and modeling, *J. Vac. Sci. Technol. A*, **26**, 44.
8. Chu W. K., Mayer J. W., and Nicholet M. A. (1978). *Backscattering Spectrometry*, Academic Press, USA.

9. Cuenat A. and Aziz M. J. (2002). Spontaneous pattern formation from focused and unfocused ion beam irradiation, *Mat. Res. Soc. Symp. Proc.*, **696**, N2.8.1–N2.8.6.
10. Cuerno R., Vázquez L., Gago R., and Castro M. (2009). Surface nanopatterns induced by ion-beam sputtering. *J. Phys.: Condens. Matter*, **21**, 220301.
11. Dahmen K., Giesen M., Ikonomov J., Starbova K., and Ibach H. (2003). Steady-state surface stress induced in noble gas sputtering, *Thin Solid Films*, **428,** 6.
12. Facsko S., Dekorsy T., Koerdt C., Trappe C., Kurz H., Vogt A., and Hartnagel H. L. (1999). Formation of ordered nanoscale semiconductor dots by ion sputtering, *Science*, **285**, 1551–1553.
13. Facsko S., Bobek T., Stahl A., Kurz H., and Dekorsy T. (2004). Dissipative continuum model for self-organized pattern formation during ion-beam erosion, *Phys. Rev. B*, **69**, 153412.
14. Feldman L. C., Mayer J. W., and Picraux S. T. (1982). *Materials Analysis by Ion Channelling*, Academic Press, USA.
15. Gago R., Vázquez L., Cuerno R., Varela M., Ballesteros C., and Albella J. M. (2001). Production of ordered silicon nanocrystals by low-energy ion sputtering, *Appl. Phys. Lett.*, **78**, 3316–3318.
16. Gago R., Vázquez L., Cuerno R., Varela M., Ballesteros C., and Albella J. M. (2002). Nanopatterning of silicon surfaces by low-energy ion-beam sputtering: Dependence on the angle of ion incidence, *Nanotechnology*, **13**, 304–308.
17. Gago R., Vázquez L., Plantevin O., Metzger T. H., Muñoz-García J., Cuerno R., and Castro M. (2006). Order enhancement and coarsening of self-organized silicon nanodot patterns induced by ion-beam sputtering, *Appl. Phys. Lett.*, **89**, 233101.
18. Gago R., Vázquez L., Sánchez-García J. A., Varela M., Ballesteros M. C., Plantevin O., Albella J. M., and Metzger T. H. (2006). Temperature influence on the production of nanodot patterns by ion beam sputtering of Si(001), *Phys. Rev. B*, **73**, 0155414.
19. Hofsass H. and Zhang K. (2008). Surfactant sputtering, *Appl. Phys. A*, **92**, 517.
20. Kahng B., Jeong H., and Barabási A. L. (2001). Quantum dot and hole formation in sputter erosion, *Appl. Phys. Lett.*, **78**, 805–807.
21. Kalyanasundaram N., Moore M. C., Freund J. B., and Johnson H. T. (2006). Stress evolution due to medium-energy ion bombardment of silicon, *Acta Mater*, **54,** 483.

22. Kótai, E. (1994). Computer methods for analysis and simulation of RBS and ERDA spectra. *Nucl. Instrum. Methods B*, **85**, 588–596.
23. Le Roy S., Søndergård E., Nerbø I. S., Kildemo M., and Plapp M. (2010). Diffuse-interface model for nanopatterning induced by self-sustained ion-etch masking, *Phys. Rev. B*, **81**, 161401.
24. Li R.-S. (1993). Influence of bombardment-induced Gibbsian segregation on alloy sputtering, *Nucl. Instrum. Methods B*, **82**, 283–290.
25. Ludwig F., Eddy C. R., Malis O., and Headrick R. L. (2002). Si(100) surface morphology evolution during normal-incidence sputtering with 100–500 eV Ar^+ ions, *Appl. Phys. Lett.*, **81**, 2770–2772.
26. Macko S., Frost F., Ziberi B., Förster D. F., and Michely M. (2010). Is keV ion-induced pattern formation on Si(001) caused by metal impurities?, *Nanotechnology*, **21**, 085301.
27. Madi C. S., Davidovich B., George H. B., Norris S. A., Brenner M. P., and Aziz M. J. (2008). Multiple bifurcation types and the linear dynamics of ion sputtered surfaces. *Phys. Rev. Lett.*, **101**, 246102.
28. Madi C. S., George H. -B., and Aziz M. J. (2009). Linear stability and instability patterns in ion-sputtered silicon, *J. Phys.: Condens. Matter*, **21**, 224010.
29. Makeev M. A., Cuerno R., and Barabási A. -L. (2002). Morphology of ion-sputtered surfaces, *Nucl. Instrum. Method B*, **197**, 185–227.
30. Muñoz-García J., Vázquez L., Cuerno R., Sánchez-García J. A., Castro M., and Gago R. (2009). Self-organized surface nanopatterning by ion beam sputtering, in *Towards Functional Nanomaterials: Lecture Notes on Nanoscale Science and Technology 5* (Wang Z., ed), Springer, New York, pp. 323–398.
31. Ozaydin G., Özcan A. S., Wang Y., Ludwig K. F., Zhou H., Headrick R. L., and Siddons D. P. (2005). Real-time x-ray studies of Mo-seeded Si nanodot formation during ion bombardment, *Appl. Phys. Lett.*, **87**, 163104.
32. Ozaydin G., Ludwig K. F., Zhou H., and Headrick R. L. (2008). Effects of Mo seeding on the formation of Si nanodots during low-energy ion bombardment, *J. Vac. Sci. Technol. B*, **26**, 551.
33. Plantevin O., Gago R., Vázquez L., Biermanns A., and Metzger T. H. (2007). In situ X-ray scattering study of self-organized nanodot pattern formation on GaSb(001) by ion beam sputtering, *Appl. Phys. Lett.*, **91**, 113105.
34. Sánchez-García J. A., Vázquez L., Gago R., Redondo-Cubero A., Albella J. M., and Czigány Zs. (2008). Tuning surface morphology in self-organized

ion beam nanopatterning of Si(001) via metal incorporation: From holes to dots, *Nanotechnology*, **19**, 355306.

35. Sánchez-García J. A., Gago R., Caillard R., Redondo-Cubero A., Martin-Gago J. A., Palomares F. J., Fernández M., and Vázquez L. (2009). Production of nanohole/nanodot patterns on Si(001) by ion beam sputtering with simultaneous metal incorporation, *J. Phys.: Condens. Matter*, **21**, 224009.
36. Shenoy V. B., Chan W. L., and Chason E. (2007). Compositionally modulated ripples induced by sputtering of alloy surfaces, *Phys. Rev. Lett.*, **98**, 256101.
37. Tanemura M., Yamauchi H., Yamane Y., Okita T., and Tanemura S. (2004). Controlled fabrication of Mo-seeded Si microcones by Ar^{+}-ion bombardment, *Nucl. Instrum. Methods B*, **215**, 137–142.
38. Tanemura M., Kobayashi M., Kudo M., Yamauchi H., Okita T., Miao L., and Tanemura S. (2006). Fabrication of densely distributed Si nanorods by Ar+-ion bombardment, *Surf. Sci.*, **600**, 3668–3672.
39. Tsamouras D., Palasantzas G., and De Hosson J.Th.M. (2001). Growth front roughening of room-temperature deposited oligomer films, *Appl. Phys. Lett.*, **79,** 1801–1803.
40. Valbusa U., Boragno C., and Buatier de Mongeot F. (2002). Nanostructuring surfaces by ion sputtering, *J. Phys.: Condens. Matter*, **14**, 8153–8175.
41. Wehner G. K. (1985). Cone formation as a result of whisker growth on ion bombarded metal-surfaces, *J. Vac. Sci. Technol. A*, **3,** 1281.
42. Zaporozchenko V. I. and Stepanova M. G. (1995). Preferential sputtering in binary targets, *Prog. Surf. Sci.*, **2,** 155.
43. Zhou J. and Lu M. (2010). Mechanism of Fe impurity motivated ion-nanopatterning of Si (100) surfaces, *Phys. Rev. B*, **82**, 125404.
44. Ziberi B., Frost F., Tartz M., Neumann H., and Rauschenbach B. (2004). Importance of ion beam parameters on self-organized pattern formation on semiconductor surfaces by ion beam erosion, *Thin Sol Films*, **459**, 106–110.
45. Ziberi B., Frost F., Rauschenbach B., and Höche Th. (2005). Highly ordered self-organized dot patterns on Si surfaces by low-energy ion-beam erosion, *Appl. Phys. Lett.*, **87**, 033113.

Chapter 7

Kinetic Monte Carlo Simulations of Low-Energy Ion-Induced Surface Patterning

Wai-Lun Chan[a] **and Eric Chason**[b]

[a]*Department of Chemistry and Biochemistry, University of Texas at Austin, Austin, TX 78712, USA*
[b]*School of Engineering, Brown University, Providence, RI 02912, USA*
wlchan@austin.utexas.edu, Eric_Chason@brown.edu

In this chapter, we review some of our recent works in using kinetic Monte Carlo (KMC) simulations to model the pattern formation behavior on crystalline surfaces during low-energy ion sputtering. The KMC simulation allows us to control the rate of different atomistic processes (e.g., diffusion, defects creation, clustering) independently. Thereby, it allows us to study how each of these processes controls the pattern formation behavior. We found that by using Sigmund's sputtering mechanism, we can quantitatively reproduce many of the results predicted by the continuum models. One of the exceptions is in the high-temperature and low-ion-flux regime, where roughening is limited by the nucleation of new steps on the surface. These nucleation processes are often neglected in the continuum models. Furthermore, many recent experiments and simulations show that surface craters and adatoms created during

Nanofabrication by Ion-Beam Sputtering: Fundamentals and Applications
Edited by Tapobrata Som and Dinakar Kanjilal

ISBN 978-981-4303-75-0 (Hardcover), 978-981-4303-76-7 (eBook)
www.panstanford.com

an ion impact can play a more important role in the patterning behavior than the sputter vacancies. In the last part of this chapter, we demonstrate the effects of multiple defect creation on the pattern formation kinetics.

7.1 Introduction

It is well known that low-energy ion irradiation can lead to the formation of highly ordered, nanoscaled, self-organized patterns on surfaces [10, 37, 42, 57]. This phenomenon has also been demonstrated to occur in a wide range of materials such as metals, semiconductors, and oxide surfaces. Although a significant amount of work has been devoted to understanding the mechanisms that control the formation of these surface patterns, we still lack a comprehensive theory that can predict the dynamics of the pattern formation behavior quantitatively, even in relatively simple materials systems such as Si or noble metals. The complexity arises from the interaction of multiple atomistic processes during the ion irradiation. These basic surface processes include morphology-dependent sputtering yield; defect generation, clustering, and annihilation; surface diffusion, and so on. Furthermore, different kinetic mechanisms can dominate the pattern formation behavior in different processing regimes, e.g., ion flux, ion energy, incident angle, and irradiation temperature. The surface relaxation mechanisms are also very different in various material systems. These complications together make quantitative prediction of the pattering behavior very difficult. On the other hand, if the effects of each of these mechanisms are known, a variety of surface patterns such as highly ordered quantum dots, one-dimensional or two-dimensional patterns that align with the ion beam or with the surface crystallographic orientation can be produced in a controlled way by varying the sputtering conditions. This has many potential technological applications since a large surface area of nanopatterns can be produced in a single ion sputtering step.

Our current understanding of the morphology evolution is primarily based on continuum theories that have been developed over the last two decades. The most well-known theory perhaps is the Bradley–Harper theory [5], which states that pattern formation is due to a competition between the surface-curvature-dependent sputter yield and the surface diffusion. Although there is an increase in experimental and simulation evidences [31, 34, 35, 55], showing that the assumption employed by Bradley and Harper is probably too simplistic, and the curvature-dependent sputter yield can barely account for the observed pattern formation behavior [10], nevertheless, it serves as a starting point for us to understand the patterning phenomenon and a basis for the development of more sophisticated continuum models [8, 16, 19, 36, 37, 45].

A major challenge of the continuum approach is developing differential equations that account for the surface evolution resulting from a large number of interacting atomistic processes. An alternative approach is to simulate the kinetic behavior at the atomic level rather than capturing the overall evolution with analytical expressions. This approach can be implemented in a number of ways, including KMC methods [14, 15, 17, 26, 29, 32, 44, 47, 52, 54, 59, 60] such as the one described in this chapter. The evolution of the surface then arises from allowing the individual particles to act according to these rules without an *a priori* understanding of how this behavior will occur. In this respect, continuum models and simulations are complementary in that they are both ways of predicting the surface evolution under the action of different physical mechanisms. Note that the approach we are employing here is not to use the KMC simulation to model the surface evolution of a specific material system, but to let the simulation serve as a generic model to study how different mechanisms can interact with each other to change the pattern formation kinetics.

In this chapter, we will start by discussing our KMC results when we apply Sigmund's sputtering mechanism [50, 51] (the same sputtering mechanism is used in the Bradley–Harper model and many other continuum models) to a crystalline surface. The use of the KMC model is necessary in this case because most of the continuum models are formulated based on amorphous surfaces. The thermal or ion-induced relaxation of a crystalline surface can be quite complicated compared to that of an amorphous surface, and to describe it using a continuum model can be theoretically challenging [12, 30, 38, 48, 49]. Our KMC model shows that while many features of the pattern formation predicted by the Bradley–Harper model (and similar continuum model using Sigmund's mechanism) are reproduced in the KMC simulation, the wavelength of the patterns depends heavily on the concentration of mobile defects on the surface. How the defect concentration depends on the temperature, ion flux, and surface morphology is difficult to be modeled by a continuum approach in some cases, but it can be determined explicitly in KMC simulations.

Furthermore, we notice that there are many recent findings suggesting that Sigmund's sputtering mechanism is insufficient to explain many of the observed dynamics. Molecular dynamics (MD) simulations [23, 24, 31] and scanning tunneling microscopy (STM) studies [33, 40] show that craters and multiple defects are created around the impact point of the ion. While Sigmund's sputtering mechanism only accounts for the sputter vacancies resulting from sputtering, its accuracy in predicting the pattern formation behaviors becomes questionable. In the last part of this chapter, we will describe our recent efforts in incorporating the formation of multiple defects per incident ion into our simulation and discuss the possible effects of the inclusion of these defects. The purpose of the chapter is not to give a comprehensive review of KMC simulations of ion-beam-induced surface patterning, but to highlight some of our recent findings using KMC simulations [13, 14] that give insight into the underlying fundamental mechanisms.

7.2 KMC Model

Before we discuss our simulation results, we will first describe the principles behind our KMC modeling. In the KMC approach, different surface processes, e.g., sputtering, adatoms diffusion, defect recombination (some of these processes are shown in Fig. 7.1a), are assigned to have a specific rate of execution. Then different events are executed with a probability proportional to their relative rates. In our simulation, the surface is modeled as a square lattice and we use a solid-on-solid (SOS) model so that there is only one surface site associated with each lattice site (no overhangs are allowed) [14, 15].

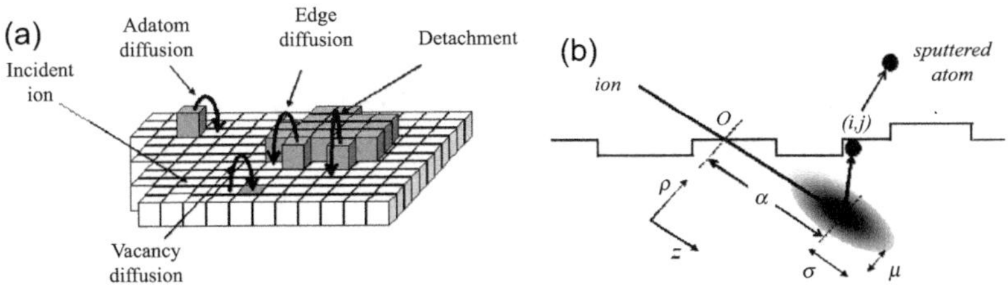

Figure 7.1 (a) Example of surface kinetic processes that can be happened during sputtering. Each process is assigned with its transition rate according to Eq. 7.2. (b) Schematics showing Sigmund's sputtering mechanism. Reprinted from reference [14].

7.2.1 *The Sputtering Process*

The sputtering process is incorporated into our simulation using Sigmund's model [50, 51] for sputtering. In this model, Sigmund assumed that the amount of material sputtered at a particular surface site is proportional to the amount of energy deposited at that site by an incoming ion through a series of atomic collisions. The energy distribution averaged over a large number of ion impacts can be approximated by a Gaussian distribution, centered at a point below the surface.

A schematic representation of the deposited energy density proposed in Sigmund's model is shown in Fig. 7.1b; the ion enters the surface at point O and the resulting energy distribution is centered at distance a from the entry point along the same direction as the incident trajectory. Using coordinates where z is the distance from O along the ion trajectory and ρ is the distance normal to the trajectory, the energy per unit volume deposited by collisions at each point in the material is described by

$$E\left(\rho, z\right) = \varepsilon_0 \exp\left\{-\left[\frac{(z-a)^2}{2\sigma^2} + \frac{\rho^2}{2\mu^2}\right]\right\} \tag{7.1}$$

where ε_0 is a normalization factor and the parameters σ and μ describe the width of the energy deposition profile in the directions parallel and perpendicular to the initial trajectory, respectively. The contours of uniform energy deposited by the ion form ellipsoids around the average stopping point of the ion. For a given incident ion trajectory, Eq. 7.1 can be used to calculate the average amount of energy deposited at each site on the surface [for example, the site labeled as (i, j) in Fig. 7.1b].

For continuum theories such as the Bradley–Harper model, the amount of material sputtered by a single ion from each position on the surface is taken to be proportional to the energy deposited there. Because it is a continuum description, the change in the surface height at each position does not have to be quantized in units of atomic step heights. In contrast, the KMC simulations are atomistic in nature, so we require that only whole atoms can be removed from a surface site due to the sputtering process. To accomplish this, we use the Sigmund form for the energy deposition to determine the *probability* of sputtering a single atom from a site on the surface rather than to determine the continuum amount of material removed. From this probability distribution, we use a Monte Carlo scheme to determine how many discrete atoms are removed for each incident ion and from which sites they are removed.

The average total sputter yield Y (number of atom removed per incident ion) can be adjusted by tuning the normalization factor ε_0 in Eq. 7.1. In our case, Y is set to be 2 for a flat surface. For the sputtering of a rough surface, the sputter yield can be either smaller or larger than 2, and it depends on the local surface morphology. Indeed, one of the major consequences resulting from Sigmund's mechanism is that the sputter yield at a valley is higher than the sputter yield at a crest. This causes a surface instability, which can lead, in some cases, to the formation of periodic surface patterns.

Removal of atoms by sputtering in the simulation leads to the creation of only surface vacancies. This ignores the creation of other defects such as subsurface interstitials and vacancies that can be created by the ion-induced displacements. This can be true at high temperature where most of the defects recombine [21] so that only the vacancies created by the sputtering of atoms will remain. Similarly, the continuum theories also only model the removal of atoms from the surface by sputtering and ignore the creation of other defects. Still it is worth noting that the defect formation in real systems may be different than the approximation made by Sigmund's mechanism. It has also been pointed out that there may be significant rearrangement of atoms on the surface due to the incident ion, which may lead to different behavior than predicted by Sigmund's model [19, 31, 34, 40]. These issues will be discussed more in Section 7.6.

7.2.2 *Simulation of Surface Transport*

In addition to sputtering, we simulate the surface transport by thermally activated diffusion mechanisms. The motion of atoms on the surface is determined by the probability of making a thermally activated hop over the energetic barrier between sites. For an atom at an initial site denoted by i and a final neighboring site denoted by j, the rate of making a transition between them is

$$r(n_i, n_j) = \Gamma \exp(-E(n_i, n_j)/kT) \tag{7.2}$$

where Γ is a prefactor equal to the attempt frequency, $E(n_i, n_j)$ is an activation energy for the transition, and n_i and n_j refer to the number of bonding on the initial and final sites. Because the lattice is square, there are four neighboring sites j around each atom i that must be considered, with a potentially different transition rate associated with each jump direction.

The activation energy $[E(n_i, n_j)]$ depends on the local environment as well as on the bonding around the initial and final sites. We define the activation energy for transitions between any two neighboring sites as:

$$E(n_i, n_j) = E_o(n_i, n_j) \quad \text{for } n_j \geq n_i \text{ (no bonds broken)} \tag{7.3a}$$

$$E(n_i, n_j) = E_o(n_i, n_j) + (n_i - n_j)E_b \quad \text{for } n_j < n_i \text{ (bonds broken)} \tag{7.3b}$$

In these equations, $E_o(n_i, n_j)$ is the activation energy for transitions, which do not break any bonds (usually referred to as diffusion barriers). This energy is defined to be symmetric over the switching of the indexes i and j. Therefore, the only difference between a forward and a reverse process is due to the change in bond energy. E_b is the energy per bond, and the term $(n_i - n_j)E_b$ denotes the extra energy barrier that is needed when the transition is toward a lower coordination site. In our simulation, $E_o(n_i, n_j)$ is taken to be 0.8 eV for most of the transitions, and E_b is equal to 0.2 eV. Further details on the simulation parameters can be found in reference [14].

From the above description, we can assign a transition rate to each configuration of atoms based on the coordination of the initial (n_i) and final (n_f) sites. By summing over all the sites on the surface, we determine $N(n_i, n_f)$, which is the total number of configurations at which atoms can make a transition of this type. The total rate $R(n_i, n_f)$ at which transitions of each type can happen is then given by:

$$R(n_i, n_f) = N(n_i, n_f)r(n_i, n_f) \tag{7.4}$$

In addition to the transitions for different atomic configurations, there is also a rate for ion impingement Φ (the site at which an atom is sputtered is described above). Summing over all the possible transitions that can occur, we obtain the total rate R_{tot} for all events happening on the surface as:

$$R_{tot} = \sum_{n_i, n_f} R(n_i, n_f) + \Phi \tag{7.5}$$

where the sum over n_i and n_f is from 0 to 4 to cover all the possible surface configurations.

In a single iteration of the simulation, an event is chosen probabilistically based on its relative rate, i.e., the probability of choosing a transition from a site with n_i coordination to one with n_f is$R(n_o, n_f)/R_{\text{tot}}$ and the probability of choosing an ion impingement event is Φ/R_{tot}. The actual site for executing the transition is chosen by picking randomly from a list of all surface sites that have this starting and final coordination (such lists are maintained in the simulation). Once the site is chosen, the surface morphology is modified by moving an atom from its initial position to its new position. If the type of event is ion impingement instead of atom hopping, we use the algorithm described in the previous section to remove atoms from the surface and create surface vacancies according to Sigmund's mechanism. After the morphology is modified, the time duration of the simulation is incremented by $dt = 1/R$ and any transition rates or lists of configurations that have been affected by this event are recalculated. The process is then repeated in order to execute subsequent events on the surface.

Although we consider the hopping of atoms in all possible configurations on the surface, the surface evolution is dominated by the diffusion of what we refer to as mobile surface defects, i.e., adatoms and surface vacancies. The motion of mobile defects is central to surface transport, so it is important to be clear about what we mean by these terms. The transitions corresponding to adatom and surface vacancy diffusion are shown schematically in Fig. 7.1a. Adatom diffusion refers to the hopping of an atom that sits on top of the surface so that it has no nearest neighbors bonding to it in the plane of the surface (its only bond is to the surface below). Since the initial and final configurations have zero coordination ($n_i = 0$ and $n_j = 0$), the activation energy for adatom hopping is $E_o(0,0)$ using the notation defined above.

Vacancy diffusion corresponds to the hopping of a surface atom into an adjacent empty surface site, which we refer to as a surface vacancy. Although the vacancy is not in itself a surface species (being the absence of an atom), the hopping of an atom into the vacant site results in the apparent motion of the vacancy. We refer to this as surface vacancy diffusion in a manner that is analogous to the use of the term vacancy diffusion for bulk material. In both cases, the actual motion corresponds to the hopping of atoms, but it is useful to think of the vacancy as an entity that moves in the process of exchanging places with an atom. In the configuration labeled vacancy diffusion in the figure, the atoms jump from one highly coordinated site to another ($n_i = 3$ and $n_j = 3$); so the barrier for diffusion in this case is equal to E_o(3,3).

Adjustments of these diffusion barriers allow us to independently control the diffusion rates of the adatoms and surface vacancies. Since vacancy diffusion does not require a net decrease in the number of bonds, the barrier for vacancy diffusion can be comparable to the diffusion of an adatom. Indeed, the rapid diffusion of vacancies has been well established by measurements and calculations [1, 4, 22, 25, 53] on a number of surfaces. Mobile surface vacancies are especially important for the simulation of ion-induced processes since vacancy-type surface defects can be more numerous than adatoms.

In the following sections, we will present our simulation results. Our simulation results can be compared directly with continuum model formulated based on Sigmund's sputtering mechanism (e.g., Bradley–Harper model and its variations). We will not go into the details of the continuum models and readers are suggested to refer to references [10, 37] for details.

7.3 Ripples Formation in the Linear Instability Regime

One of the benchmarks for the instability model is the formation of ripples with a wavelength that does not change with time as the sputtering continues. Furthermore, the ripple amplitude grows exponentially with time from an initially flat surface (at least in the early period before saturation) [5, 37]. These characteristics are indeed observed in our simulations. In order to make a quantitative comparison between our KMC models and the continuum theories, we also compare our ripple growth rate and ripple travelling velocity with continuum models and we find quantitative agreements between the two.

7.3.1 *Time Evolution of the Ripple Wavelength and Amplitude*

Figure 7.2 shows some of the typical surface morphologies from the simulations. The ion beam parameters used are $a = 20$, $\sigma = \mu = 10$ (Fig. 7.1), and $Y = 2$, with all the lengths measured in units of the lattice spacing a_0. The angle of incidence is 45° and the ion beam direction projected onto the surface is indicated as arrows in the figure. The images represent surfaces after simulated sputtering at temperatures of 187°C for removal of 30, 60, and 90 monolayers (Fig. 7.2a–c) and 250°C for removal of 20, 40, and 60 monolayers (Fig. 7.2d–f); the ion flux f is 2 monolayers s^{-1}. The grey scale corresponds to the height at each surface position as shown on the accompanying marker. The ripples that form have their wavevector parallel to the ion beam direction. Adjacent to each image is the power spectral density (PSD) of the surface height; the peak in the graph confirms the presence of a preferred periodicity on the surface. The wavelength and amplitude of the ripple are determined from the centroid and total area under the peak in the PSD as a function of simulation time.

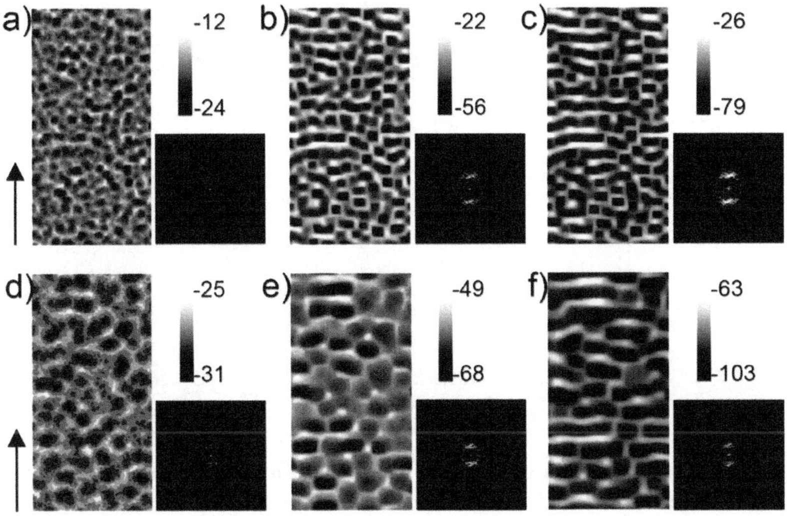

Figure 7.2 Morphology of the surface at different times during the sputtering simulation. (a–c) images correspond to removal of 20, 40, and 60 monolayers at temperature 187°C; (d-f) images correspond to removal of 30, 60, and 90 monolayers at temperature of 250°C. The scale next to each figure relates the grey level to the surface height in units of the simulation lattice parameter. The power spectral density of the surface is shown below each image of the surface. Reprinted from reference [13].

The evolution of the characteristic ripple wavelength and amplitude is shown in Fig. 7.3a,b for the same conditions used in Fig. 7.2. In the early stages, the ripple amplitude is seen to grow exponentially with sputtering time. During this growth, the corresponding wavelength is essentially constant with a well-defined peak in the PSD. In addition, the orientation of these ripples is defined by the ion beam direction; when the azimuthal orientation of the beam is changed in the simulation, the orientation of the pattern changes as well [3]. The exponential growth, fixed wavelength, and sample orientation are all features of the linear instability mechanism as predicted by the Bradley–Harper theory. After extended periods, the amplitude saturates, which signals the end of the region over which the behavior can be compared with the linear model and indicates the onset of nonlinear behavior. Note that the ripples formed at high temperature have a longer wavelength

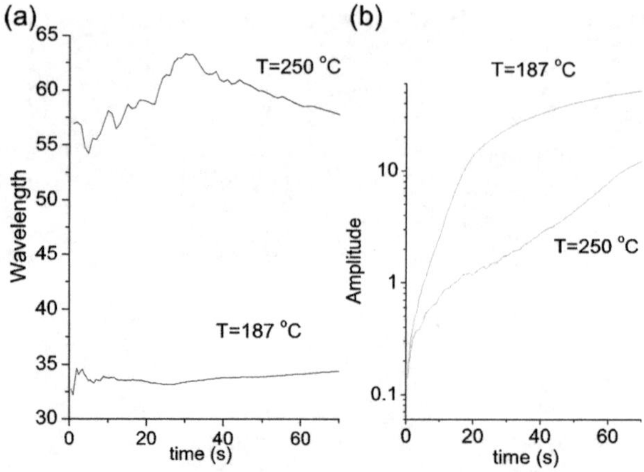

Figure 7.3 Evolution of (a) wavelength and (b) amplitude of the ripple pattern during sputtering simulation. The sputtering conditions are same as Fig. 7.2. Reprinted from reference [13].

and a slower growth rate than those formed at low temperature. This corresponds to the faster surface relaxation rate at high temperatures. As discussed below, the temperature dependences of the growth rate and wavelength are well explained by the continuum model.

7.3.2 *Ripple Growth Rate*

The Bradley–Harper model predicts that the exponential growth r^* of the ripples is proportional to the square of the wavevector k^* of the ripples. Mathematically, they can be related as:

$$r^* = \frac{1}{2}\nu_{\max}k^{*2} \tag{7.6}$$

In this equation, $\nu_{\max}$ is a parameter that depends linearly on the flux f but is independent of the temperature. Therefore, this relationship predicts that r^*/f is proportional to k^{*2} without any explicit dependence on the temperature and flux (for details, see references [10, 14]). When expressed this way, the relationship between r^* and k^* is independent of any kinetic processes. In Fig. 7.4, we plot r^*/f as a function of k^*. In agreement with Eq. 7.6, all the data from the different flux and temperature measurements collapse

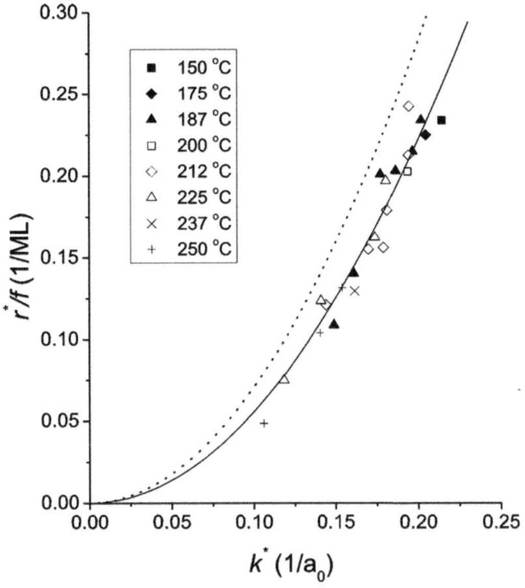

Figure 7.4 The ripple growth rate as a function of the wavevector of the ripples. The solid curve is a fit to Eq. 7.6, and the dashed line is the prediction by continuum theory. Reprinted from reference [14].

into a single curve and we can extract the parameter $\nu_{\max}$ from the simulation. Since $\nu_{\max}$ depends only on the sputtering parameters, it can serve as a quantitative comparison between the simulation and continuum models. The fit is shown as the solid line in the figure with a value of $\nu_{\max}$ equal to 5.56 ± 0.09. The same constant predicted by the Bradley–Harper theory [5] using the ion beam parameters we used in the simulation is 7.07, which is shown as the dashed solid line in the figure.

The reasonably good quantitative agreement between the simulation and the Bradley–Harper theory is reassuring since both approaches use Sigmund's model for calculation of the sputter yield. It further confirms that the linear instability theory is an appropriate continuum description of morphology evolution during sputtering. This appears to be true even though the surface relaxation mechanism used in the Bradley–Harper theory is very different from the simulations, i.e., the calculations of relaxation in the Bradley–Harper theory are appropriate for an amorphous surface, while the simulation is based on transport on a crystalline stepped surface. Despite this difference, the roughening rate from the KMC simulations is only decreased by about 20% comparing to the predictions on an amorphous surface of the Bradley–Harper theory.

7.3.3 *The Ripples Velocity*

Under the sputtering conditions we used in the simulation, the instability theory predicts that the ripple will travel across the surface. In order to quantify the effect of the curvature on ripple velocity, we measured the velocity of the simulated ripples at different wavevectors, which were obtained by using different conditions of temperature and flux. In Fig. 7.5, we show the measured velocity, normalized by the ion flux, versus the square of the characteristic wavevector (k^*) of the ripples. From the figure, we see that the velocity decreases dramatically as the ripple wavevector increases.

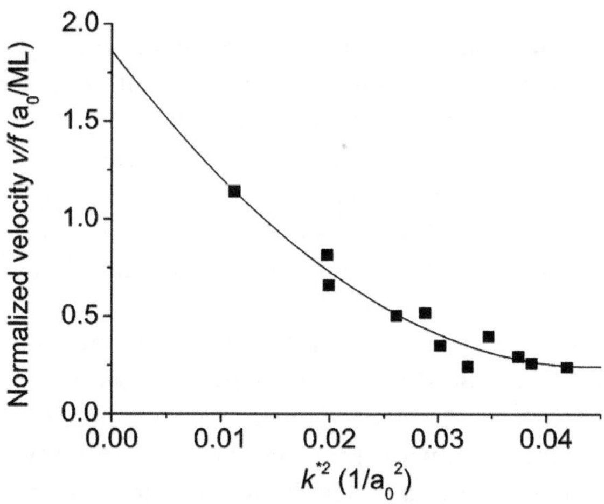

Figure 7.5 The lateral velocity of the ripples on the surface depends on the wavevector due to dispersion. The velocity from the simulation, normalized by the ion flux, is presented vs. the square of the wavevector. The solid line represents the fit to Eq. 7.7. Reprinted from reference [14].

The velocities we observed in Fig. 7.5 disagree with the original Bradley–Harper model. For example, at a temperature 250°C and a flux of 1 monolayer s^{-1}, the ripple pattern is observed to travel in the negative direction at a velocity 0.57 lattice spacing per monolayer of atoms removed. For sputtering at lower temperature ($T = 187$°C) and the same flux, the ripple pattern travels in the same direction at a velocity ~0.2 lattice spacing per monolayer of atoms removed. In comparison, calculations from the linear Bradley–Harper model [5] predict a ripple velocity of 1 lattice spacing per monolayer in the negative direction for both cases.

These discrepancies are a consequence of the fact that the ripple velocity depends on the ripple wavelength, which in turn depends on the sputtering temperature and flux. The linear Bradley–Harper theory (Eq. 7.5) assumes that the surface curvature is small relative to the range of the ions used and, therefore, the calculated velocity is independent of the wavelength. For the simulated ripples, however, the surface curvature can be large so that higher order terms must be included to predict the actual velocity (by symmetry, only even powers of the wavevector are allowed). The lowest order correction to the velocity is calculated by Makeev *et al.* [37] and depends on the square of the wavevector but otherwise not on any of the kinetic parameters. The correction term depends linearly on the flux and in a complex way on the other sputtering parameters.

We can model the effect of dispersion by expanding the velocity in powers of k^{*2}:

$$\frac{v}{f} = \gamma - \Omega_1 k^{*2} + \xi k^{*4} + \text{(higher order terms)} \qquad (7.7)$$

where γ and Ω_1 are calculated in reference [37]. The solid line represents the fit of the velocity data to this form, keeping terms only up to order k^4. The values of the parameters from the fit are $\gamma = 1.86$, $\Omega_1 = 73.3$, and $\xi = 829.7$. For the first two terms (the only ones calculated in reference [37]), these values are in very good agreement with the calculated values of $\gamma = 2$ and $\Omega_1 = 100$. These results strongly support the interpretation that the dependency of the velocity on the wavevector is due to dispersion and can be explained by continuum models that include higher order terms.

7.4 Dependence of Ripple Wavelength on Temperature and Ion Flux

When the ripple growth rate and travelling velocity are expressed in terms of the characteristic wavevector k^*, the remaining coefficients (Eqs. 7.6 and 7.7) depend only on the ion beam parameters (e.g., ion beam energy, incident angle, and range), but not on the processing conditions such as temperature and flux. From the above discussion, we can see that the growth kinetics of a ripple structure at a given wavelength follows closely with the continuum model. On the other hand, the ripple wavelength itself is very sensitive to parameters such as ion flux f and temperature T. In this section, we explore the temperature and flux dependences of the ripple wavelength and show that these dependences mainly originate from the relaxation process on the surface and the concentration of the mobile defects on the surface.

Figure 7.6 shows the temperature dependence of the ripple wavelength and the flux dependences at several different temperatures. The temperature dependence clearly cannot be described by an Arrhenius form with a single activation energy. Different temperatures also show very different flux dependences. At low temperatures, the wavelength is weakly dependent on the temperature,

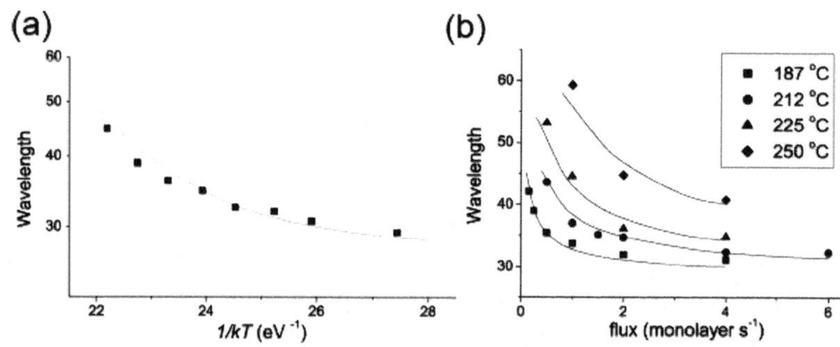

Figure 7.6 Dependence of the simulated ripple wavelength on (a) temperature at a flux of 2 monolayers s^{-1} and (b) flux at different temperatures as indicated in figure. Solid lines represent fit to linear instability theory evaluated using measured value of surface defects. Reprinted from reference [14].

while at elevated temperatures, the ripple wavelength increases rapidly with temperature. As the flux is increased, the ripple wavelength decreases, eventually saturating at high flux rates. For higher temperatures, the wavelength increases. Similar trends have been observed in different material systems (e.g., Cu [9, 11], Si [6, 20]). In the following, we use the simulation results together with the continuum model to explain the origin of these complicated dependences.

In the continuum model, the wavelength can be written as a function of concentration C of the surface mobile defects, and their diffusivity D as [14]:

$$\lambda^*(T, f) = \frac{2\pi}{k^*} \propto \sqrt{\frac{D(T)C(f, T)}{fT} + A_I} \tag{7.8}$$

where A_I is a constant independent of temperature and ion flux [36], and its value can be calculated from the ion beam parameters we used in the simulation. $D(T)$ refers to the diffusivity of the mobile defects, which is proportional to $\exp(-E_0/kT)$ in the simulations. The dependence of C on the temperature and flux is more complicated and depends on the details of the surface defect formation, clustering, and annihilation mechanisms [9, 20]. Because of the difficulty in formulating a universal analytically expression for C, it is often difficult to compare experimental results with continuum models. One of the advantages of the KMC model is that we can determine the number of mobile defects on the surface by direct counting. As a result, we can determine the role of the defect concentration and surface relaxation mechanisms on the ripple wavelength.

In our simulation, the majority of mobile defects are surface vacancies. This is because the mechanism that we use for sputtering (by Sigmund) only creates vacancy-like defects. In addition, we find that the average surface defect concentration remains fairly constant and uniform over the surface. A typical evolution of the adatom and vacancy concentration as a function of time is shown in Fig. 7.7a. These data are taken at $T = 187°\text{C}$ and $f = 1$ monolayer s^{-1}. Although the processes controlling the defect concentration are complex, the data over all the simulations studied is found to scale as $C \sim (f/D)^{1/2}$ over the range of parameters studied. The total time-averaged defect concentration as a function of f/D is shown in Fig. 7.7b.

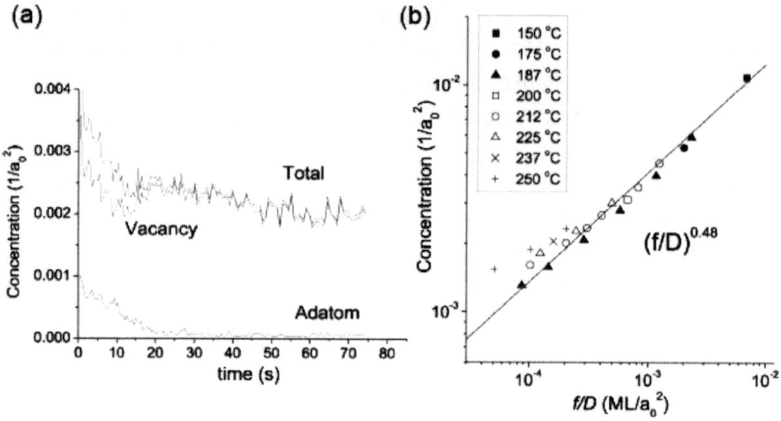

Figure 7.7 (a) Evolution of surface defect concentration (adatoms, vacancies) with sputtering time in KMC simulations. (b) Dependence of average simulated total defect concentration on f/D from simulations conducted at different temperatures and fluxes. Reprinted from reference [14].

Using the defect concentration we found, we fit our simulation results with Eq. 7.8. The values of the fitting are shown as solid lines in Fig. 7.7b. The only free parameters are A_I and E_0. The value for the activation energy E_0 determined from the fitting procedure is 0.86 eV, in good agreement with the value of 0.8 eV used as input to the simulation. The value for A_I is equal to 709 a_0^2, which corresponds to a wavelength of 26.6 a_0 in the absence of thermal diffusion. Based on the sputtering parameters used in our simulation, the saturated wavelength predicted by the continuum theory is 25.7 a_0 when thermal diffusion is absent. Again the KMC simulations show a very good agreement to the continuum model, with the defect concentration we found in the simulation as an input parameter to the continuum model.

The good agreement between the continuum model and simulation is rather surprising, considering that the continuum model assumes the surface to be amorphous. For an amorphous surface, the rate of surface relaxation is simply proportional to the divergence of the surface curvature [27, 41]. On the other hand, many studies have shown that the thermal relaxation kinetics on a crystalline surface are much more complicated and there is no simple analytical solution [12, 30, 38, 48, 49]. Our present results show that during sputtering (where the defect concentration is much higher than the equilibrium defect concentration), the surface relaxation kinetics of a crystalline surface are indeed very similar to those of an amorphous surface. Further, we would like to point out that an equation similar to Eq. 7.8 can also fit reasonably well with experimental data taken on Cu(001) surfaces [9]. However, in case of the experiment, we cannot count the defect concentration directly. Instead, we must estimate the defect concentration using a simple model for the defect kinetics [9].

From Eq. 7.8, we can explain the physical origin of the complicated dependences shown in Fig. 7.7. At high temperatures, the surface relaxation is thermally activated so that the temperature dependence of the wavelength is dominated by the thermally activated diffusivity (with an apparent activation energy comparable to E_0). At low temperatures, the wavelength approaches a constant determined by the value of A_I so that the slope of the Arrhenius plot decreases. In this regime, the surface relaxation is dominant by an ion-induced mechanism [36] so that the ripple wavelength depends on the ion beam parameters rather than the temperature and flux. The trends for the simulated wavelength with flux are equally well modeled. At lower fluxes or higher temperatures ($DC/fT >> A_I$), the thermally controlled processes dominate and the enhanced smoothing leads to an increase in the ripple wavelength. At high fluxes ($DC/fT << A_I$), the wavelength approaches the asymptotic value determined by the ion-induced smoothing. The good agreement seen here between the simulations and the linear instability model suggests that the continuum model can adequately describe many features of ripples formed by Sigmund's sputtering mechanism, even on a crystalline surface.

7.5 Barrier for Ripple Formation in the Low-Flux and High-Temperature Regime

In the last two sections, we have shown many similarities between results from the simulation and the continuum model. In this

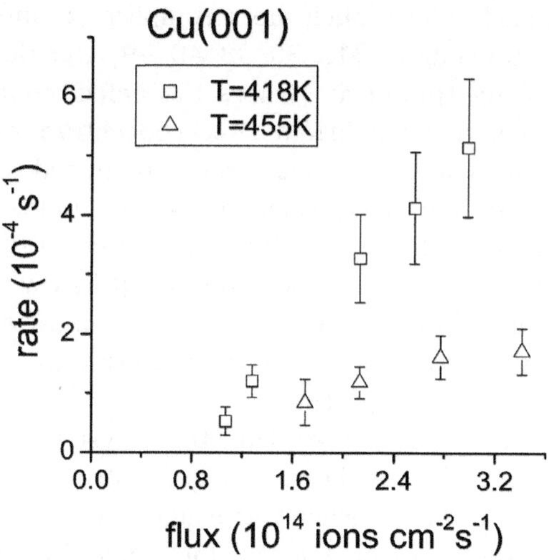

Figure 7.8 Ripples growth rate as a function of flux at temperatures 418 K and 455 K on Cu(001) surfaces. Reprinted from reference [11].

section, we focus on the simulation results that are not predicted by the continuum model, even though similar results are observed in experiments. In the linear instability model, ripples can be formed even at very small ion fluxes, i.e., there is no threshold flux for ripple formation. In experiments conducted on crystalline surfaces (e.g., metals [43, 46], semiconductors [2, 18, 58]), however, it is often reported that at a low-ion-flux and high-temperature regime, the surface is eroded through a layer-by-layer mechanism and no observable periodic pattern is formed.

This threshold behavior for ripple formation can be seen by studying the ripple formation rate r^* as a function of ion flux. Figure 7.8 shows the ripple growth rate as a function of ion flux [11]. If we extrapolate the rate to the x-axis, we see a non-zero intercept, which indicates that a threshold flux is present for ripple formation.

The same behavior is observed in our simulation. The results for the growth rate r^* from the simulations under different flux and temperature conditions are shown in Fig. 7.9a on a linear scale. For each temperature, the ripple growth rate increases with the ion flux. For the same flux, the growth rate is larger at low temperatures than at high temperatures. This trend in the growth rate is correlated with

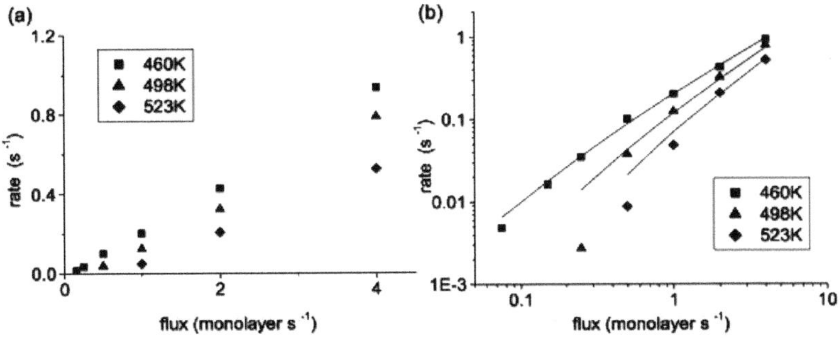

Figure 7.9 Dependence of ripple formation rate as a function of flux on (a) linear plot and (b) log plot. The solid line in (b) represents fit to the continuum model. Reprinted from reference [13].

the wavelength so that ripples with shorter wavelength grow faster than ripples with long wavelength. The growth rate appears to go to zero at a critical flux for each temperature, with the critical flux being larger at higher temperatures. Shown on a logarithmic scale in Fig. 7.9b, there is a rapid drop in the growth rate as the flux is decreased.

The Bradley–Harper model predicts that the dependence of the growth rate on flux and temperature can be written as:

$$r^* \propto \frac{f}{(D(T)C(f,T)/fT) + A_I} \tag{7.9}$$

By using the same fitting parameters as in Eq. 7.8, we can calculate the ripple amplification rate as a function of flux, which is shown by the solid lines in Fig. 7.9b. The model and simulations agree well when $DC/fT \approx A_I$ (at high fluxes or low temperatures). At lower fluxes or higher temperatures ($DC/fT >> A_I$), however, the model and simulation results diverge. The value of r^* found in the simulations is significantly lower than the value predicted by Eq. 7.9 in the low flux regime. We attribute this to the fact that in the simulation at low fluxes, there is a transition to non-roughening behavior so that the ripples grow much slower than predicted by the Bradley–Harper theory or not at all within the simulated time. Figure 7.10 shows the surface roughness as a function of time in the low flux regime. The oscillations in the *rms* roughness show that the surface undergoes layer-by-layer erosion in the low flux regime.

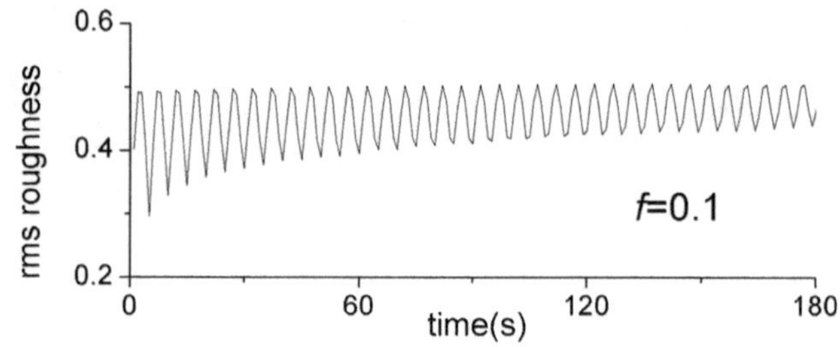

Figure 7.10 The evolution of *rms* roughness of the surface as a function of time, with $T = 187°\text{C}$ and $f = 0.1$ monolayer s^{-1}.

The rapid decline in the growth rate at low fluxes is seen in both the simulations and the experiments but is not consistent with the linear instability mechanism, which predicts that the ripple growth rate should go smoothly to zero with decreasing flux. We attribute this discrepancy to a significant difference between the assumptions of the continuum model and the actual system. In the continuum model, the transport on the surface is modeled by the Mullins–Herring theory [27, 41], which ignores the effect of steps on the surface. The success of the Bradley–Harper model in predicting the wavelength and the growth rate on flux and temperature indicates that this is an acceptable simplification over a wide range of conditions. However, when the flux is low or the temperature is high, this approximation becomes most problematic. For a crystal surface under the roughening-transition temperature to roughen, new terraces must be nucleated on the surface [56]. In the low-flux and high-temperature regime, where the ion-induced damage is quickly recovered by thermal diffusion, new terraces cannot be created. Instead of forming ripples, the surface will be eroded by layer-by-layer removal. Indeed, as we have shown, a transition to oscillation in the *rms* roughness (a signature of layer-by-layer erosion) is observed for the simulations at the point where the roughening rate deviates from the continuum model prediction.

7.6 Effects of Multiple Defects

The simulation results discussed above are obtained by using Sigmund's model of sputtering. However, as mentioned earlier, increasing numbers of experimental evidences show that Sigmund's sputtering mechanism alone cannot account for many of the experimental results. For example, the ripple growth rate observed in experiments is generally $\approx$10–100 times larger than the prediction of the continuum theory. The discrepancy is widely observed in a number of surfaces such as Ge, Si, and Cu (for a summary, see reference [10]), which indicates that additional roughening mechanisms need to be included in the continuum model. Furthermore, recent studies on sub-keV sputtering of Si [34, 35] and Ge [61] surfaces show that patterns with different symmetries or even surface smoothening can be observed on the same surface by varying the ion incident angle and ion energy. Many of these transitions between patterning behavior with respect to the ion energy and incident angle are not predicted by the simple Bradley–Harper model. Again these experimental results call for a more sophisticated model for ion-induced pattern formation.

Recently, there have been a number of new models proposed for ion-beam-induced surface patterning, which include different ion-induced mechanisms [8, 19, 31, 35, 39, 42]. One of the mechanisms that have received much attention is the redistribution of atoms, creating craters and adatom clusters on the surface during a single-ion impact [7, 31, 55]. Indeed, the observation of these defect features has been previously reported in a number of simulations [23, 24] and experiments [33, 40]. Only recently, they have been incorporated into the continuum model to explain the pattern formation behavior in the high fluence regime [19, 31, 35]. In the following, we will demonstrate the effects of multiple defects formation on the patterning behavior using our KMC model. In particular, we will focus on the ripple growth rate. Note that the inclusion of multiple defects into KMC models has been done in some earlier studies [54]. One of the distinct features in our KMC model is that the locations of these defects depend on the local surface curvature. This is especially important in modeling the kinetics of pattern formation since the surface would not remain atomically flat once patterns have started to develop on the surface.

7.6.1 Simulation Method

In the multiple-defect model, single vacancies and adatoms are introduced on the surface. In the simulation that we are showing, eight pairs of adatoms and vacancies are generated on the surface per ion incidence. To determine the location of these defects, a bulk vacancy and interstitial pair is first generated, with their positions determined randomly using a Gaussian profile centered at point O (similar to the energy contours in Sigmund's mechanism — Eq. 7.1). The parameters used are $a = 15$, $\sigma = \mu = 7.5$ for vacancies, and $a = 15$, $\sigma = \mu = 15$ for interstitials. The Gaussian width of the interstitial profile is two times larger than the width of the vacancy profile to account for the observation that interstitials are usually created further away from the stopping point of the ion. Note that the center of the energy distribution (the parameter a) is taken to be the same for both types of defects. The processes for generating multiple defects are summarized in Fig. 7.11a.

The defects are produced underneath the surface. To propagate the bulk defects onto the surface, a random walk probability is used. For a bulk defect at a depth d away from the surface, the probability for it to propagate onto a certain surface site is:

$$P(r) = \exp\left(-\frac{r^2}{2d^2}\right) \qquad (7.10)$$

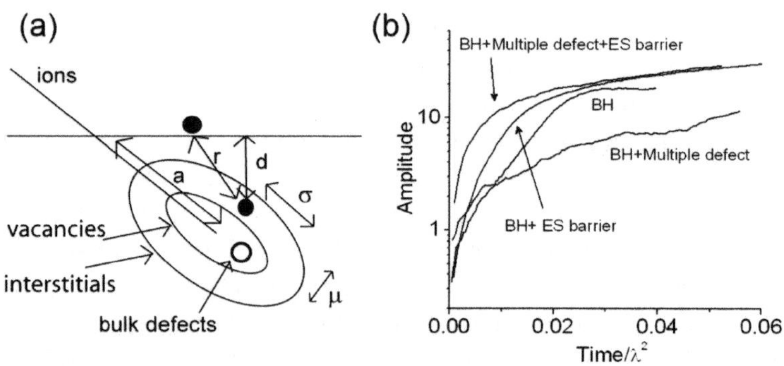

Figure 7.11 (a) Schematic diagram for multiple defects generation. (b) Evolution of ripple amplitude with time showing effects of including multiple defects per ion and adding an ES barrier to the interlayer transport. Reprinted from reference [13].

where r is the distance between the bulk defect and the surface site. This makes the location of the surface defect depend on the surface morphology. Each individual subsurface defect is propagated onto the surface according to Eq. 7.10 once it is created. In the simulation, we also include the sputter vacancies using the same method as described in Section 7.2 with $a = 15$, $\sigma = \mu = 7.5$, and sputter yield $= 2$. Therefore, even though the total number of surface defects generated per ion incidence is 18, only two atoms are removed from the surface.

7.6.2 *Simulation Results*

All the simulations discussed in this section are run at $T = 225°\text{C}$ and $f = 1$ monolayer s^{-1}. Regardless of the addition of multiple defects, the ripples that are formed on the surface have wavelengths that do not change with time. The evolution of ripple amplitude is shown in Fig. 7.11b, with the time axis normalized by the square of the ripple wavelength (Eq. 7.6). Under the same temperature and flux, we can see that adding the multiple defects alone actually lowers the roughening rate (Bradley–Harper + multiple defects) relative to the Bradley–Harper mechanism alone. This is possibly due to the rapid recombination of the single adatoms and vacancies after their creation, which makes it so that the additional defects do not contribute to increasing the roughening.

In order to lower the defect recombination rate, we can add a barrier to defects that diffuse across a step. This barrier is known as the Ehrlich–Schwoebel barrier, and it is present in many metal surfaces. If we add an Ehrlich–Schwoebel barrier with a barrier height of 0.1 eV, the wavelength of the pattern is reduced and there is an increase in the apparent roughening rate, which is shown in Fig. 7.11b (Bradley–Harper + multiple defect + Ehrlich–Schwoebel barrier). For comparison, we also show the amplitude evolution with Ehrlich–Schwoebel barrier alone (Bradley–Harper + Ehrlich–Schwoebel barrier). The above results suggest that multiple defects can enhance roughening if they are not readily recombined. The detailed temperature and flux dependences of these results are reported in reference [28]. Finally, we note that we do not include the creation of craters and adatom clusters. Since clusters are immobile as compared to the single surface defects and they are more difficult to recombine, we expect that adding defect clusters will enhance the roughening rate. We have also not considered the possibility of using a different parameter a for each type of defect, which will create a redistribution of surface atoms similar to the crate functions calculated by [31].

7.7 Conclusion

The KMC simulation is a useful method for determining the relative importance of different atomistic processes on the ion-induced surface patterning behavior. In the first part of this chapter, we described a KMC model that includes Sigmund's sputtering mechanisms on a crystalline surface. The same sputtering mechanism is used in many of the continuum models. However, most of these continuum models are formulated based on amorphous surfaces, which have very different relaxation kinetics compared to that of a crystalline surface. In spite of these differences, we find that our KMC model quantitatively reproduces many of the ripple formation characteristics found in the continuum model. Furthermore, from the KMC model, we can determine the concentration of mobile surface defects explicitly. Surface defect concentration is important in determining the pattern relaxation rate but cannot be readily obtained from experiments. We find that the complicated temperature and flux dependences of ripple wavelength are mainly originated from the temperature and flux dependences of the defect concentration. Once we determine the surface defect concentration, we can predict the ripple wavelength based on the continuum models with high accuracy.

In spite of the success described above, the continuum model deviates from the results of the KMC simulation at the very-low-flux or high-temperature regime. In this regime, the ripple formation is limited by the nucleation of new terraces, but these mechanisms are often missing in continuum models. Finally, we demonstrate the effect of inclusion of multiple defect formation per ion impact on the ripple formation rate. We find that the ripple growth rate is increased only if both the multiple defects and Ehrlich–Schwoebel barrier are spontaneously included in the simulation. The Ehrlich–Schwoebel barrier limits the recombination of surface defects formed by the multiple defect mechanism. The two mechanisms couple together and enhance the surface roughening rate.

Acknowledgement

We gratefully acknowledge many helpful discussions and input from Vivek Shenoy, M. S. Bharathi, and Y. Ishii. The work is supported by U. S. Department of Energy under Contact DE-FG02-01ER45913.

References

1. Ala-Nissila, T., Ferrando, R., and Ying, S. C. (2002). *Adv. Phys.*, **51**, 949.
2. Bedrossian, P., Houston, J. E., Tsao, J. Y., Chason, E., and Picraux, S. T. (1991). *Phys. Rev. Lett.*, **67**, 124.
3. Bharathi, M. S., Chan, W. L., and Chason, E. (2006). *Phys. Rev. B*, **74**, 224103.
4. Boisvert, G., and Lewis, L. J. (1997). *Phys. Rev. B*, **56**, 7643.
5. Bradley, R. M., and Harper, J. M. E. (1988). *J. Vac. Sci. Technol. A*, **6**, 2390.
6. Brown, A.-D., and Erlebacher, J. (2005). *Phys. Rev. B*, **72**, 075350.
7. Carter, G., and Vishnyakov, V. (1996). *Phys. Rev. B*, **54**, 17647.
8. Castro, M., Cuerno, R., Wázquez, L., and Gago, R. (2005). *Phys. Rev. Lett.*, **94**, 016102.
9. Chan, W. L., and Chason, E. (2005). *Phys. Rev. B*, **72**, 165418.
10. Chan, W. L., and Chason, E. (2007). *J. Appl. Phys.*, **101**, 121301.
11. Chan, W. L., Pavenayotin, N., and Chason, E. (2004). *Phys. Rev. B*, **69**, 245413.
12. Chan, W. L., Ramasubramaniam, A., Shenoy, V., and Chason, E. (2004). *Phys. Rev. B*, **70**, 245403.
13. Chason, E., and Chan, W. L. (2009). *J Phys.: Condens. Matter*, **21**, 224016.
14. Chason, E., Chan, W. L., and Bharathi, M. S. (2006). *Phys. Rev. B*, **74**, 224103.
15. Chason, E., and Dodson, B. W. (1991). *J. Vac. Sci. Technol. A*, **9**, 1545.
16. Cuerno, R., and Barabasi, A. L. (1995). *Phys. Rev. Lett.*, **74**, 4746.

17. Cuerno, R., Makse, H. A., Tomassone, S., Harrington, S. T., and Stanley, H. E. (1995). *Phys. Rev. Lett.*, **75**, 4464.
18. Cvetko, D., de Renzi, V., Floreano, L., Morgante, A., Peloi, M., Tommasini, F., Chab, V., and Prince, K. C. (1995). *Surf. Sci. Lett.*, **323**, L305.
19. Davidovitch, B., Aziz, M. J., and Brenner, M. P. (2007). *Phys. Rev. B*, **76**, 205420.
20. Erlebacher, J., Aziz, M. J., Chason, E., Sinclair, M. B., and Floro, J. A. (1999). *Phys. Rev. Lett.*, **82**, 2330.
21. Floro, J. A., Kellerman, B. K., Chason, E., Picraux, S. T., Brice, D. K., and Horn, K. M. (1995). *J. Appl. Phys.*, **77**, 2351.
22. van Gastel, R., Somfai, E., van Albada, S. B., van Saarloos, W., and Frenken, J. W. M. (2001). *Phys. Rev. Lett.*, **86**, 1562.
23. Ghaly, M., and Averback, R. S. (1994). *Phys. Rev. Lett.*, **72**, 364.
24. Ghaly, M., Nordlund, K., and Averback, R. S. (1999). *Philo. Mag. A*, **79**, 795.
25. Hannon, J. B., Klunker, C., Giesen, M., Ibach, H., Bartelt, N. C., and Hamilton, J. C. (1997). *Phys. Rev. Lett.*, **79**, 2506.
26. Hartmann, K., Kree, R., Geyer, U., and Kölbel, M. (2001). *Phys. Rev. B*, **65**, 193403.
27. Herring, C. (1950). *J. Appl. Phys.*, **21**, 301.
28. Ishii, Y., Chan, W. L., and Chason, E. (2012). *Nucl. Instrum. Methods B*, **272**, 188.
29. Jacobsen, J., Cooper, B. H., and Sethna, J. P. (1998). *Phys. Rev. B*, **58**, 15847.
30. Jeong, H.-C., and Williams, E. D. (1999). *Surf. Sci. Rep.*, **34**, 177.
31. Kalyanasundaram, N., Ghazisaeidi, M., Freund, J. B., and Johnson, H. T. (2008). *Appl. Phys. Lett.*, **92**, 131909.
32. Koponen, I., Hautala, M., and Sievänen, O.-P. (1997). *Phys. Rev. Lett.*, **78**, 2612.
33. Kyuno, K., Cahill, D. G., Averback, R. S., Tarus, J., and Nordlund, K. (2001). *Phys. Rev. Lett.*, **83**, 4788.
34. Madi, C. S., Davidovitch, B., George, H. B., Norris, S. A., Brenner, M. P., and Aziz M. J. (2008). *Phys. Rev. Lett.*, **101**, 246102.
35. Madi, C. S., George, H. B., and Aziz, M. J. (2009). *J. Phys.: Condens. Matter*, **21**, 224010.
36. Makeev, M. A., and Barabasi, A. L. (1997). *Appl. Phys. Lett.*, **71**, 2800.
37. Makeev, M. A., Cuerno, R., and Barabasi, A. L. (2002). *Nucl. Instrum. Methods B*, **197**, 185.
38. Margetis, D., Aziz, M. J., and Stone, H. A. (2005). *Phys. Rev. B*, **71**, 165432.

39. Medhekar, N. V., Chan, W. L., Shenoy, V. B., and Chason, E. (2009). *J. Phys.: Condens. Matter*, **21**, 224021.
40. Morgenstern, M., Michely, T., and Cosma, G. (1999). *Philo. Mag. A*, **79**, 775
41. Mullins, W. W. (1959). *J. Appl. Phys.*, **30**, 77.
42. Munoz-Garcıa, J., Vazquez, L., Cuerno, R., Sanchez-Garcıa, J. A., Castro, M., and Gago, R. (2009). Self-organized surface nanopatterning by ion beam sputtering, in *Lecture Notes on Nanoscale Science and Technology* (Wang, Z., ed), Springer, Heidelberg.
43. Murty, M. V. R., Couture, A. J., Cooper, B. H., Woll, A. R., Brock, J. D., and Headrick, R. L. (2000). *J. Appl. Phys.*, **88**, 597.
44. Murty, M. V. R., Cowles, B., and Cooper, B. H. (1998). *Surf. Sci.*, **415**, 328.
45. Park, S., Kahng, B., Jeong, H., and Barabási, A.-L. (1999). *Phys. Rev. Lett.*, **83**, 3486.
46. Poelsema, B., Verheij, L. K., and Comsa, G. (1984). *Phys. Rev. Lett.*, **53**, 2500.
47. Pomeroy, J. M., Jacobsen, J., Hill, C. C., Cooper, B. H., and Sethna, J. P. (2002). *Phys. Rev. B*, **66**, 235412.
48. Shenoy, V. B., Ramasubramaniam, A., and Freund, L. B. (2003). *Surf. Sci.*, **529**, 365.
49. Shenoy, V. B., Ramasubramaniam, A., Ramanarayan, H., Tambe, D. T., Chan, W. L., and Chason, E. (2004). *Phys. Rev. Lett.*, **92**, 256101.
50. Sigmund, P. (1969). *Phys. Rev.*, **184**, 383.
51. Sigmund, P. (1973). *J. Mater. Sci.*, **8**, 1545.
52. Stepanova, M., and Dew, S. K. (2004). *Appl. Phys. Lett.*, **84**, 1374.
53. Stoltze, P. (1994). *J. Phys.: Condens. Matter*, **6**, 9495.
54. Strobel, M., Heinig, K.-H., and Michely, T. (2001). *Surf. Sci.*, **486**, 136.
55. Sule, P., and Heinig, K. H. (2009). *J. Chem. Phys.*, **131**, 204704.
56. Tersoff, J., Denier van der Gon, A. W., and Tromp, R. M. (1994). *Phys. Rev. Lett.*, **72**, 266.
57. Valbusa, U., Borangno, C., and de Mongeot, F. R. (2002). *J. Phys.: Condens. Matter*, **14**, 8153.
58. Watanabe, H., and Ichikawa, M. (1996). *Appl. Phys. Lett.*, **68**, 2514.
59. Yewande, E. O., Hartmann, A. K., and Kree, R. (2005). *Phys. Rev. B*, **71**, 195405.
60. Yewande, E. O., Kree, R., and Hartmann, A. K. (2006). *Phys. Rev. B*, **73**, 115434.
61. Ziberi, B., Frost, F., and Rauschenbach, B. (2006). *Appl. Phys. Lett.*, **88**, 173115.

Chapter 8

From Cascades to Patterns: A Monte Carlo Approach

Reiner Kree and Taha Yasseri
Institut für Theoretische Physik, Georg-August University, Friedrich-Hund Pl. 1, D-37077 Goettingen, Germany
kree@theorie.physik.uni-goettingen.de

This chapter dscribes a top-down Monte Carlo modeling approach to self-organized pattern formation by ion beam erosion. After a discussion of the general topic and its present status, we briefly introduce two of the more microscopic theories — the binary collision approximation (BCA) molecular dynamics and kinetic transport theory — to highlight some results, which should be taken into account when constructing Monte Carlo models. Then we describe a basic simulation model in detail and very briefly discuss its range of applications. The basic model ignores several of the results from the more detailed theories. We show that it is very easy to include the disregarded findings and exemplify the impact of the refinements and extensions on ion-induced pattern formation.

Nanofabrication by Ion-Beam Sputtering: Fundamentals and Applications
Edited by Tapobrata Som and Dinakar Kanjilal

ISBN 978-981-4303-75-0 (Hardcover), 978-981-4303-76-7 (eBook)
www.panstanford.com

8.1 Introduction

It is a fascinating observation that surfaces of metals and semiconductors spontaneously develop ordered morphological patterns at nanometer scales if they are bombarded by keV or MeV ions. It is a challenge for theoretical physicists to derive this nonequilibrium, self-organized pattern formation from basic laws of physics. But scientific curiosity is not the only motivation to study patterns emerging from ion beam erosion. Once our understanding of the physical mechanisms behind this phenomenon has become good enough to control it at a technological level, we will have an extremely cheap and versatile "bottom-up" fabrication method for surface structures in the $10^1 - 10^3$ nm range at hand, which is material independent and simply scalable by ion energy. During the last years, prototypical structures have been prepared on a variety of materials, among them many technologically relevant materials like Si, Ge, III-V semiconductors, SiO_2 and TiO_2. Typical patterns are fields of ripples with a high degree of periodicity and ordered (as well as disordered) arrays of regular nanodot structures. Until recently it was a common belief that the basic physical mechanisms leading to ripple formation were understood well on the basis of seminal work by Sigmund,[1,2] Carter Nobes and Webb,[3] and Bradley and Harper.[4] There was hope that more refined versions of the same theory could also explain the appearance and the features of other types of patterns. So the overall situation seemed quite satisfying at first sight. More recently, however, a number of studies[5-7] have seriously shattered our belief that we have already understood all the relevant pattern-forming physical mechanisms and the antagonistic smoothing mechanisms, which are at work.

Our present understanding of self-organized pattern formation by ion beam erosion is founded upon the experimental fact that the mean sputtering yield Y (the number of sputtered-off atoms per ion impact) depends on the morphology of the surface (for a broader review of the field see reference 8 and reference 9). In particular, Y of a flat surface depends on the angle of incidence θ of incoming ions, measured with respect to the surface normal. $Y(\theta)$ can be approximated as $\sim 1/\cos(\theta)$ for not too large incidence angles (up to $\sim 60^\circ$) for a variety of materials.[10] If the surface is already structured on larger ($\mu m \cdots mm$) scales, the angle of incidence varies smoothly and becomes $\theta - \gamma(\vec{r})$, where $\gamma(\vec{r})$ is the angle of local slope at surface point $\vec{r}$ in the plane of ion incidence direction containing $\vec{r}$. Under stationary irradiation conditions, the sputtering yield $Y(\theta - \gamma(\vec{r}))$, and thus the local erosion rate, gives rise to a surface motion with morphology-dependent normal velocity $c(\gamma(\vec{r}))$. The evolution of these surfaces is described by an eikonal equation (see reference 8 for a review, and references therein). Many larger scale patterns have been discussed within this deterministic framework. On top of these larger scale structures, the observed nanoscale structures require more refined explanations.

Early nanoscale patterns were roughly periodic, ripple-type morphologies, developing from erosion of a flat surface caused by ion beams at oblique angles of incidence θ. The average orientation of these ripples was often found to depend on θ, such that for small θ the ripple wave vector is directed parallel to the projection of the ion beam direction $\vec{n}_{ion}$ onto the substrate surface, whereas at large θ (near grazing incidence), it is directed perpendicular. We will call these patterns *parallel ripples* and *perpendicular ripples*, respectively.[a] These findings were explained by an intrinsic curvature instability of the erosion process, as depicted in Fig. 8.1. The instability leads to a roughening of the surface, i.e., it becomes unstable at all length scales, but diffusion smoothes a rough surface, thus counteracting the curvature instability and leading to a unique periodic height modulation, which exhibits the largest growth rate and thus will dominate the morphology.

A continuum theory based on this mechanism was worked out by Bradley and Harper,[4] which could explain the rotation of ripple orientation with increasing angle θ. This theory has been extended to include nonlinear terms, resulting in an evolution equation for the surface height profile, which is an anisotropic variant of the Kuramoto–Sivashinsky equation.[11–13] These models allow to follow the evolution of surface morphologies into regimes, which are dominated by nonlinear terms. Further extensions of the continuum theories have been proposed, which are inspired by the models of sand dune dynamics.[14,15] Within these models,

[a]Be warned that other authors will use the exactly reversed naming. They describe the ripple orientation by the direction of crests and valleys.

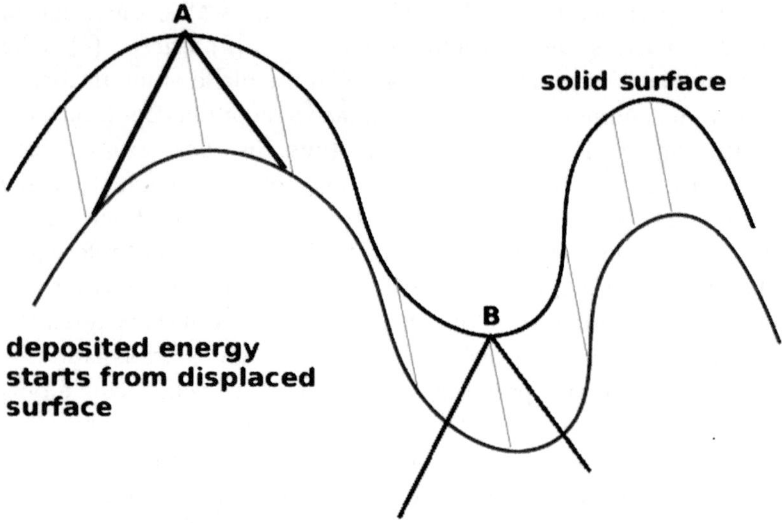

Figure 8.1 Generic curvature instability: Point A on the surface is near a maximum, point B near a minimum. Both get energy from cascades, which start from surface points shifted along ion trajectories by penetration depth. The cascade energy decays as it spreads, so the energy reaching A or B decreases with distance. Depicted are the distances of two starting points of cascades from A. Shifting these distances rigidly along the displaced surface, it becomes clear that B gets more energy than A as the corresponding starting points of cascades are closer to B. Thus matter at B is eroded faster than at A.

it is also possible to obtain a self-organized formation of regular dots, forming a hexagonal lattice.[15] In more recent experimental work on ion beam eroded Si and Ge surfaces (for a review, see reference 16), highly regular ripple patterns, as well as ordered arrays of dots and other highly regular patterns have been reported, but the essential mechanisms, which lead to the improved and diversified patterns, remained unclear. In the last years, it became more and more obvious that ion beam erosion of very pure surfaces irradiated by a mass-filtered ion beam may not lead to self-organized pattern formation, at least in a wide range of θ from normal incidence to $\sim 45°$. Carter and Vishnyakov[17] proposed an additional θ-dependent smoothing mechanism, which emerges from the transport of momentum of the stopped ion to the surface via the cascade. Cascade particles will induce a material transport directed preferentially parallel to $\vec{n}_{ion}$, which induces a material flux $\sim \cos(\theta - \gamma) d \sin(\theta - \gamma)$ parallel to the surface, called a *downhill current*, because for small and slowly varying slopes, it is proportional to the slope and parallel to the direction of steepest descent. Here γ is the local surface slope and $d \sin(\theta - \gamma)$ is the projection of the typical displacement d of recoil atoms in the direction of the ion beam onto the surface. In a continuum description, this mechanism will indeed show up as an additional smoothing of surface structures, but its absolute value is hard to calculate, even though a rough estimate in reference 17 claimed that it is of a relavant order of magnitude. The downhill current has been identified as the major mode of induced transport on amorphous, diamond-like carbon (ta-C) irradiated by 100 eV C atoms in a quantum and classical molecular dynamics simulation by Moseler *et al.*[18] So there is at least a possibility to reconcile the Bradley–Harper theory with the experimental finding of smooth surfaces. Other experimental results produced more doubts about our understanding of the physical mechanisms of the self-organized pattern formation. Madi *et al.*[6] found a bifurcation scenario for 250 eV Ar ion-sputtered Si at low θ, within which a finite wavelength parallel ripple pattern vanishes (type I bifurcation), whereas all the variants of continuum theories like that of Bradley and Harper predict only bifurcations with diverging wavelengths (type II bifurcation) at the bifurcation point. Most embarrassing of all, it was found that extremely small amounts of surface contaminations by cosputtered or codeposited atomic species profoundly influence the pattern formation, and that many, if not all, of the highly ordered patterns are due to such "surfactant effects,"[7] mostly due to uncontrolled contaminations with Fe or Mo within the experimental setup. A main lesson to draw from these recent results is that we should carefully reexamine physical mechanisms, which can play a role in the self-organized pattern formation.

Computer simulations are a versatile theoretical tool, which in many cases can be used profitably to identify and discriminate relevant physical mechanisms in a complex system. However, a direct molecular dynamics simulation of the pattern formation from ion sputtering is currently still out of reach (for a recent attempt, see reference 19), because the energy scales of relevant processes range from keV (stopping of the ion) to meV (surface diffusion) and corresponding timescales from femtoseconds to minutes or hours. Therefore, one resorts to modeling strategies of two different types, both containing Monte Carlo methods. In a bottom-up approach, one simulates the short time dynamics of single-collision cascades (up to some $\sim 10^{-15}$ sec) in an approximate molecular dynamics algorithm, called binary collision approximation (BCA), and then uses Monte Carlo rules for the further evolution of the resulting displaced atoms. The first demonstration that BCA cascades indeed lead to ripple patterns has been given by Koponen, Hautala, and Sievänen.[20,21] A combination of the BCA-based TRIM (transport of ions in matter) package with kinetic Monte Carlo dynamics has been developed and applied by Heinig and coworkers.[22] Here we will follow a top-down approach, which implements all the physical mechanisms as Monte Carlo rules, starting from a simple basic model and then studying variants and refinements of this model. The Monte Carlo rules are taken from or inspired by results of more microscopic theories or from experiments. Both approaches are to a certain extent complementary, but it should be noted that the distinction between the two strategies is more of a conceptual type, as both approaches have to construct Monte Carlo rules from additional information, which cannot be obtained within the models. Ideally, both efforts will end up in a computer simulation model containing all the relevant physical mechanisms, which can be used to control and predict ion-beam-induced pattern formation processes.

In the following, we first discuss results from BCA and from a kinetic theory of sputtering, which are relevant for the construction of Monte Carlo rules describing the erosion process. Smoothing mechanisms of thermal surface diffusion are not discussed in depth in this chapter, as there is an extensive literature available, and the physical mechanisms do not differ from those used in models of molecular beam epitaxy and crystal growth. For a detailed and general introduction of the subject, see reference 23. In Sec. 8.3, we introduce our basic Monte Carlo model. To keep the discussion self-contained, we will also present a simple Monte Carlo model of thermal surface diffusion, which will be used in the subsequent examples. The basic model has been used frequently in the literature, although it does not respect all of the lessons to be learned from BCA and kinetic theory. In Sec. 8.4, we will, therefore, introduce and study variants and refinements of the basic model. We will find that some physical mechanisms, which have been neglected so far, have a profound impact on pattern formation. The discussion will be limited to the presentation of a few examples, without a detailed quantitative analysis. This is not only due to limited space, but it is our belief that we should currently explore qualitative impacts of physical mechanisms and not enter a detailed analysis, before we can be convinced about the relevance and validity of the mechanism.

8.2 Lessons from BCA and Kinetic Theory

As mentioned in the previous section, a Monte Carlo approach uses a set of stochastic updating rules, which have to be constructed from more fundamental theories or from experiments. One could, for example, use data from a many-body molecular dynamics simulation to set up the Monte Carlo rules. Recently, one indeed begins to approach simple aspects of ripple formation by molecular dynamics simulations.[19] For most practical cases, the construction of Monte Carlo rules for ion-beam erosion still starts from an approximate molecular dynamics approach, called BCA or from the kinetic level of description. In the following, we briefly describe these two approaches, and in particular discuss the main lessons to be learned from their results for the construction of Monte Carlo rules. Both approaches essentially cut down the dynamical many-body problem of the stopping of ions within the sample, including the induced displacements of sample atoms, to two-body collisions. Thus the general picture is that of a collision cascade: an incident ion transfers its momentum and energy to substrate atoms, which in turn may collide with other substrate atoms, which again collide with substrate atoms until atoms of this cascade reach the surface of the sample or get stuck and become structural defects in the substrate. If they can overcome the surface potential barrier, they are ultimately sputtered off. All our discussions here are limited to *linear cascades*, which means that the incident ion and all the moving recoil atoms only scatter from resting atoms (thermal motion neglected), whereas collisions between moving particles are not taken into account. These collisions become more and more important with increasing ion energy and will ultimately lead to various kinds of nonlinear cascades, like displacement cascades, thermal spikes, and shock wave fronts.

8.2.1 *Binary Collision Approximation*

At high energies, interaction processes between the incident ion or recoil atoms and the resting atoms can be approximated as two-particle collisions. The BCA follows all the trajectories generated from the collisions until the energy of a moving particle is smaller than some threshold, which is approximately the energy necessary to produce a Frenkel pair. The two-particle collisions are treated as classical, elastic collisions. The interaction potentials are of the form of screened Coulomb potentials. Additional inelastic processes due to electronic stopping can be included. Detailed description of such programs and the underlying physics can be found in reference 24 for the TRIM program and references 25,26 for the MARLOWE program. What are the main lessons to be learned from BCA single-ion impact simulations for the construction of Monte Carlo rules? We have studied the statistics of 5 keV Cu single-ion impacts on Cu surfaces[27] in detail to draw the following conclusions:

(1) The energy E of cascade atoms reaching the surface is a strongly fluctuating quantity, distributed according to a probability density

$$\lambda(E) \approx E^{-2}, \text{ for } E > E_0 \tag{8.1}$$

(2) The joint probability density $p(\vec{r}, E)$ to find a cascade particle at a point $\vec{r}$ on the surface (provided the ion was stopped at $\vec{r}_0 = (x, y, -a)$ within the substrate) with energy E factorizes, i.e., $p(\vec{r}, E) \approx \nu(\vec{r})\lambda(E)$.

(3) The probability to find a cascade particle at a surface point can be fitted to

$$\nu(x, y, z) \approx \nu_0 \rho^2 \exp(-\rho/\mu - \sqrt{(z+a)^2}/\sigma) \tag{8.2}$$

with $\rho^2 = x^2 + y^2$.

(4) The spatial distribution of atoms, which are not sputtered off, is not significantly different from that of eroded atoms. The energy of these adatoms is essentially cut off by the surface barrier (details depend on the detailed form of the barrier).
(5) A downhill current is clearly visible.
(6) The penetration depth of an ion is a statistical quantity with considerable fluctuations around its mean.

8.2.2 *Kinetic Theory*

Collision cascades are stochastic events, triggered by the slowing down of an ion arriving at a random surface point. BCA treats the evolution of the cascade as a sequence of two-particle collisions, typically of a moving recoil atom and a resting substrate atom. An analytic stochastic theory may be based on the same assumption. If collisions between moving particles are neglected, the cascade trajectories form a tree, i.e., a graph without loops. At the stochastic kinetic level of description, all the information about the cascades is contained in the probabilities $P_n(\Gamma, t|\vec{r}_0, E_0, \vec{e}_0)$ to find n atoms of a cascade in region Γ of $\{\vec{r}, E, \vec{e}\}$ phase space at time t, provided the cascade was generated at time $t = 0$ by a single particle at $\vec{r}_0$, moving with energy E_0 in direction $\vec{e}_0$. If all the scattering probabilities are homogeneous, a nonlinear integro-differential equation for the P_n, or rather for the generating function $Z(\mu) = \sum_n \exp(\mu n) P_n$ has been constructed.[28] But the overwhelming majority of results of kinetic theory of sputtering have been obtained from evolution equations for the average phase space numbers $N(\Gamma, t|\vec{r}_0, E_0, \vec{e}_0) = \sum_n n P_n = \partial Z(\mu)/\partial\mu|_{\mu=0}$, or the phase space densities $n(\vec{r}, E, \vec{e}|\vec{r}_0, E_0, \vec{e}_0)\, d^3r\, dE\, d^2\vec{e}$, respectively. They obey Boltzmann-like transport equations, which become linear integro-differential equations for homogeneous materials. One should keep in mind that this level of averaged kinetic description already neglects intrinsic fluctuations of collision cascades, which are fully kept in BCA simulations. Many results about sputtering have been obtained from kinetic theory, either by elementary arguments on the cascade statistics[29] or by approximate solutions of the transport equation.[1] What are the main results that are useful in constructing Monte Carlo rules?

(1) Findings 1) and 6) of BCA are also obtained in kinetic theories[29]
(2) The average cascade volume is of ellipsoidal shape, with the larger axis parallel to ion beam direction. The average energy deposited at $\vec{r}$ is a function, which can in some cases be approximated as[1]

$$E(\rho, z) \approx E_0 \exp(-\rho^2/2\mu^2 - (z+a)^2/2\sigma^2). \tag{8.3}$$

(3) The average rate at which atoms are sputtered off is proportional to the average deposited energy.[1]
(4) The momentum relaxation can be calculated, but it has not been used for detailed evaluations of downhill currents up to now.

8.3 Basic Monte Carlo Model

A basic model used in many Monte Carlo simulations of pattern formation under ion beam erosion treats the sample as a lattice (in most cases simple cubic) composed of material points or blocks, which can be roughly identified with sample atoms. The lattice constant is chosen as the unit of length, so that lattice points have integer coordinates. To uniquely construct surfaces and volumes from the lattice points, let us agree upon the convention that each cubic block is addressed by its corner with smallest integer coordinates (i, j) on the top facette. The material surface is characterized by a time-dependent, single-valued discrete height function $k = h(i, j)$ and is thus composed of horizontal and vertical square facettes. Often a flat reference surface $h(i, j) = 0$ is chosen as initial condition, so that h takes on negative values due to erosion. This description of surfaces leads to very efficient code, but it is not ideally suited to map physical erosion processes, because erosion of blocks from vertical parts of the surface creates overhangs, which violate the condition of a single-valued height profile $h(i, j)$ (the solid-on-solid (SOS) constraint). Some rules have to be constructed to deal with these situations. In our simulations, we allow for the erosion of blocks from vertical parts, but they are followed by an immediate relaxation of the column to restore the SOS constraint (see Fig. 8.2).

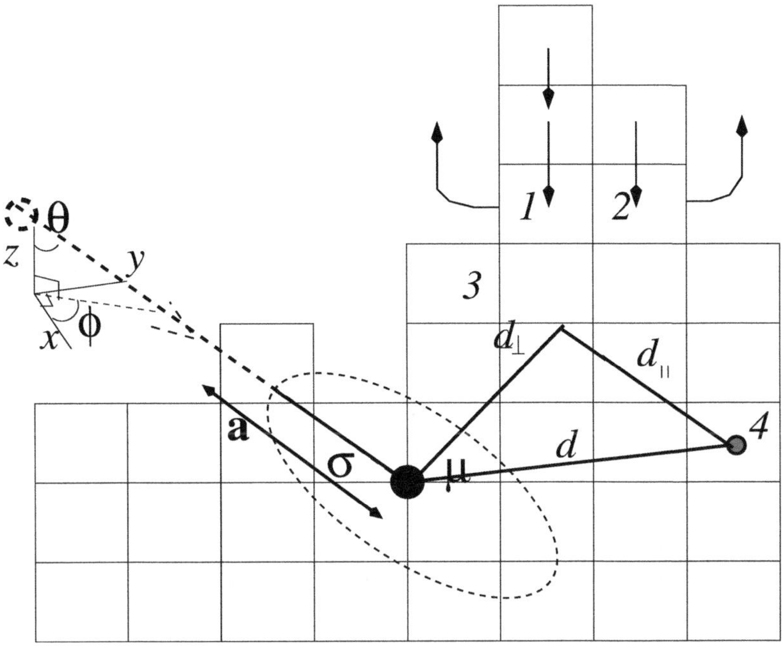

Figure 8.2 Elements of erosion in basic Monte Carlo model. Ions are approaching from a direction characterized by the polar angle of incidence θ and the azimuthal angle ϕ. It penetrates the substrate without being deflected until it is stopped after penetration depth a. The cascade of recoil atoms emerging from the stopped ion transports energy. The distribution of deposited energy has ellipsoidal isosurfaces and decays as a Gaussian with half widths σ and μ. Therefore the mean energy reaching a surface point (4) depends on the lengths $d_{||}$ and $d_{\perp}$. Eroded surface atoms can be part of either a horizontal (3) or a vertical piece of surface (1,2). Erosion of (1) and (2) must be followed by an immediate relaxation process, which restores the SOS constraint.

A basic Monte Carlo simulation run consists of sequences of erosion steps and surface diffusion steps. The erosion creates surface defects by sputtering off single blocks, which leave vacancies behind. Recombination of vacancies with adatoms smoothes the surface morphology. In the following subsections, we will introduce some detailed rules for these steps. Then we will discuss how to adapt sequences of Monte Carlo steps to physical time.

8.3.1 *Erosion*

In the basic model, it is assumed that the ion beam profile is homogeneous over the sample and that all ions follow straight trajectories parallel to a fixed direction of incidence $\vec{n} = -(\sin\theta\cos\phi, \sin\theta\sin\phi, \cos\theta)$. An erosion step starts by picking a random point (x, y, z) of an ion trajectory and then following this trajectory until it is stopped within the substrate. Although this sounds like a very simple procedure, one has to note that it requires some decisions on seemingly finer points, which can severely influence the pattern formation. In BCA, the stopping is modeled as a sequence of two-particle collisions, which take the current position of the resting particles into account. In the kinetic approach, the scattering is described by a free path length l between two scattering events. l may be either random or fixed. This modeling is adequate for amorphous materials, but it already neglects special effects of crystal structures, like, for example, channeling. In the basic Monte Carlo model, the stopping of the ion takes place at exactly one point. One assumes that after a surface impact, the ion penetrates the material along its undisturbed trajectory for a fixed penetration depth a and then initiates a cascade emerging from that point. This implies a number of idealizations, which one should be aware of. Here we mention two important ones.

(i) There is only one primary collision of the ion, and there is no electronic stopping along the path of the ion. This implies that surface structures with lateral length scales smaller than $\sim a$ are crossed undisturbed. Thus sharp edges or columnar or tower-like structures appear stable under this type of Monte Carlo erosion.
(ii) It suppresses fluctuations of the penetration depth.

The validity of these idealization can be tested be refining the Monte Carlo model. This will be discussed in Sec. 8.4. As long as no suspicious tower- or edge-like structures appear, one may also judge by inspection that (i) has not artificially distorted the surface morphology. The fluctuations of the penetration depth will always turn out to be an important factor influencing pattern formation.

Energy and momentum of the stopped ion are distributed by a cascade of recoil atoms. In the basic Monte Carlo model, only the energy deposition is taken into account and is modeled by the average energy $E(x, y, z)$ transported to (x, y, z) by a cascade emerging from the $(0, 0, 0)$. Finally, surface atoms at $i, j, h(i, j)$ are eroded with probabilities proportional to $E(i, j, h(i, j))$, which completes a single erosion step. Even by superficial inspection it should be clear that this is a very crude caricature of the physical erosion process. Not only does it neglect a number of important effects like a surface barrier, the creation of adatoms along with sputtering, or the redeposition of eroded atoms, but also it seems to be at odds with basic cascade theory. We have stated above that both BCA and kinetic theory find the energy brought to the surface to be a highly fluctuating quantity with a very broad, approximately $1/E^2$ distribution. Then what is its average? And why should the average be the basis for a Monte Carlo decision upon sputtering? In the next section, we will reconcile the findings of BCA and kinetic theory with the Monte Carlo modeling.

8.3.2 *Surface Diffusion*

Modeling surface diffusion is a very complex topic of its own, but it has been discussed extensively in the past, so we can be brief here. Let us stress that we only refer to thermal surface diffusion and not to mechanisms of surface transport induced by the collision cascade. The Monte Carlo modeling of surface diffusion is based upon moves that are local updating (hopping) rules of the current surface configuration S to some configuration S', which are accepted or rejected in a stochastic Monte Carlo decision based on an energy change $E(S \to S')$ assigned to this move. The idea is that the move requires the crossing of an energy barrier $E_B = E(S \to S')$, which has to be overcome by thermal fluctuations, and thus will appear at a rate

$$k = k_0 \exp(-E_B/kT), \tag{8.4}$$

the Arrhenius hopping rate. The prefactor k_0 is the trial rate, which is of the order of magnitude of the vibrational frequency of a surface adatom, i.e., $k_0 \sim 10^{12} \cdots 10^{13}\text{sec}^{-1}$. Local configurational changes at a surface can be studied by molecular dynamics simulations, which in principle could provide valid Monte Carlo updating rules. But in practice, this turns out to be very complicated, as barriers may involve the concerted motion of many particles. So, before adapting to molecular dynamics or other results, one has to choose some simplifying modeling principles. Usually one considers *one-particle hops*, that is S and S' differ by the position of just one surface atom. The simplest modeling strategy is *bond breaking*, which assumes that there is a fixed amount of energy necessary to change the number of nearest neighbors of an atom by 1 (bond energy ϵ_b). Then the barrier that an atom at site $(i, j, h(i, j))$ has to cross to leave this place is just ϵ_b multiplied by the number of nearest neighbors of this atom. Note that this barrier $E_B(i, j)$ only depends on the position from which the atom starts. The rule is completed by specifying the range of a single particle hop, usually to a nearest neighbor site. The advantage of these models is that they contain just one adjustable parameter, ϵ_b. Often, however, bond-breaking models cannot be fitted to molecular dynamics or experimental data. The models can be slightly refined by distinguishing between substrate atoms and in-plane nearest neighbors $n(i, j)$ and postulating $E_B(i, j) = \epsilon_s + \epsilon_b * nn(i, j)$. Still they often are in conflict with basic observations. For example, the barrier seen by an isolated adatom is just ϵ_s, whereas for an isolated vacancy it is $\epsilon_s + 3\epsilon_b$, which implies that adatom mobility should always be much larger than vacancy mobility, which is not true for many materials.

In the framework of single-particle hops, the most general barriers may depend on the initial and the final position of the hopping atom. Such barriers are also called *Kawasaki-type barriers* in analogy to the Kawasaki dynamics of Ising models, as the Monte Carlo updating rule requires the state of the surface at 2 sites. For our simulations, we use two such Kawasaki barriers: one (*the net bond breaking barriers*) to avoid the mentioned oversimplifications of the bond-breaking model and one (*the Ehrlich–Schwoebel barrier*) to generate a pattern-forming instability by surface diffusion. In the net bond-breaking model, the barrier is just

$$E_B(i, j \rightarrow k, l) = \epsilon_s + \max\{0, \epsilon_b \cdot [nn(i, j) - nn(k, l)]\}, \quad (8.5)$$

i.e., it counts the net change of the number of nearest neighbor bonds for the whole move. If this change is smaller than zero, the move gains energy and there is no extra barrier. Models of this type have been successfully calibrated to real materials. The Ehrlich–Schwoebel barrier is the asymmetric barrier, which an atom experiences at the upper edge of a step due to the asymmetric atomic structure. The barrier for hopping downward (inter-layer transport) is different from hopping on the flat terrace, i.e., ϵ_s and either larger (positive Ehrlich–Schwoebel barrier) or smaller (negative Ehrlich–Schwoebel barrier) than ϵ_s. If adatoms are created randomly, a positive Ehrlich–Schwoebel barrier causes an effective uphill current and thus induces a pattern-forming instability.[23] We implement the Ehrlich–Schwoebel barrier by the following prescription: $E_{ES}(ij \to kl) = \epsilon_{ES}$ if (k, l) is at the upper edge of a step and (i, j) and (k, l) are in-plane nearest neighbors. In all other cases, E_{ES} vanishes.

The total barrier, which enters the Arrhenius rate is $E = \epsilon_s + \Delta E = \epsilon_s + E_B + E_{ES}$. Note that the factor $\exp(-\epsilon_s/kT)$ can be absorbed in a redefined attempt rate

$$k_{diff} = k_0 e^{-\epsilon_s/kT} \ll k_0. \tag{8.6}$$

For typical values ($\epsilon_s = 0.75\text{eV}$, $kT \approx 0.025\text{eV}$) one gets $k_1 \approx 10^2 \text{sec}^{-1}$, which is a basic scale to connect Monte Carlo steps to physical time.

8.3.3 *Timescales*

If we accept a trial with probability $\exp(-\Delta E/kT)$, there should be one Monte Carlo sweep (one Monte Carlo step per surface atom) of hopping trials every $k_{diff}^{-1} \simeq 10^{-2}$ sec. On the other hand, let us estimate the number of ions that reach each surface atom per second. If we have a nearest neighbor distance of $2 \ldots 3\mathring{A}$, i.e., $\simeq 10^{15}$ atoms/cm^2 and an ion current density of $\sim 7.5 \times 10^{14}$ ions/(cm^2 sec), we get $k_{er} \simeq 0.75$ ions/(atom sec), or, as a rule of thumb, one ion per atom per second. Thus there will be $\sim 10^2$ diffusion steps of a surface atom before an ion hits a nearby surface point again. This estimate could be slightly improved by taking into account that an ion influences surface atoms via the cascade, but this is less important, because the critical parameter is the diffusion rate, which depends sensitively on the temperature and the barriers. In order to simulate slight changes in temperature, the attempt rate

k_1 has to be changed drastically. For the example barrier ϵ_s = 0.75 eV, a temperature change from 350 K to 700 K increases k_1 by a factor $\simeq 3 \times 10^4$. Thus there should, in principle, exist two distinct physical regimes: an *erosion-dominated regime,* where the number of diffusion steps between two ion impacts near the same surface point is not large enough to override the structure due to the previous impact and a *diffusion-dominated regime,* where the erosion structure is wiped out by diffusion in between two impacts. Note that in the diffusion-dominated regime, erosion constitutes a constant source of vacancies and (at least on a somewhat rough surface) adatoms, so that Ehrlich–Schwoebel currents may drive a kind of pattern formation, which is completely different from the pattern formation in erosion-dominated regimes.

The connection to physical time may be implemented in two different ways. In the continuous time Monte Carlo approach, one first generates the next (random) instance of time, at which any one out of all the possible events (i.e., diffusion or erosion) will happen. As we know the rates for each of these events and we know that they are independent, the total rate for any event to happen is just the sum Γ over all the rates of possible events, and the waiting time τ until the next event is distributed exponentially $p(\tau) = \Gamma \exp(-\Gamma\tau)$. After having drawn τ, we make a random decision among all the possible events, as event e may happen with probability Γ_e/Γ (see reference 30 for more details). A second way to implement the kinetic timescales is to work with a fixed time step, corresponding to the highest rate. In our example case, the time of 100 diffusion sweeps equals $0.75^* L^2$ ion impacts (in a system with L^2 surface sites.) Thus there will be one diffusion sweep every $0.0075^* L^2 > 1$ impacts, provided L is large enough, otherwise there will be one impact every $100/(0.75^*L^2)$ diffusion sweeps. We have implemented the kinetic Monte Carlo model in both ways and up to now have not found any statistically significant differences. The second implementation is usually faster.

8.3.4 *Results from the Basic Model*

The basic model has been used to study a number of salient features of the self-organized pattern formation. These will not be discussed here in depth, we rather refer the reader to the reviews reference 31 and reference 9 for most details. The main purpose of the present subsection is to indicate useful applications as well as limitations of the basic model.

Figure 8.3 depicts prototypical results of morphologies with parallel ripples (at $\theta = 50°$) and perpendicular ripples (at $\theta = 80°$)

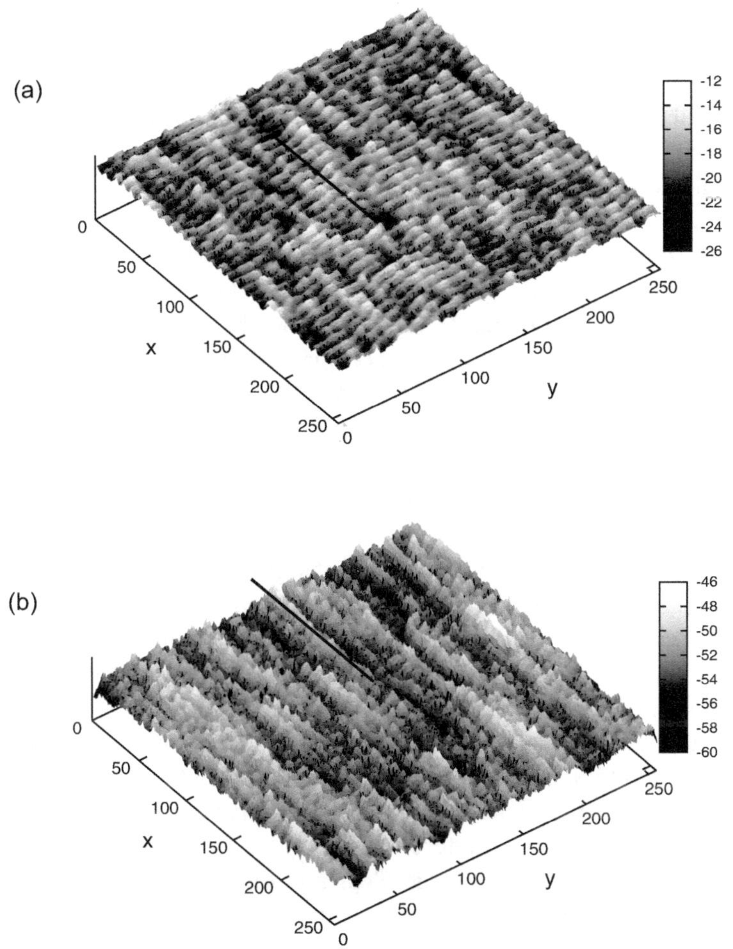

Figure 8.3 Typical morphologies of ripples obtained from the basic Monte Carlo model. (a) For smaller θ, ripples are parallel; (b) for larger θ, ripples are perpendicular. Parameters of simulation are given in the main text.

in a 256 $\times$ 256 cutout of a $L \times L$ system with $L = 512$. The penetration depth has been set to $d = 4$ and the other erosion parameters are $\sigma = 3.5$ and $\mu = 1.0$. The surface diffusion is of net bond-breaking type at room temperature with a hopping barrier of 0.18 eV. Timescales are set as explained in Sec. 8.3.3, with the ratio between diffusion and erosion steps as mentioned there. The morphology shown emerges after $6 \times L^2$ ion impacts (corresponding to $\sim$10 sec of irradiation with 7.5×10^{14} ions/sec). Although the pictures might suggest that the basic model essentially reproduces the results of Bradley–Harper theory, a more detailed analysis reveals important differences. Note that the patterns shown are far from perfect regularity. The only noise kept in the basic model is the shot noise of ion impacts and the intrinsic noise of the random hopping processes, which is not fully included even in the noisy version of the Bradley–Harper theory and its extensions. In simulations, it turns out that the impact of these types of noise on pattern regularity depends sensitively on the erosion parameters. The standard Bradley–Harper scenario of ripples, which change their orientation from parallel to perpendicular, is not as generic as it is sometimes suggested. There are large areas in the space of erosion parameters with quite different scenarios.[32]

The model has been used for three main purposes:

(1) To study the temporal evolution of surface morphology from the early regime, where it should be comparable to linear Bradley–Harper theory up to very late times to find quasi-asymptotic behavior.[30,33,34] These studies revealed the important influence of the diffusion mechanisms on surface morphology.
(2) To study the dependence of surface patterns upon the erosion parameters,[32] i.e., penetration depth, σ and μ. Note that all details of the complicated cascade processes are summed up in just three parameters. Nevertheless, a variety of pattern-forming scenarios emerge. Typically, only irregular patterns or surface roughening is observed at small θ. At larger θ, longer or shorter ripples as well as dots and pits may occur. Near grazing incidence, perpendicular ripples, but also parallel ripples, and irregular patterns may occur.

(3) To study more complex experimental setups, like rotating samples, dual beam irradiation, etc.[35,36] Several such setups have been proposed and experimentally tested.[37] They can be easily implemented and studied in the Monte Carlo models.

8.4 Variants and Refinements

Certainly, the reader has noticed that the basic model ignores several of the lessons from BCA and kinetic theory. Let us list some important cases of disregard:

(1) The basic model completely neglects noise in the penetration depth.
(2) It uses an erosion mechanism based on averaged energy, whereas we know that the energy of cascade particles is strongly fluctuating.
(3) Even the expression for the deposited energy is only approximate.
(4) The erosion model produces only vacancies (sputtered-off atoms) but no adatoms.
(5) It does not include a downhill current.
(6) It neglects redeposition.

In the following subsections, we show that it is not difficult to refine or vary the basic model in ways, which take these effects into account. The reader will notice that this flexibility is one of the major advantages of a top-down Monte Carlo modeling approach. Although easily implemented, most of the refinements and extensions bring along new parameters, which considerably enlarge the parameter space and make it more difficult to study all the possible impacts on pattern formation in a systematic fashion. Here we only give some examples of significant qualitative changes in the standard scenarios of the basic model caused by the refinements. In all our examples in this section, we have used simple bond-breaking surface diffusion at room temperature with a bond energy of $\epsilon_b = 0.18$ eV.

8.4.1 *Noisy Stopping*

For each ion impact, we may draw a different, random penetration depth a. The kinetic theory of amorphous materials assumed that the free path lengths are distributed exponentially, i.e., there is a fixed probability per length for a moving atom to undergo a collision. This assumption from kinetic theory of gases has been criticized and a kinetic theory has been proposed, which is based on fixed free path lengths.[38] In Monte Carlo simulations, we may test the consequences of any of these assumptions for pattern formation. In Fig. 8.4, you see that a regular ripple pattern (Fig. 8.4(a)) at $\theta = 50°$ is completely destroyed, if the penetration depth is distributed exponentially as depicted in Fig. 8.4. In fact, even a small amount of noise in the penetration depth has profound impact. We have tested noise of the form $a = (1 - \lambda)\bar{a} + \lambda\delta a$, where $0 < \lambda < 1$ and δa is random and evenly distributed from 0 to $2\bar{a}$. Fig. 8.4(b) shows the limiting case $\lambda = 1/2$), beyond which the ripple pattern vanishes. Fig. 8.5 demonstrates that noise in penetration depth can change the type of surface morphology from parallel to perpendicular ripples. In this example, erosion parameters were chosen such that without noise in penetration depth, parallel ripples persist up to grazing incidence. In Fig. 8.5(b), noise has been added, which changes the ripple orientation to perpendicular. We have checked that the distribution of noise is less important than its variance, so this scenario can be obtained from exponentially distributed noise (as shown in Fig. 8.5(b)) as well as evenly distributed fluctuations.

It has been argued[39] that the noise in erosion can be subsumed by a simple additive Gaussian white noise in a continuum model of the Kuramoto–Sivashinsky type. But our results show more pronounced and surprising effects than one would expect from the noisy continuum theory.

8.4.2 *Cascade Noise and Surface Barrier*

The simple erosion rule of the basic model can be replaced by the following rule, which is taken from the results of BCA simulations and cascade kinetic theory. First, it is decided stochastically whether a cascade particle will arrive at a certain point of the surface

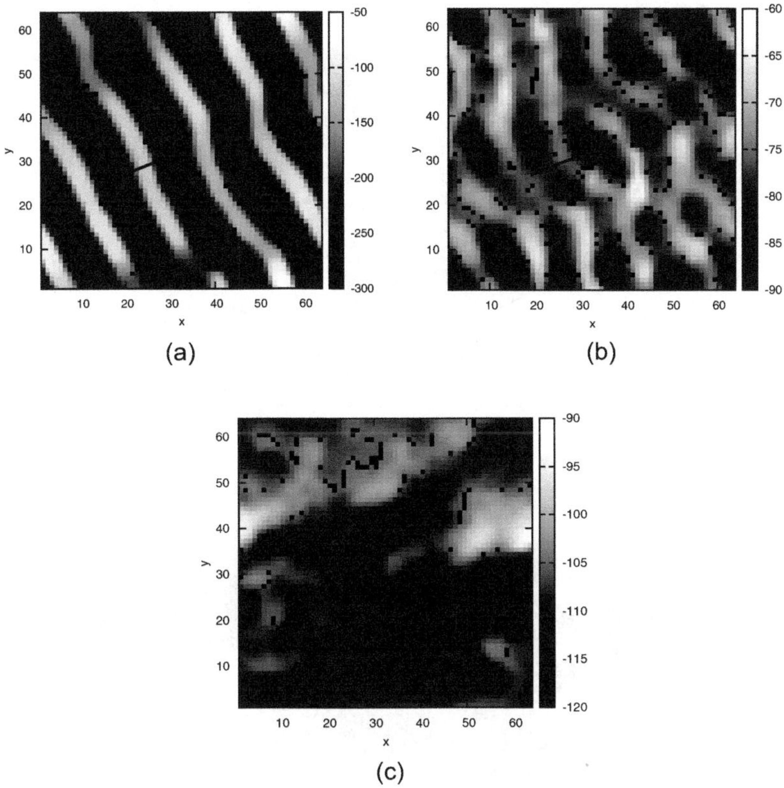

Figure 8.4 Noise in penetration depth: Simulation of a 256×256 system (64×64 cutout displayed) with average penetration depth $\bar{a} = 9.3, \sigma = 6.0$, and $\mu = 2.0$ for $t = 2 \times 256^2$ ion impacts. Morphologies (a) as obtained from basic model, without any noise in penetration depth; (b) with $a = \bar{a}/2 + \delta a$, where δa is evenly distributed from 0 to $\bar{a}$; and (c) with exponentially distributed a (with average $\bar{a}$), as assumed in kinetic theory.

on the basis of a formula like Eq. (8.3). Note that this implies a reinterpretation. Instead of the average deposited energy, we use an arrival probability for cascade particles. Although these two quantities are related, they are not identical. But our quantitative knowledge about this probability from BCA and kinetic theory is at least as good as it is for the average deposited energy.

For each surface point at which a cascade particle arrives, a random energy is drawn from the distribution $\lambda(E) = E_0\Theta(E -$

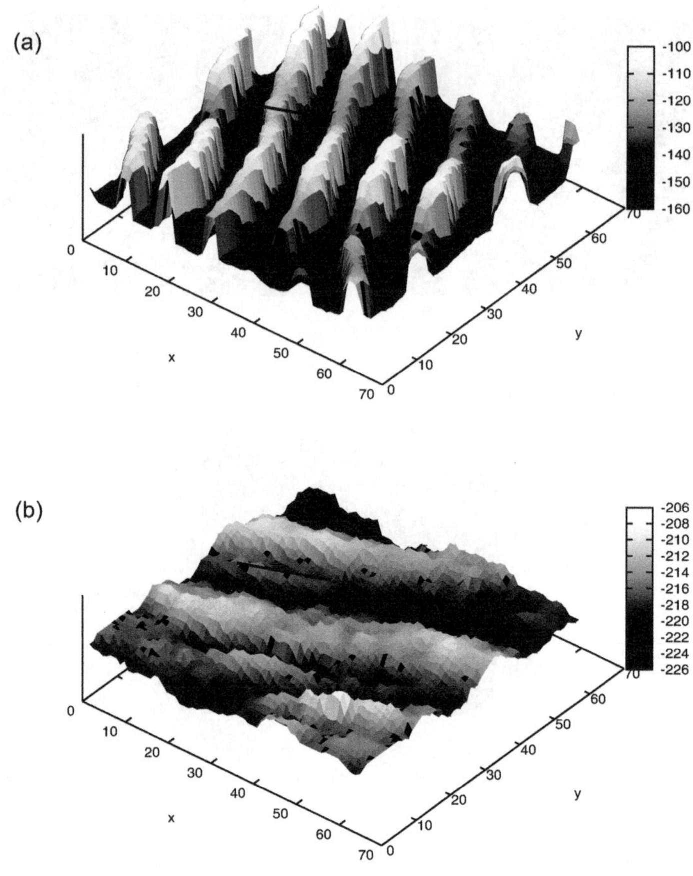

Figure 8.5 Noise in penetration depth may change the type of pattern. Morphology after two ions per surface atom have hit the substrate from $\theta = 80^\circ$ (average $a = 6.0$ and $\sigma = 4.0$, $\mu = 2.0$) (a) from the basic model, parallel ripples persist; (b) with exponentially distributed a, perpendicular ripple structures appear.

$E_0)E^{-2}$ and the surface atom is sputtered off, if this energy is larger than a surface barrier E_{surf}. The erosion rule adds more energy scales, i.e., the surface barrier and the cutoff energy E_0. The latter may be interpreted as the minimum energy necessary to create a cascade atom at the surface. Surprisingly, we have found that the basic pattern-forming scenarios, including ripples with rotating

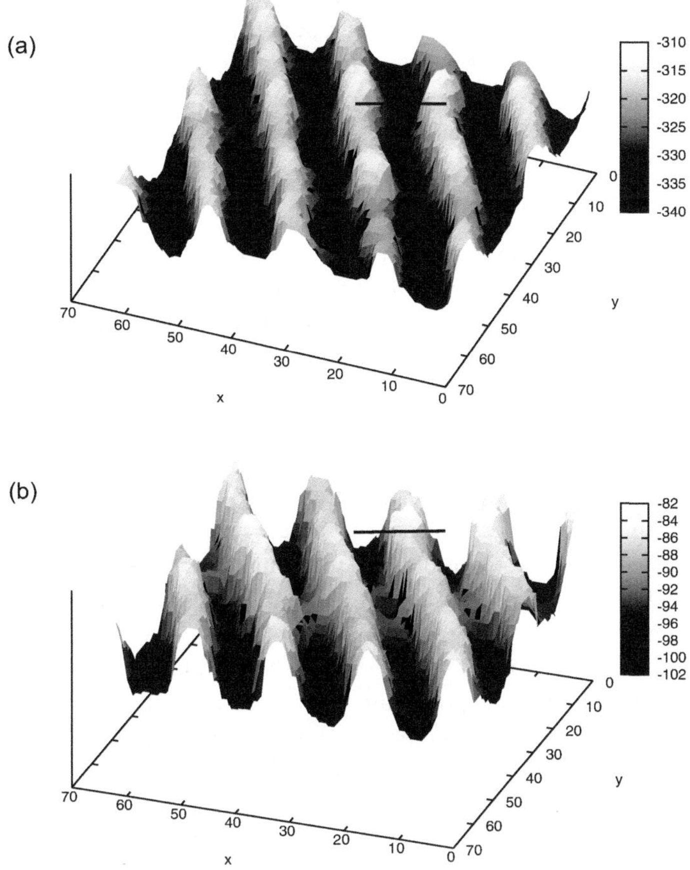

Figure 8.6 Basic and refined erosion model may lead to similar patterns. Morphologies with penetration depth $a = 7.0, \sigma = 3.0, \ \mu = 1.5$, (a) ripple pattern from the basic model; (b) ripple patterns from the refined model, including cascade noise and adatom creation (see Sec. 8.4.4).

orientations, are rather robust against this type of cascade noise, despite the fact that the additional energy fluctuations are so strong that an average no longer exists. In Fig. 8.6, you can compare the morphologies obtained from the basic model and from the model including the cascade noise under otherwise same conditions. The surface barrier has been set to 5 eV and the cutoff $E_0 = 3.2$ eV. This simple example should not, however, raise the belief that the cascade noise can simply be neglected. The larger parameter space is much more difficult to scan systematically, but we found that for increasing surface barriers and decreasing E_0, the effects of this type of noise become more and more pronounced. Furthermore, this erosion rule is necessary to generate adatoms, as discussed in Sec. 8.4.4.

8.4.3 *Modified Cascade Statistics*

We can replace the Gaussian form of the arrival probability of cascade atoms Eq. (8.3) by a distribution obtained from experiments or from BCA simulations, for example by Eq. (8.2). Although this replacement has been shown to conserve the main features of the Bradley–Harper scenario in a continuum theory derived from,[27,40] we find that it may profoundly change scenarios of surface morphologies in Monte Carlo simulations. In Fig. 8.7(a), a ripple pattern is shown, which emerges from ion bombardment under $\theta = 50°$, with erosion parameters $a = 10, \sigma = 5.0$ and $\mu = 3.0$, using the basic model. The completely different morphology of chains of dots in Fig. 8.7(b) is obtained using Eq. (8.2) instead of Eq. (8.3) with the same parameters a, σ, and μ.

Although one might argue that a more detailed comparison should consider the possibility of a nontrivial map between equivalent erosion parameters of the two models, this exemplary result should at least warn against the belief, that the Monte Carlo results are stable against the functional form of the energy deposition function. An extreme case of the dependence of patterns on the form of the arrival probability is shown in Fig. 8.8. A slight change in length scales (less than 10%) changes a standard ripple pattern to an ordered chain of dots. For the parameters given, the chain of dots will at later times (i.e., larger ion fluence) return to ripples again, but this is no longer observed for parameters beyond $\sigma = 4.0$, $\mu = 2.0$. As a general conclusion, the pattern formation does depend on cascade statistics in subtle ways, which cannot be completely captured in local continuum theories.

8.4.4 *Adatom Creation*

Given the refined erosion rule of the Sec. 8.4.2, it is now easy to include adatom creation. If the random energy of a cascade particle

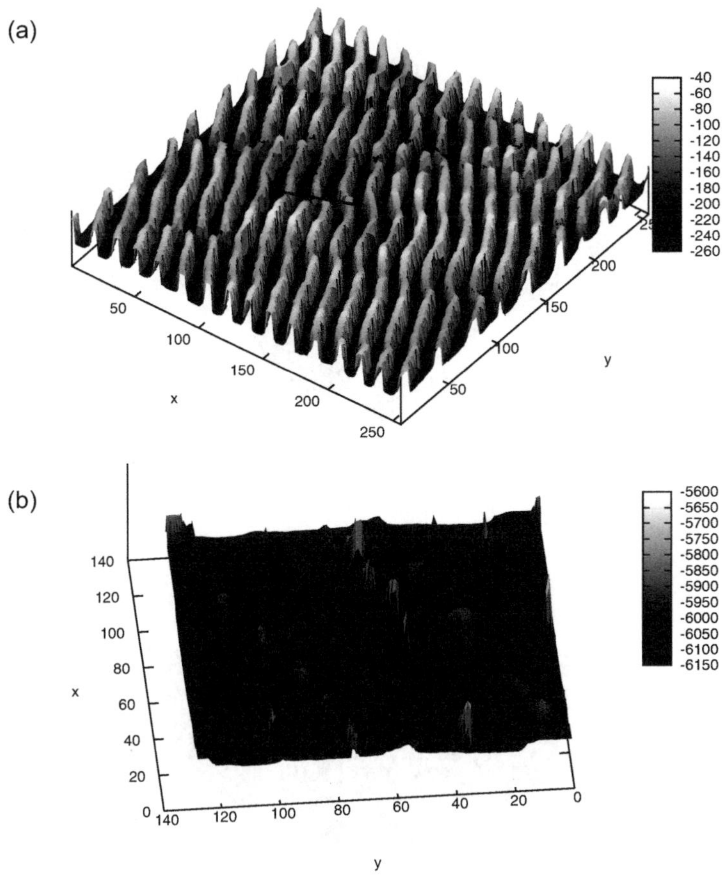

Figure 8.7 Different cascade statistics lead to different patterns: Morphologies obtained from (a) the basic model using Eq. (8.3), (b) using Eq. (8.2), instead of Eq. (8.3) with same parameters.

at the surface does not overcome the surface barrier, but is larger than a barrier for adatom creation E_{ad}, it is placed on the top of a randomly chosen nearest neighbor column. Extending the range of adatom sites, it is also possible to include *cratering* in this way, if there is information on the distribution of crater atoms (which will further blow up the parameter space, of course). As soon as adatom creation is taken into account, the basic argument of a generic curvature instability can be invalidated, if the number of

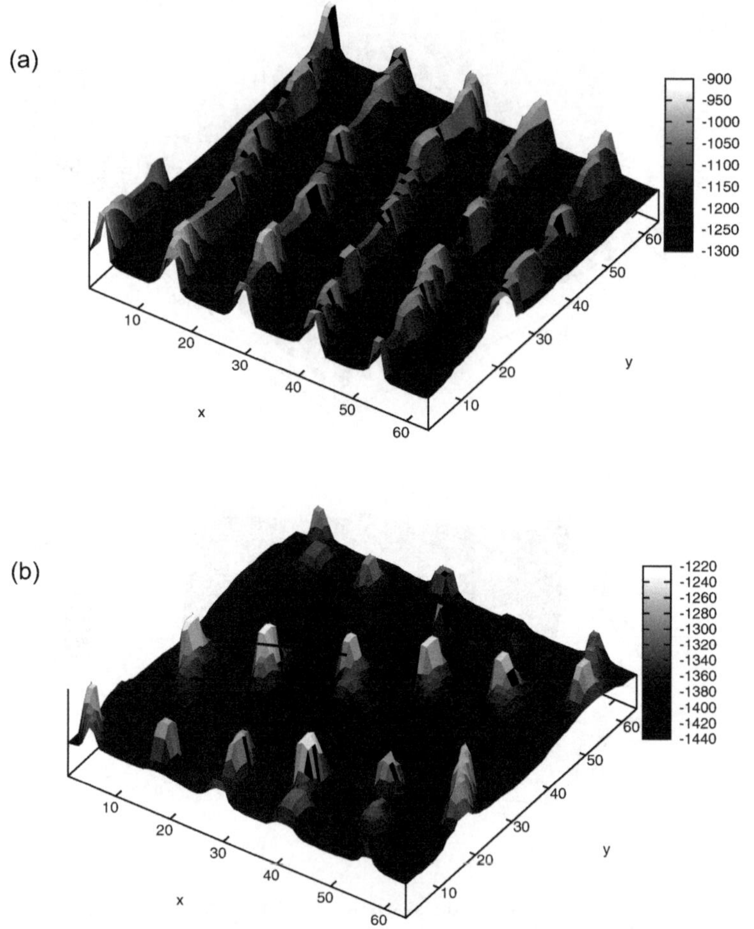

Figure 8.8 Sensitive dependence of patterns on the extension of the cascades. Note the slight changes in erosion parameters. Morphologies obtained using Eq. (8.2) with penetration depth $a = 7.0$ (a) $\sigma = 2.9$, $\mu = 1.45$; (b) $\sigma = 3.0$, $\mu = 1.5$.

locally created adatoms exceeds the number of sputtered-off atoms. Then valleys can become filled up with adatoms instead of being preferentially eroded. Furthermore, the created adatoms may by themselves induce growth instabilities. In Monte Carlo simulations, it is easy to suppress the sputtering (for example, by chosing

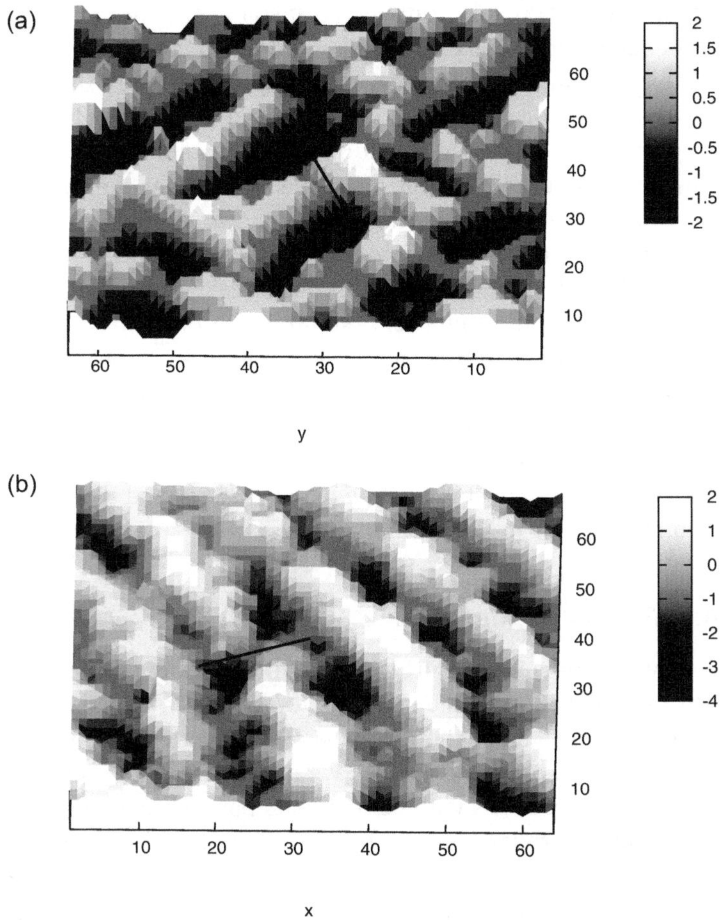

Figure 8.9 Adatom dynamics with suppressed sputtering. Morphologies obtained using Eq. (8.2) with penetration depth a $= 7.0$, $\sigma = 4.0$, and $\mu = 2.0$. (a) At $t = 4$ ions/atom, a diffusional instability starts in (110) directions, but (b) at $t = 20$ ions/atom, it is overcome by the anisotropic adatom generation and ripple-like structures orient parallel to the ion beam.

a very large surface barrier). In this way, the dynamics of the adatoms created by the ion impacts can be studied *in silico*, which is not possible in experiments. In Fig. 8.9, the adatom pattern is shown as it emerges from $\theta = 50^\circ$ ion impact, if sputtering is largely suppressed by setting the surface barrier to 1 keV. First, the adatoms show signs of a diffusional instability, which is due to a Schwoebel barrier (0.15 eV), which we added to the diffusion model. Note that the patterns follow crystallographic axes (110 directions) instead of being oriented with respect to the ion beam. If adatoms were randomly deposited, the diffusional instability will lead to persistent structures. For large ion fluences, however, the anisotropy of the adatom-generating mechanism may overcome the diffusional instability and adatoms tend to form parallel ripples as depicted in Fig. 8.9(b). Note that in the basic model, where no adatoms are generated by ion impacts, a suppression of sputtering would leave the flat surface untouched forever. Whether or not the created adatoms play an important role for the pattern formation during ion beam erosion depends on other parameters, like the diffusional hopping barriers. Many different scenarios from slight changes of the shape of ripples to stabilization of flat surfaces and to diffusional instabilities can be found in simulations.

8.4.5 *Further Refinements*

We may also include a surface downhill current of the created adatoms by biasing the choice of the target site in the adatom creation rule. So instead of choosing one of four nearest neighbor sites with equal probabilities $p = 1/4$, we may enhance the probability for hops into directions with components parallel to ion beam, and correspondingly reduce probabilities for antiparallel directions.

A refinement, which extends the model by an important physical mechanism, is the inclusion of a second atomic species of codeposited "surfactant atoms." As long as these atoms stay strictly at the surface, this is easily done by marking each surface site by a two-valued flag variable (A or B) corresponding to the substrate (A) or the surfactant (B). All the energetic barriers may be different for A and B particles and in surface diffusion, there are now A-A, B-B, and A-B bonds, so another energy barrier parameter enters. To ensure a stationary coverage, every sputtered-off B particle is replaced in the next step.[41] Finally, the inclusion of redeposition requires to trace the trajectories of sputtered-off particles. Although this can be done in the same way the incoming ion trajectory is traced, it requires detailed knowledge about the directions of sputtered-off particles and their energies. Furthermore, the secondary impacts on the surface have to be modeled. This brings lots of new and largely unknown parameters into play, which is why we have not yet undertaken such a study.

8.5 Conclusion

We have studied qualitative examples of the impact of variations of Monte Carlo rules, as derived from BCA and kinetic theory, on the self-organized pattern formation at solid surfaces under ion beam irradiation. The presented top-down modeling strategy is easily implemented, very flexible, and capable of providing insight into physical mechanisms of pattern formation by ion beam erosion. It constitutes a sort of transmission shaft between continuum theories, more microscopic approaches like BCA or kinetic theory, and experiments. Results from experiments or any microscopic theory may be used to construct Monte Carlo rules, but the same is true for principles of continuum theories, which can be reformulated as Monte Carlo algorithms. Therefore, the impact of experimental results on model building may be checked as effectively as the impact of a new hypothesized physical mechanism on observable patterns. Although mainly qualitative examples have been presented, they have already revealed that the pattern formation depends very sensitively on some details of the erosion mechanism. In particular, we have shown that noise in the penetration depth is much more important than one expects from arguments in reference 39. Exponentially distributed penetration depths, as used in kinetic theory of amorphous materials, usually wipe out patterns in most cases. Even smaller amounts of noise can change the type of pattern qualitatively. Changes in the cascade statistics, in particular in the arrival probabilities of cascade particles at the surface, also have pronounced effects, which were not expected from a continuum theory. On the other hand, we found that the noise in the energy of arriving cascade particles has less pronounced impact, although the fluctuations in energy are strong. Adatom creation from collision cascades may act as an additional smoothing mechanism, but may also induce additional diffusional instabilities.

The dependence on details of the erosion process may turn out to limit our approach in an important way, because a more and more detailed modeling of the collision cascade ultimately leads back to BCA. At present it is not clear how far this road has to be traveled and whether an effective intermediate description of erosion processes exists, which captures the important features that are relevant for pattern formation. But the basic model is definitely not sufficient in this respect. The examples also constitute a kind of warning against quick, hand-waving arguments, which try to pinpoint the important physical mechanisms for pattern formation. After a period of belief in generic explanations from simple continuum models, one now has to reconsider the underlying concepts critically to identify the generic ways in which the pattern forming mechanisms, like the Bradley–Harper curvature instability, fail and how other mechanisms show up, which depend on new factors like codeposited atomic species. We firmly believe that Monte Carlo modeling will stay a valuable tool in these studies.

References

1. P. Sigmund (1969) Theory of sputtering. I. Sputtering yield of amorphous and polycrystalline targets, *Phys. Rev.*, **184**(2), 383–416. doi: 10.1103/PhysRev.184.383.
2. P. Sigmund (1973) A mechanism of surface micro-roughening by ion bombardment, *J. Mater. Sci.*, **8**, 1545–1553.
3. G. Carter, M. Nobes, and R. Webb (1981) A second-order erosion slowness theory of the development of surface topography by ion-induced sputtering, *J. Mater. Sci.*, **16**, 2091–2102.
4. R. M. Bradley and J. M. E. Harper (1988) Theory of ripple topography induced by ion-bombardment, *J. Vac. Sci. Technol. A*, **6**(4), 2390–2395.
5. H. Hofsäss and K. Zhang (2008) Surfactant sputtering, *Appl. Phys. A*, **92**(3), 517–524. doi: 10.1007/s00339-008-4678-9.
6. C. Madi, H. George, and M. Aziz (2009) Linear stability and instability patterns in ion-sputtered silicon, *J. Phys.: Condens. Matter*, **21**, 224010.
7. S. Macko, F. Frost, B. Ziberi, D. Förster, and T. Michely (2010) Is kev ion-induced pattern formation on si(001) caused by metal impurities? *Nanotechnology*, **21**, 085301.

8. G. Carter (2001) The physics and applications of ion beam erosion, *J. Phys. D: Appl. Phys.*, **34**, R1–R22.
9. W.-L. Chan and E. Chason (2007) Making waves: Kinetic processes controlling surface evolution during low energy ion sputtering, *J. Appl. Phys.*, **101**(12):121301, (2007).
10. H. Oechsner (1975) Sputtering: a review of some recent experimental and theoretical aspects, *Appl. Phys. A*, **8**, 185–198.
11. Y. Kuramoto and T. Tsuzuki (1977) Persistent propagation of concentration waves in dissipative media far from thermal equilibrium, *Prog. Theor. Phys.*, **55**(2), 356–369.
12. R. Cuerno and A.-L. Barabási (1995) Dynamic scaling of ion-sputtered surfaces, *Phys. Rev. Lett.*, **74**(23), 4746–4749. doi: 10.1103/PhysRevLett.74.4746.
13. S. Park, B. Kahng, H. Jeong, and A.-L. Barabási (1999) Dynamics of ripple formation in sputter erosion: Nonlinear phenomena, *Phys. Rev. Lett.*, **83**(17), 3486–3489. doi: 10.1103/PhysRevLett.83.3486.
14. T. Aste and U. Valbusa (2005) Ripples and ripples: From sandy deserts to ion-sputtered surfaces, *New J. Phys.*, **7**, 122–142.
15. J. Muñoz-García, M. Castro, and R. Cuerno (2006) Nonlinear ripple dynamics on amorphous surfaces patterned by ion beam sputtering, *Phys. Rev. Lett.*, **96**(8), 086101.
16. B. Ziberi, M. Cornejo, F. Frost, and B. Rauschenbach (2009) Highly ordered nanopatterns on Ge and Si surfaces by ion beam sputtering, *J. Phys.: Condens. Matter*, **21**(22), 224003.
17. G. Carter and V. Vishnyakov (1996) Roughening and ripple instabilities on ion-bombarded Si, *Phys. Rev. B.*, **54**(24), 17647–17653.
18. M. Moseler, P. Gumsch, C. Casiraghi, A. Ferrari, and J. Robertson (2005) The ultrasmoothness of diamond-like carbon surfaces, *Science*, **309**(5740).
19. P. Süle and K.-H. Heinig (2009) The molecular dynamics simulation of ion-induced ripple growth, *J. Chem. Phys.*, **131**(20), 204704.
20. I. Koponen, M. Hautala, and O.-P. Sievänen (1996) Simulations of submicrometer-scale roughening on ion-bombarded solid surfaces, *Phys. Rev. B*, **54**(19), 13502–13505. doi: 10.1103/PhysRevB.54.13502.
21. I. Koponen, M. Hautala, and O.-P. Sievänen (1997) Simulations of ripple formation on ion-bombarded solid surfaces, *Phys. Rev. Lett.*, **78**(13), 2612–2615. doi: 10.1103/PhysRevLett.78.2612.
22. B. Liedke, K.-H. Heinig, S. Facsko, and W. Möller (2010) Ion-induced surface pattern evolution described by combined trim and kinetic

Monte Carlo simulation, in *Workshop Ion Beam Physics*, (D. G. FZD, ed.) 29–31.03.2010, Dresden, Germany.

23. T. Michely and J. Krug (2004) *Islands, Mounds, and Atoms. Patterns and Processes in Crystal Growth Far from Equilibrium*, Springer, Berlin, Germany; *Surface Sciences.* ISBN 3-540-40728-6.
24. J. F. Ziegeler, J. P. Biersack, and K. Littmark (1985) *The Stopping and Range of Ions in Matter*, Pergamon, New York. see also http://www.srim.org/
25. M. Robinson and I. Torrens (1974) Computer simulation of atomic-displacement cascades in solids in the binary-collision approximation, *Phys. Rev. B*, **9**(2), 5008–5024.
26. M. Robinson (1994) The binary collision approximation: background and introduction, *Rad. Eff. Def. Sol.*, **130**, 3–20.
27. M. Feix, A. K. Hartmann, R. Kree, J. Muñoz-García, and R. Cuerno (2005) Influence of collision cascade statistics on pattern formation of ion-sputtered surfaces, *Phys. Rev. B:Condens. Matter*, **71**(12), 125407.
28. M. Williams (1977) A stochastic theory of particle transport, *Proc. Roy. Soc. Lond. A*, **358**, 105–120.
29. M. Thompson (1968) The energy spectrum of ejected atoms during the high energy sputtering of gold pt 2, *Philos. Mag.*, **18**, 377–414.
30. E. Chason, W. L. Chan, and M. S. Bharathi (2006) Kinetic Monte Carlo simulations of ion-induced ripple formation: Dependence on flux, temperature, and defect concentration in the linear regime, *Phys. Rev. B: Condens. Matter*, **74**(22), 224103.
31. A. Hartmann, R. Kree, and T. Yasseri (2009) Simulating discrete models of pattern formation by ion beam sputtering, *J. Phys.: Condens. Matter*, **21**, 224015.
32. E. O. Yewande, R. Kree, and A. K. Hartmann (2006) Morphological regions and oblique-incidence dot formation in a model of surface sputtering, *Phys. Rev. B: Condens. Matter*, **73**(11), 115434.
33. A. K. Hartmann, R. Kree, U. Geyer, and M. Kölbel (2002) Long-time effects in a simulation model of sputter erosion, *Phys. Rev. B*, **65**(19), 193403. doi: 10.1103/PhysRevB.65.193403.
34. E. O. Yewande, A. K. Hartmann, and R. Kree (2005) Propagation of ripples in Monte Carlo models of sputter-induced surface morphology, *Phys. Rev. B: Condens. Matter*, **71**(19), 195405.
35. E. O. Yewande, R. Kree, and A. K. Hartmann (2007) Numerical analysis of quantum dots on off-normal incidence ion sputtered surfaces, *Phys. Rev. B: Condens. Matter*, **75**(15), 155325.

36. T. Yasseri and R. Kree (2010) A Monte Carlo study of surface sputtering by dual and rotated ion beams, *Nucl. Instrum. Methods*, **268**(16), 2496–2503.
37. M. Joe, J.-H. Kim, C. Choi, B. Kahng, and J.-S. Kim (2009) Nanopatterning by multiple-ion-beam sputtering, *J. Phys.: Condens. Matter*, **21**(22), 224011.
38. M. Stepanova (1995) Master equation for atomic collision cascades with fixed path length between collisions, *Nucl. Instrum. Methods*, **103**, 33–37.
39. K. B. Lauritsen, R. Cuerno, and H. A. Makse (1996) Noisy kuramoto-sivashinsky equation for an erosion model, *Phys. Rev. E*, **54**(4), 3577–3580. doi: 10.1103/PhysRevE.54.3577.
40. S. Norris, M. Brenner, and M. Aziz (2009) From crater functions to partial differential equations: A new approach to ion bombardment induced nonequilibrium pattern formation, *J. Phys.: Condens. Matter*, **21**, 224017.
41. R. Kree, T. Yasseri, and A. Hartmann (2009) Surfactant sputtering: Theory of a new method of surface nanostructuring by ion beams, *Nucl. Instrum. Methods*, **267**, 1403–1406.

Chapter 9

Understanding Surface Patterning by Lattice Gas Models

Géza Ódor,[a,*] Bartosz Liedke,[b] and Karl-Heinz Heinig[b]

[a] *Research Institute for Technical Physics and Materials Science, P.O. Box 49, H-1525 Budapest, Hungary*
[b] *Institute of Ion Beam Physics and Materials Research, Helmholtz-Zentrum Dresden-Rossendorf, P.O. Box 51 01 19, 01314 Dresden, Germany*
*odor@mfa.kfki.hu

In present day nanotechnology, surface structures are effectively made via natural processes instead of conventional carving methods. We show that the emergence of different surface patterns can be understood by suitable mapping onto the simple nonequilibrium lattice gases. Surface adsorption/desorption processes correspond to migration of oriented dimers representing local heights. The surface diffusion can be mapped onto attracting or repelling moves of dimers of the lattice. While attracting dimer moves describe roughening surfaces, repulsive ones realize smoothing processes. The competition of these different reactions leads to nontrivial surface pattern formation. With the help of this effective approach, difficult unanswered questions of surface growth and scaling can be investigated and have been resolved. Besides, the mapping onto binary variables facilitates effective simulations and enables

Nanofabrication by Ion-Beam Sputtering: Fundamentals and Applications
Edited by Tapobrata Som and Dinakar Kanjilal

ISBN 978-981-4303-75-0 (Hardcover), 978-981-4303-76-7 (eBook)
www.panstanford.com

us to consider very large system sizes. We have shown that the fundamental Kardar–Parisi–Zhang universality class is stable against a competing roughening diffusion. A strong smoothing diffusion leads to logarithmic growth and mean-field class scaling behavior in two dimensions. These lattice gas simulations result in ripple coarsening, if parallel surface currents are present, otherwise logarithmic behavior can be observed.

9.1 Introduction

In nanotechnologies, large areas of nanopatterns are needed, which can be fabricated today only by expensive techniques, e.g., electron-beam lithography or direct writing with electron and ion beams. Besides the conventional top-down technologies, which use masks, photoresists, etc., to create structures on the surfaces, nowadays bottom-up approaches are getting close to achieve the same results more efficiently. In that case, the self-assembly of patterns of large areas is facilitated in a cost-effective way [1]. This has led reopening of the research for fundamental theoretical understanding of the ion-beam-induced surface patterning and scaling [2], which was flourishing at the end of the previous century [3, 4]. Although the basic universality classes and important models have been explored, many notoriously difficult fundamental questions have been unanswered. Perturbative renormalization-group methods and analytical tools have limited applicability and precise numerical simulations, approaching asymptotic scaling regimes, were feasible in one dimension mainly.

One of the simplest nonlinear stochastic differential equations set up by Kardar, Parisi, and Zhang [5] describes the dynamics of growth processes in the thermodynamic limit. It specifies the evolution of the height function $h(\mathbf{x}, t)$ in the d-dimensional space

$$\partial_t h(\mathbf{x}, t) = v + \sigma \nabla^2 h(\mathbf{x}, t) + \lambda_2 (\nabla h(\mathbf{x}, t))^2 + \eta(\mathbf{x}, t) . \qquad (9.1)$$

Here v and λ_2 are the amplitudes of the mean and local growth velocity, σ is a smoothing surface tension coefficient, and η roughens the surface by a zero-average Gaussian noise field exhibiting the variance $\langle \eta(\mathbf{x}, t)\eta(\mathbf{x}', t')\rangle = 2D\delta^d(\mathbf{x}\text{-}\mathbf{x}')(t - t')$. The notation D is used for the noise amplitude and $\langle\rangle$ means the distribution average.

The Kardar–Parisi–Zhang (KPZ) equation was inspired in part by the stochastic Burgers equation [6], which belongs to the same universality class [7], and it became the subject of many theoretical studies [3, 4, 8]. Besides, it models other important physical phenomena such as directed polymers [9], randomly stirred fluid [7], dissipative transport [10, 11], and the magnetic flux lines in superconductors [12]. The equation is solvable in $(1+1)d$ [9], but in higher dimensions approximations are available only. As the result of the competition of roughening and smoothing terms, models described by the KPZ equation exhibit a roughening phase transition between a weak coupling regime ($\lambda_2 < \lambda_2^*$), governed by the Edwards–Wilkinson fixed point at $\lambda_2 = 0$ [13], and a strong coupling phase. The strong coupling fixed point is inaccessible by perturbative renormalization group method. Therefore, the KPZ phase space has been the subject of controversies and the value of the upper critical dimension has been debated for a long time [14–22].

Surface diffusion is a much studied basic process [8]. Several atomistic models have been constructed and investigated with the aim of realizing Mullins–Herring diffusion [23, 24] and scaling (for a recent review see [25]). The Langevin equation of Mullins and Herring is a linear one, with a ∇^4 lowest order gradient term

$$\partial_t h(\mathbf{x}, t) = \nu_4 \nabla^4 h(\mathbf{x}, t) + \eta(\mathbf{x}, t)\,. \tag{9.2}$$

emerging as the result of a curvature-driven surface current $j(\mathbf{x}, t) \propto \nabla(\nabla^2 h(\mathbf{x}, t))$, which obeys the conservation law

$$\partial_t h(\mathbf{x}, t) + \nabla j(\mathbf{x}, t) = \eta(\mathbf{x}, t)\,. \tag{9.3}$$

Here the noise $\eta(\mathbf{x}, t)$ is a nonconserved, Gaussian white one, which can be the result of fluctuations in the ion-beam intensity directed against the surface. This equation is exactly solvable and exhibits a scaling invariance of the roughness characterized by the exponents:

$$\alpha = (4-d)/2, \quad \beta = (4-d)/8, \quad z = 4 \tag{9.4}$$

below the upper critical dimension $d_c = 4$.

Microscopic models realizing this behavior are mainly unrestricted solid-on-solid (SOS) type, that can provide steep slopes and strong curvatures necessary for the $\alpha \geq 1$ roughness exponent in one and two dimensions. However, it turned out that the asymptotic universality class of the various limited mobility growth models is a surprisingly subtle issue. Many of the earlier findings proved to be incorrect due to pathologically slow crossover and extremely long transient effects [26]. Anomalous scaling, by which the local and global behavior is different, has been found to be relevant in such "super-rough" models, where large local slopes are present [27–32]. In fact, according to our knowledge, only the "larger curvature (or Kim–Das Sarma) model" [33, 34] and the $n = 2$ model of reference [35] exhibits Mullins–Herring universality class scaling asymptotically.

The scaling in other "atomistic" models crosses over to behavior dictated by more relevant terms in the sense of renormalization group. That can be the Edwards–Wilkinson class in case of $\nabla^2(h)$ [36] or the molecular beam epitaxy (MBE) class in case of the fourth-order nonlinearity $\nabla^2[\nabla(h)^2]$ [37–40].[a] The nonlinear MBE equation:

$$\partial_t h = \nu_4 \nabla^4 h + \lambda_{22} \nabla^2 [\nabla(h)^2] + \eta \tag{9.5}$$

with nonconserved Gaussian noise η is just the conserved version of KPZ (CKPZ) and exhibits the following scaling exponents:

$$\alpha = (4-d)/3, \quad \beta = (4-d)/(8+d), \quad z = (8+d)/3\,. \tag{9.6}$$

As we can see, all exponents are smaller than those of the linear Mullins–Herring (9.4) class values.

Here we present restricted solid-on-solid (RSOS) models with $\Delta h = \pm 1$ height restriction within the framework of our previous approach [41], which in the limit of weak external noise exhibit Mullins–Herring or MBE scaling. Due to simple construction, these can be mapped onto lattice gases of diffusing dimers, allowing an easier way to study the effects of Mullins–Herring and MBE subprocesses of more complex system. Earlier less restrictive SOS models were used to describe these classes especially in one dimension (for a review see reference [25]).

A further step in generalizing our surface models is the combination of different subprocesses resulting in a nonequilibrium system.

---

[a]Cubic nonlinearity is also possible by the same symmetries; However, not much realization or occurrence has been found.

For example by adding a competing Mullins–Herring diffusion to the KPZ updates, one can model the noisy Kuramoto–Sivashinsky equation [42, 43].

$$\partial_t h = v + \sigma \nabla^2 h - \nu_4 \nabla^4 h + \lambda (\nabla h)^2 + \eta \, . \qquad (9.7)$$

Here the surface tension coefficient σ is negative (in contrast with the KPZ), whereas ν_4 is a positive surface diffusion coefficient. However, in our simulations, we realize a gauge transformed situation: normal KPZ (with positive σ) competing with an inverse Mullins–Herring (iMH) with negative ν_4. In Kuramoto–Sivashinsky, even the deterministic variant ($\eta = 0$) exhibits spatio-temporal chaos and is useful to describe pattern formation, such as chemical turbulence and flame-front propagation [42]. There are other physical systems, including ion sputtering, where the noisy version of the Kuramoto–Sivashinsky equation is used [44]. In one-dimensional field, theory has proved that Kuramoto–Sivashinsky belongs to the $1 + 1$-dimensional KPZ universality class [44]. However, in higher dimensions, perturbative field theory cannot access the strong coupling fixed point. Numerical studies [45–47] have provided controversial results in $2 + 1$ dimensions, hence it remained a controversial and challenging problem to clarify the asymptotic scaling behavior of Kuramoto–Sivashinsky [48–51].

It was pointed out [52, 53] that grooved phases and growth instabilities may emerge as the consequence of broken detailed balance condition:

$$P(\{i\}) w_{i \to i'} = P(\{i'\}) w_{i' \to i} \qquad (9.8)$$

where $P(\{i'\})$ denotes the probability of the state $\{i\}$ and $w_{i \to i'}$ is the transition rate between states $\{i\})$ and $\{i'\}$. This means that complex structures and patterns can emerge in a nonequilibrium system. In one dimension, a model of massive particles exhibiting momentum has been shown to exhibit Kuramoto–Sivashinsky scaling behavior [54]. Later another one-dimensional RSOS growth model was constructed [55], in which deposition and diffusion of single adatoms were competing. It was suggested that large-scale behavior could be described by the noisy Kuramoto–Sivashinsky equation.

First, let us consider the simple adsorption/desorption model and later we will turn toward a more complex system.

9.2 Mapping of Surface Adsorption/Desorption onto Lattice Gases

Mapping of surface growth onto reaction-diffusion system allows effective numerical simulations [56, 57]. As a generalization of the 1 + 1-dimensional rooftop model [58, 59] and the 2 + 1-dimensional octahedron model [41], we consider the deposition and removal processes of higher-dimensional objects on $d \geq 2$-dimensional surfaces. We remind that in 1 + 1 dimensions, a continuous surface line having no overhangs can be approximated by 45° up/down slope elements (after appropriate length rescaling). A process with KPZ scaling can be realized by deposition at local minima or removal of local maxima (rooftop). If we associate the up slopes with "particles" and down slopes with "holes" of the base lattice (see Fig. 9.1 a), the adsorption/desorption corresponds

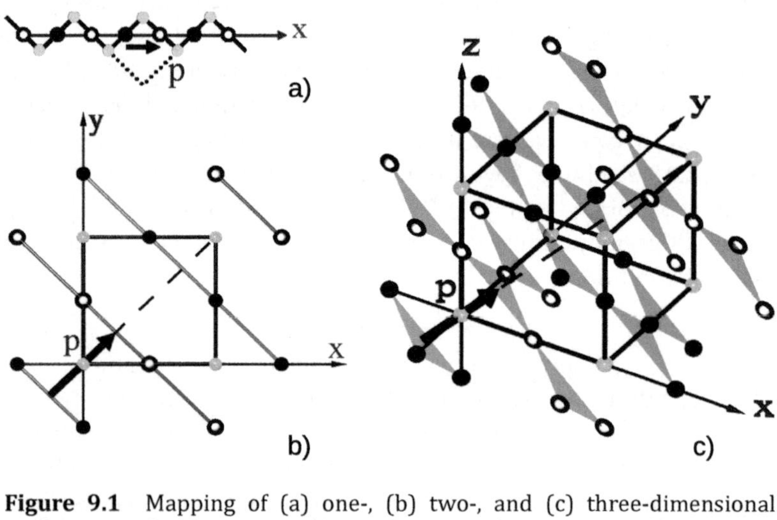

Figure 9.1 Mapping of (a) one-, (b) two-, and (c) three-dimensional surface growth (discrete up/down derivatives) onto 1, 2, 3-dimensional oriented motion of d-mers. In general, a deposition event corresponds to a d-Kawasaki exchange, or a d-mer hopping in the bisectrix direction of axes. Up/down slopes between lattice points (pale green dots at the line corners) are denoted by full/empty circles, corresponding to particles/holes on the projection space. The surface continuity translates into particle exclusion, which makes the model nontrivial in higher dimensions. Please change this globally. Rewrite such constructions as Reprinted from reference [63].

to the asymmetric simple exclusion process (ASEP) [60]. This is a lattice gas [61, 62] where particles can hop on adjacent sites with asymmetric rates and hard-core exclusion. Its behavior is well known, and variations of ASEP (disorder, Interactions, etc.) correspond to variations of $1+1$-dimensional KPZ growth models.

We have extended the rooftop construction to $(2+1)$ dimensions [41] by the introduction of octahedra having four slopes. The up edges in the x or y directions can be represented by '$+1$', while the down ones by '-1', and a surface element update is a generalized (Kawasaki) exchange (Eq. (3) of reference [41]). The translation of up edges to "particles" and the down ones to "holes" of the base lattice maps particle deposition/removal processes onto two simultaneous particle moves (one in the x and one in the y direction). One can also consider it a dimer move in the bisectrix direction of x and y (see Fig. 9.1 b). Therefore, the $(2+1)$ surface dynamics can be mapped onto a "two-dimensional ASEP" of oriented dimers exhibiting hard-core exclusion. The asymmetric drift corresponds to an evolving surface exhibiting KPZ scaling, while the symmetric dimer diffusion is related to the Edwards–Wilkinson behavior.

Now we proceed with this kind of construction, considering the discrete slope variables in higher dimensions and generalize the simultaneous $+1 \leftrightarrow -1$ (Kawasaki) exchange rule of them (Eq. (3) of reference [41]) to d-dimensional updates

$$\begin{pmatrix} -1 & 1 \\ -1 & 1 \\ -1 & 1 \\ \ldots & \end{pmatrix} \underset{q}{\overset{p}{\rightleftharpoons}} \begin{pmatrix} 1 & -1 \\ 1 & -1 \\ 1 & -1 \\ \ldots & \end{pmatrix}, \tag{9.9}$$

with probability p for attachment and probability q for detachment (see Fig. 9.1 c for the 3d case). It is well known [4] that the surface evolution of the deterministic KPZ growth are described also by the Burgers equation [6] for growth velocities $\mathbf{v}(\mathbf{x},t)$ in the surface normal obeying

$$\partial_t \mathbf{v}(\mathbf{x}, t) = \sigma \nabla^2 \mathbf{v}(\mathbf{x}, t) + \lambda_2 \mathbf{v}(\mathbf{x}, t) \nabla \mathbf{v}(\mathbf{x}, t) \tag{9.10}$$

due to the transformation $\mathbf{v}(\mathbf{x}, t) = \nabla h(\mathbf{x}, t)$.

In the forthcoming part, we will prove that our microscopic model for d-mers in the continuum limit can be mapped onto the anisotropic version of Eq. (9.10), similarly as shown in lower dimensions [41, 58]. The derivation is based on the formulation of the reduction of possible updates. Our surface model is represented by the discrete derivative elements: $\delta_x, \delta_y, \delta_z \ldots (\in \pm 1)$ at every lattice points. A generalized Kawasaki update (9.9) is defined by a matrix

$$\begin{pmatrix} \delta_x(i-1, j, k, \ldots) \; \delta_x(i, j, k, \ldots) \\ \delta_y(i, j-1, k, \ldots) \; \delta_y(i, j, k, \ldots) \\ \delta_z(i, j, k-1, \ldots) \; \delta_z(i, j, k, \ldots) \\ \ldots \end{pmatrix}. \tag{9.11}$$

In d dimensions, we define vectors of the slopes, the columns of (9.11), analogously to one and two dimensions: $\overline{\sigma}_{i,j,k,\ldots} = (\delta_x(i-1, j, k, \ldots), \delta_y(i, j-1, k, \ldots), \ldots)$, around the lattice point, which we select for deposition/removal update and set up a microscopic master equation

$$\begin{aligned} \partial_t P(\{\overline{\sigma}\}, t) = & \sum_{i,j,k,\ldots} w'_{i,j,k,\ldots}(\{\overline{\sigma}\}) P(\{\overline{\sigma}'\}, t) \\ & - \sum_{i,j,k,\ldots} w_{i,j,k,\ldots}(\{\overline{\sigma}\}) P(\{\overline{\sigma}\}, t) \end{aligned} \tag{9.12}$$

with the probability distribution $P(\{\overline{\sigma}\}, t)$. Here the prime index denotes the state of $\overline{\sigma}$ following the update (9.9). The transition probability of $\overline{\sigma}$-s can be expressed as

$$\begin{aligned} w_{i,j,k,\ldots}(\{\overline{\sigma}\}) = & A[1 - \overline{\sigma}_{i+1,j+1,k+1,\ldots} \overline{\sigma}_{i,j,k,\ldots} \\ & + \lambda_2(\overline{\sigma}_{i+1,j+1,k+1,\ldots} - \overline{\sigma}_{i,j,k,\ldots})], \end{aligned} \tag{9.13}$$

with $\lambda_2 = 2p/(p+q) - 1$ parametrization, which formally looks like the Kawasaki exchange probability in 1d, except the factor A, which is necessary to avoid surface discontinuity creation in higher dimensions. This means that we update the slope configurations only if the values of all coordinates of the vector $\overline{\sigma}$ are identical as shown by (9.9). One can allow formally these updates via the expression

$$\begin{aligned} A = & 1/2^{d+1} \det[(\overline{\sigma}_{i,j,k,\ldots} + C\overline{\sigma}_{i,j,k,\ldots}) \\ & \times (\overline{\sigma}_{i+1,j+1,k+1,\ldots} + C\overline{\sigma}_{i+1,j+1,k+1,\ldots})I], \end{aligned} \tag{9.14}$$

where I and C are the unity and the cyclic permutation matrices, respectively. The matrix C shifts each coordinate value to the next index value. Thus for $\overline{\sigma}$-s with mixed coordinate values, the vectors

$(\overline{\sigma}_{i,j,k,\ldots} + C\overline{\sigma}_{i,j,k,\ldots}) = \overline{k}$ or $(\overline{\sigma}_{i+1,j+1,k+1,\ldots} + C\overline{\sigma}_{i+1,j+1,k+1,\ldots}) = \overline{k'}$ possess zero elements. Therefore, the determinant of $\overline{kk'}I$, being the product of the diagonal elements, is zero in case of mixed coordinates and $A = 1/2^{d+1}$ in case of equal coordinates.

For example, a $d = 3$ update is prohibited when the slope vector is $\overline{\sigma} = (1, 1, -1)$, because $\overline{k}$ has one coordinate value of zero

$$\overline{k} = \begin{pmatrix} 1 \\ 1 \\ -1 \end{pmatrix} + \begin{pmatrix} 0 & 1 & 0 \\ 0 & 0 & 1 \\ 1 & 0 & 0 \end{pmatrix} \begin{pmatrix} 1 \\ 1 \\ -1 \end{pmatrix} = \begin{pmatrix} 2 \\ 0 \\ -2 \end{pmatrix}.$$

By calling "+1"-s as particles and the "−1"-s as holes of the base lattice, their synchronous update can be considered a single-step motion of an oriented d-mer in the bisectrix direction of the x, y, $z, \ldots$ coordinate axes. Thus d-mers follow one-dimensional kinetics, described by Kawasaki exchanges (9.13). To obtain a one-to-one mapping, we update neighborhoods of the lattice points denoted by the green dots of Fig. 9.1.

To derive Eq. (9.10), first we have to average over the slope vectors

$$\langle \overline{\sigma} \rangle = \sum_{\{\overline{\sigma}\}} \overline{\sigma} P(\{\overline{\sigma}\}, t). \tag{9.15}$$

By calculating its time derivative using the master equation (9.12) and the transition probabilities (9.13)

$$\partial_t \langle \overline{\sigma} \rangle = \sum_{\{\overline{\sigma}\}} \left[\overline{\sigma} \sum_{i,j,k,\ldots} w'_{i,j,k,\ldots}(\{\overline{\sigma}\}) P(\{\overline{\sigma}'\}, t) \right. \\ \left. - \overline{\sigma} \sum_{i,j,k,\ldots} w_{i,j,k,\ldots}(\{\overline{\sigma}\}) P(\{\overline{\sigma}\}, t) \right], \tag{9.16}$$

in which we filter out vectors of nonequal coordinates (9.14) (thus $w_{i,j,k,\ldots}$ is nonzero only if $\overline{\sigma}_{i,j,\ldots} \neq \overline{\sigma}_{i+1,j+1,\ldots}$) we can obtain

$$2\partial_t \langle \overline{\sigma}_{i,j,k,\ldots} \rangle = \langle \overline{\sigma}_{i-1,j-1,\ldots} \rangle - 2\langle \overline{\sigma}_{i,j,\ldots} \rangle \\ + \langle \overline{\sigma}_{i+1,j+1,\ldots} \rangle + \lambda_2 \langle \overline{\sigma}_{i,j} (\overline{\sigma}_{i+1,j+1,\ldots} - \overline{\sigma}_{i-1,j-1,\ldots}) \rangle, \tag{9.17}$$

analogously to one dimension [58]. Here one can see the discrete first and second differentials of $\overline{\sigma}_{i,j,k,\ldots}$ corresponding to the operators of Eq. (9.10) in the bisectrix direction of the axes. These differentials are one dimensional because the d-mer dynamics is one dimensional. In principle, one could derive a set of coupled Burgers equations for the particles in each direction in an isotropic way in accordance with isotropic surface model, but the coordinated movements reduce the dimensionality and we can map onto an anisotropic equation of d-mers.

Making a continuum limit in each direction and taking into account the relation of height and slope variables ($\mathbf{v}(\mathbf{x}, t) = \nabla h(\mathbf{x}, t)$), we can arrive to the deterministic KPZ equation. The nonlinear term vanishes for $p = q$ ($\lambda_2 = 0$). The sign of the coefficient λ_2 of the nonlinear term can be interpreted as follows: For $p > q$, positive nonlinearity (positive excess velocity) is a consequence of growth with voids.

Since this derivation was applied just for the first one in the hierarchy of equations for correlation functions, it does not prove the equivalence to the stochastic KPZ. Furthermore, the form of the noise term, which was not considered in our derivation, may also introduce differences. Although our surface model is spatially isotropic, we can map it onto a one-dimensional Burgers equation (of extended objects); therefore, anisotropic scaling behavior might be expected. However, by going into the continuum description, the hard-core exclusions necessary to provide continuous surfaces are lost and the resulting equation looks trivial.

9.3 Numerical Simulations for KPZ

Now we present numerical simulations results using this surface growth model via the one-dimensional directed migration of d-mers in the d-dimensional space. We have developed bit-coded algorithms for the updates (9.9) and run it with $p = 1$, such that randomness comes from the site selection only. Therefore, it is important to use a very good random number generator, which provides uniformly distributed numbers with high resolution. Otherwise, we would realize a process with quenched disorder, which for KPZ belongs to a different universality class ([57]). We used the latest Mersenne-Twister generator [64] in general, which has very good statistical properties and which is very fast, especially in the SSE2 registers. We tested our results using other random number generators as well. In practice, each update site can be characterized by the 2^{d^2} different local slope configurations. However, due to the surface continuity, we need only a few bits of a world (1 byte for $d = 2, 3, 4$ and two bytes for $d = 5$) for this purpose. This allows an efficient storage management in the computer memory and permits simulations of larger system sizes. The updates can be performed by logical operations, either on multiple samples at once, or on multiple (not overlapping) sites at once. Our bit-coded algorithm proved to be $\sim$ 40 times faster than the conventional FORTRAN 90 code we started with. It is important to note that this stochastic cellular automaton like representation of the surface growth opens the possibility for an implementation on extremely fast graphic cards with massively parallel processors [65, 66]. Furthermore, the construction permits the extension of the mapping for more complex surface reactions [67] (see later sections).

We performed dynamical simulations by starting from stripe-ordered particle distributions. This corresponds to a flat surface with a small intrinsic width. The considered lattices gases had the maximum linear sizes $L_{max} = 2^{15}, 2^{10}, 2^{8}, 2^{6}$ for $d = 2, 3, 4, 5$ dimensions, respectively, and periodic boundary conditions were applied. A single step of the lattice gas algorithm comprises a random site selection and in case of an appropriate neighborhood configuration a $p = 1$ Kawasaki d-mer update (9.9). The time is incremented by $1/L^d$ in units of Monte Carlo steps. Throughout the chapter, we will use this unit of time.

We could exceed by magnitudes of order all previous numerical system sizes and simulation times. For example, the largest five-dimensional simulations were done for $L = 30$ and $t_m = 230$ MCs [19]. Our $L = 64$ simulations, where we have the good bulk–surface ratio 5.4, required 2 GB memory size and a couple of weeks for a single realization up to $t_{max} = 5000$ MCs. Similarly, the largest sized simulations in $d = 2$ for $L = 11520$ system could achieve $t_{max} = 10^4$ MCs [68]. Our largest $L = 32768$ sized simulations reached $t_{max} = 44600$ MCs. The longest runs for $L = 4096$ passed the saturation at $t \simeq 4 \times 10^5$ MCs and the samples were followed up to $t_{max} = 10^6$ MCs.

We run the these lattice gas simulations for 10–1000 independent realizations for each dimension and size considered and calculated $h_{x,y,...}(t)$ and the second moment

$$W(L,t) = \left[\frac{1}{L^d} \sum_{x,y,..}^{L} h^2_{x,y,...}(t) - \left(\frac{1}{L^d} \sum_{x,y,...}^{L} h_{x,y,...}(t) \right)^2 \right]^{1/2} \quad (9.18)$$

from the height differences at certain sampling times. The growth is expected to follow the Family–Vicsek scaling [69] asymptotically, but due to the corrections, it can be described by a power series

$$W(t, L \to \infty) = bt^{\beta}(1 + b_0 t^{\phi_0} + b_1 t^{\phi_1} \ldots), \quad (9.19)$$

with the surface growth exponent β. For finite system, when the correlation length exceeds L, the growth crosses over to a saturation with the scaling law

$$W(t \to \infty, L) = aL^{\alpha}(1 + a_0 L^{\omega_0} + a_1 L^{\omega_1} \ldots), \quad (9.20)$$

characterized by the roughness exponent α. In our case, the intrinsic width of the initial state, which is represented by a zigzag surface of width 1/2 (see Fig. 9.1), results in a constant correction term. Thus we have $b_0 = 1/2$, $\phi_0 = -\beta$ and $a_0 = 1/2$, $\omega_0 = -\alpha$. During our scaling analysis, we dropped this contribution by subtracting $W^2(0) = 1/4$ from the raw data and consider the next leading order correction as leading one. Furthermore, we disregarded the initial time region $t < t_0 \simeq 50$, when basically an uncorrelated random deposition occurs. The dynamical exponent z can be expressed by the ratio $z = \alpha/\beta$ and in case of the Galilean invariance of an isotropic KPZ equation, the $z = 2 - \alpha$ relation should also hold.

Besides the extensive simulations, we have performed careful correction to scaling analysis by calculating the local slopes of the exponents. The effective exponent of the surface growth can be estimated similarly as in case of other scaling laws [57], as the discretized, logarithmic derivative of Eq. (9.18)

$$\beta_{eff}(t) = \frac{\ln W(t, L \to \infty) - \ln W(t', L \to \infty)}{\ln(t) - \ln(t')}. \quad (9.21)$$

It was determined numerically for different discretizations: $t/t' = 2, 3$, and we tried to fit it with the leading-order correction ansatz, which can easily be deduced from (9.19) (see [70] or [57])

$$\beta_{eff}(t) = \beta + b_1 \phi_1 t^{\phi_1}, \quad (9.22)$$

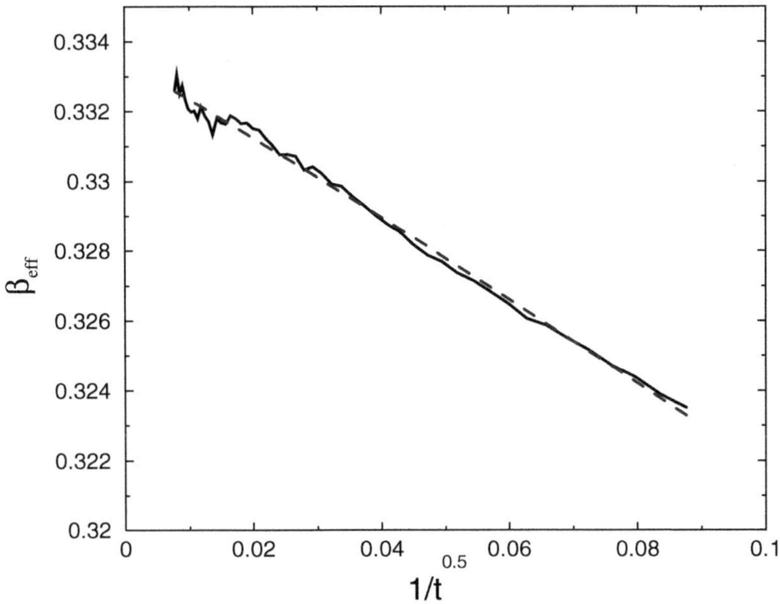

Figure 9.2 Effective exponents of the growth of the $d = 1$-dimensional model. The solid line corresponds to the simulation result. The dashed lines shows a fitting with the form (9.22). Reprinted from [63].

for $t > t_0$ and before the saturation region. In other cases, such as ballistic deposition, which has a large unknown intrinsic width, one can use another effective roughness exponent definition introduced in reference [70].

We tested our method with the *one-dimensional* exactly known case. Simulations were run on $L = 5 \times 10^5$ sized system up to $t_{max} = 16666$ MCs for 40 independent realizations. We determined the effective exponents $\beta_{eff}(t)$, which approaches $\beta = 1/3$ from below, in a perfect agreement with the leading-order correction form (see Fig. 9.2). The fitting with Eq. (9.22) on the local slopes data resulted in $\beta = 0.333(5)$ and $\phi_1 = -0.53$.

Similarly to the time dependence we can analyze the size dependence following the saturation by determining the effective exponent of the roughness, which can be defined as the logarithmic derivative of Eq. (9.18)

$$\alpha_{eff}(L) = \frac{\ln W(t \to \infty, L) - \ln W(t \to \infty, L')}{\ln(L) - \ln(L')} . \qquad (9.23)$$

The finite size scaling was done for systems of linear sizes in between L_{max} (discussed earlier) and L_{min}, which was 2^6 for 2d, 2^5 for 3d, 2^4 for 4d, and 2^3 for 5d, respectively. To handle the boundary conditions effectively, system sizes of power of 2 were simulated. To get the asymptotic values, we took into account all effective exponent points and applied a leading order, linear fitting. This method gives a better estimate for the asymptotic values than just a least-square fitting on the data points, which completely disregards corrections to scaling. We also calculated rough, but independent estimates for z by measuring the relaxation time, i.e., the time needed to reach 90% of the saturation value. The asymptotic value is extrapolated by a linear fitting: $z_{eff}(t) = z + c_1/L$.

In *two dimensions*, we estimated the growth exponent in the largest system sizes considered ($L = 2^{15}, 2^{14}, 2^{13}$) (see Fig. 9.3). Fitting in the $50 < t < 44600$ time window with the form Eq. (9.22) resulted in $b_1 = 0.83$ and $\beta = 0.245(5)$, which is somewhat bigger than what was obtained by the largest known ($L = 11520$) sized simulations: $\beta = 0.240(1)$ [68], and all other previous numerical estimates, including ours [41, 70, 71]. This value conciliates with the $\beta = 1/4$ renormalization group exponent reference [72]. One can obtain this value by the late time behavior of effective exponent, which has not been seen before, because finite size effects have screened it. On the graph one can see strong oscillations for $L = 2^{10}$ and intermediate times, which are damped before saturation. In the one-dimensional ASEP model, such oscillations are shown to be the consequence of density fluctuations being transported through a finite system by kinematic waves [73]. One can speculate that the slight final increase in β_{eff} for the largest system sizes is just a fluctuation or oscillation effect, but we could not eliminate this overall tendency by increasing the statistics. Although the statistical fluctuations grow dramatically, as $t \to \infty$ the increase in the mean value is observable for each size $L > 2^{11}$. Our error bar of β reflects this uncertainty. The width saturation values have been investigated for $L = 2^6, 2^7, \ldots, 2^{12}$. We took into account the leading order

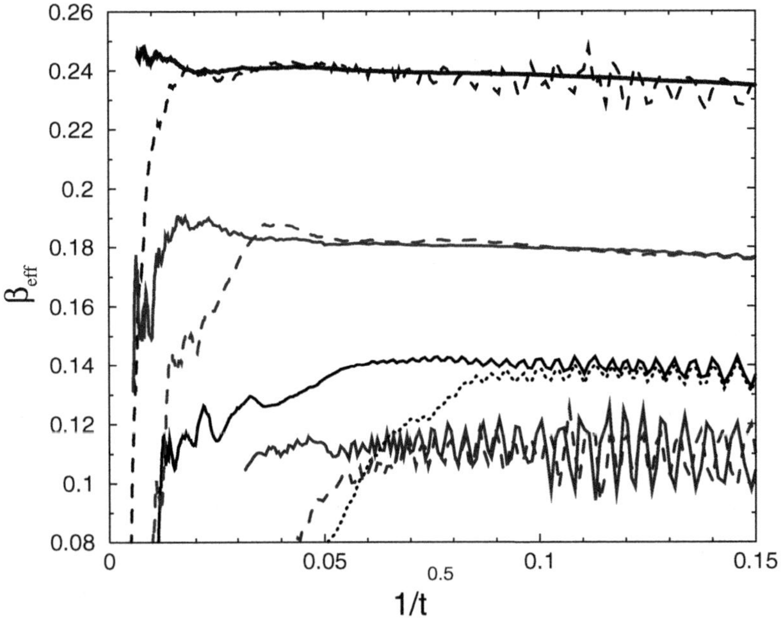

Figure 9.3 Effective exponents of the growth of the $d = 2, 3, 4, 5$-dimensional model (top to bottom). Solid lines correspond to the largest system $L = 2^{15}$, $L = 2^{10}$, $L = 2^{8}$, and $L = 2^{6}$, respectively. Dashed lines show our results of smaller sizes: $L = 2^{10}$, $L = 2^{8}$, $L = 2^{5}$, and $L = 2^{5}$, respectively, where saturation sets in earlier causing a cutoff in the scaling. Reprinted from [63]. See also Color Insert.

correction to-scaling by the following ansatz

$$\alpha_{eff}(L) = \alpha + a_1\omega_1 L^{\omega_1} \tag{9.24}$$

but due to the larger error bars, we restricted it to a linear approximation: $\omega_1 = -1$. The local slopes analysis provides $\alpha = 0.395(5)$ and $a_1 = 2.02$ for the roughness and $z_{eff} = 1.58(10)$, with the linear coefficient $c_1 = 1.83$ for the dynamical exponent. This roughness exponent is in agreement with renormalization group value [72], and somewhat bigger than the existing figures $\alpha = 0.393(3)$ [21] for $L \leq 1024$ and $\alpha = 0.385(5)$ [68] for $L \leq 128$.

Our scaling exponents results are summarized in Table 9.1. Detailed discussion of higher-dimensional results can be found in reference [63].

Table 9.1 Independent growth exponent estimates of the d-mer model in different dimensions

d	α	β	z
2	0.395(5)	0.245(5)	1.58(10)
3	0.29(1)	0.184(5)	1.60(10)
4	0.245(5)	0.15(1)	1.91(10)
5	0.22(1)	0.115(5)	1.95(15)

9.4 The Surface Diffusion Model in 2D

In two dimensions, one can map the octahedron model onto an anisotropic, but oriented migration of directed dimers. The surface built up from the octahedra can be represented by the edges meeting in the up/down middle vertexes (see Fig. 9.4). We can generate surface diffusion as the simultaneous adsorption/desorption of octahedra. Therefore, in our model after an appropriate removal site (a rooftop) selection has been done, we search for a valley bottom place in the neighborhood for deposition. The target site is chosen in the $\pm x$ or $\pm y$ direction, with the probabilities: p_{+x}, p_{-x} or p_{+y}, p_{-y}, respectively (see Fig. 9.4).[b] The maximal jump distance was fixed to be $l_m \leq 4$ lattice units following computer experiments. According to the construction, the nearest neighbor jump, corresponding to intra-layer diffusion, requires $l_m = 2$, while larger jump sizes allow intra-layer transport. To create MBE or Mullins–Herring behavior, we must allow inter-layer transport. The $l_m = 3$ or $l_m = 4$ does not make much difference, the latter one allows larger jumps on the expense of more CPU time. Having no *a priori* knowledge of the distance dependence of jump sizes and expecting insensitivity of universal scaling on the rates of short-ranged interactions, we used a constant probability for each direction (for more discussion see reference [67]).

To control this kind of surface diffusion, we impose additional constraints for accepting a move. We have tried two kinds of rules based on the local neighborhood configurations. The first one is very simple and requires that the height of a particle at the final state is

[b]Throughout our studies, we normalized the attempt probabilities.

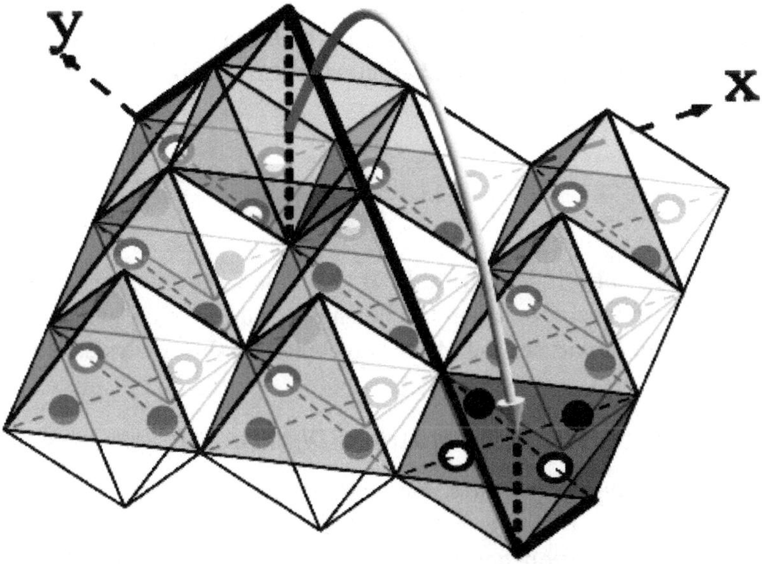

Figure 9.4 Mapping of the 2 + 1-dimensional surface diffusion onto a 2d particle model (bullets). Surface attachment (with probability p) and detachment (with probability q) correspond to Kawasaki exchanges of particles or to anisotropic migration of dimers in the bisectrix direction of the x and y axes. The crossing points of dashed lines show the base sublattice to be updated. Thick solid line on the surface shows the y cross section, reminding us to the one-dimensional rooftop model. When the shown desorption or absorption steps are executed simultaneously, they realize a surface diffusion step of size $s = 3$ along the y axis. This corresponds to a pair of dimer movement, a repulsion in case of this smoothing reaction. Reprinted from [67]. See also Color Insert.

higher than that of its initial site

$$h_{fin} - h_{ini} \geq 0 \tag{9.25}$$

which makes a surface more rough (inverse/roughening diffusion) (see Section 9.4.1). The second one is based on the local curvature at the update sites as will be discussed in Section 9.4.2.

The morphology of pattern formation in experiments is usually followed by the measurement of some characteristic size, e.g., the wavelength in periodic structures, which evolves nontrivially with time. We can easily define and measure such quantity in our model.

By following the up, or down, slopes in the x direction, we have strings of "1"-s or "-1" of length s_k in the k-th slice of the lattice gas in the y direction. We shall characterize patterns by calculating the (y) average of the longest s_k value

$$\lambda = 1/L \sum_{k=1}^{L} \max(s_k) \,. \tag{9.26}$$

This characteristic length that corresponds to the longest x slope or to the slowest x mode in the Fourier decomposition will provide information about scaling of the wavelength in our analysis.[c]

9.4.1 *Larger Height Octahedron Diffusion Model*

By the simple height difference condition, we expect to generate rough surfaces of large curvatures, since an octahedron jumping to a higher position has usually smaller number of neighbors in the lateral direction (in this model we do not allow particles to evaporate from the bottom of a V-shaped valley). In the language of lattice gases, regions of large curvature and maximal slope correspond to large regions of dense particles and holes. An attractive interaction of the dimer moves enhances such segregation. Contrary, we can realize a smoothing diffusion when we accept jumps with the condition: $h_{fin} - h_{ini} \leq 0$.

By increasing the local heights, the formation of pyramid-like structures (unstable growth) occurs similarly as in the $n = 2$ SOS model [35], but in our case the slopes are limited to 45°. However, by appropriate length rescaling, any sharp, continuous surface can be approximated. In the forthcoming, we will investigate both the inverse and the normal diffusion cases and call it the larger height octahedron diffusion model (LHOD) model.

The details of the LHOD updates and scaling results are discussed reference [67]. Due to the broken up-down and (9.8) symmetries, there are nonlinearities in LHOD, thus MBE class behavior is realized in general.

---

[c]We introduced another type of wavelength measure based on the power spectrum density calculations in [74].

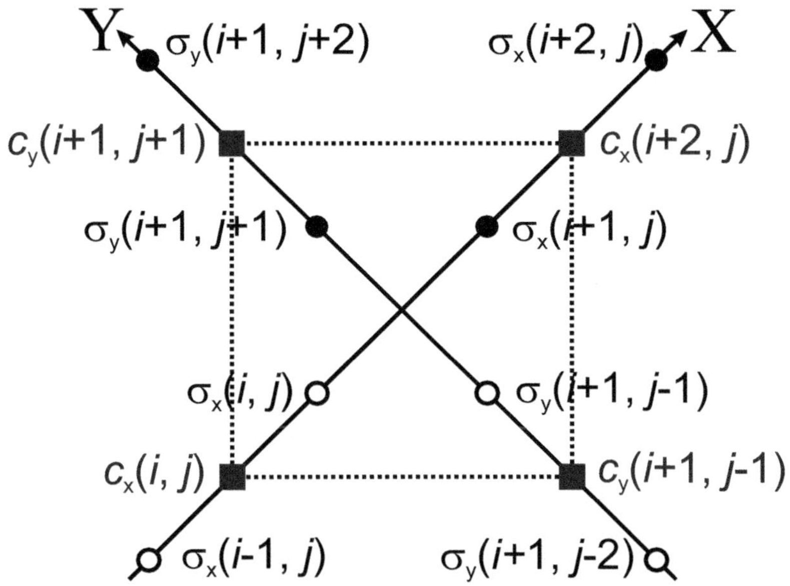

Figure 9.5 Local slope $\sigma_\chi(i, j)$ (circles) and curvature $c_\chi(i, j)$ (squares) variables at an update site. Filled circles correspond to upward, empty ones to downward slopes of the surface. This plaquette configuration models a valley bottom site, with the total curvature $H = 4$.

9.4.2 *Larger Curvature Octahedron Diffusion Model*

In this section, we describe a transition probability in addition to the octahedron surface hopping model, which satisfies the detailed balance condition (9.8), thus enables one to realize linear, equilibrium Mullins–Herring diffusion steps. The local curvature of the surface is calculated at the four edges of squares of the projected octahedra. As Fig. 9.5 shows, one can describe the curvatures $c_\chi(i, j)$ $(\chi \in (x, y))$ by the products (or differences) of the local slopes

$$c_\chi(i, j) = \sigma_\chi(i, j)\sigma_\chi(i + 1, j)\,. \qquad (9.27)$$

At each update, we calculate the sum of the change of local curvatures at the origin (i, j) and the target (i', j') sites

$$\Delta H = \Delta \sum_{\chi=x,y} \sum_{\langle i,j \rangle} c_\chi(i, j) + \Delta \sum_{\chi=x,y} \sum_{\langle i',j' \rangle} c_\chi(i', j')\,, \qquad (9.28)$$

where $\langle\rangle$ denotes the plaquette neighborhood sites as shown on Fig. 9.5. This gives maximal value: $H = 4$ for a local tops and bottoms and the minimal value: $H = -4$ for a locally flat (zigzag) configuration. Using this value, we accept the update with an Arrhenius type of probability

$$w_{i \to i'} = 1/2[1 - a \tanh(-\Delta H^2)] \tag{9.29}$$

where a is a constant. This form is very similar to what was used in case of the one-dimensional $n = 2$ model [35] and enhances (suppresses) roughening moves if $a > 0$ ($a < 0$). In reference [35], symmetry arguments were applied for the lowest order series expansion of $w_{i \to i'}$ of the model to prove a connection with the Mullins–Herring equation. Now similar derivation can be done by extending the model for dimer variables in two dimensions. In the forthcoming, we shall call this the larger curvature octahedron diffusion (LCOD) model.

We have simulated the LCOD with the parameters: $a = 0.1$, $l_m = 3$, $p = q = 0.05$. The scaling behavior has been found to agree very well with that of Mullins–Herring universality class values even in case of isotropic diffusion (see Fig. 9.6). The effective $\beta_{eff}(t)$ converges to $\beta = 0.25(1)$ before the saturation. The wavelength grows logarithmically in time (see insert of Fig. 9.6), and after a steady state value is reached, it scales logarithmically with the system size too.

For anisotropic diffusion ($p_{\pm x} = 1, p_{\pm y} = 0$), we have obtained similarly good Mullins–Herring class surface scaling, with logarithmic time and size dependence of λ again.

For $a = 0$, without any Edwards–Wilkinson noise, one can find logarithmic growth in the LCOD model in time

$$W(t, L \to \infty) \propto \ln(t) \tag{9.30}$$

and logarithmic surface roughness dependence of

$$W(t \to \infty, L) \propto \ln(L)\,. \tag{9.31}$$

Data collapse fitting for the dynamical exponent on the other hand shows $z = 4$ space-time anisotropy. This means that for the $a = 0$ noiseless case, we have realized the universality class behavior of Mullins–Herring with conserved (purely diffusive) noise, characterized by the exponents: $\alpha = \beta = 0, z = 4$ (see [75]).

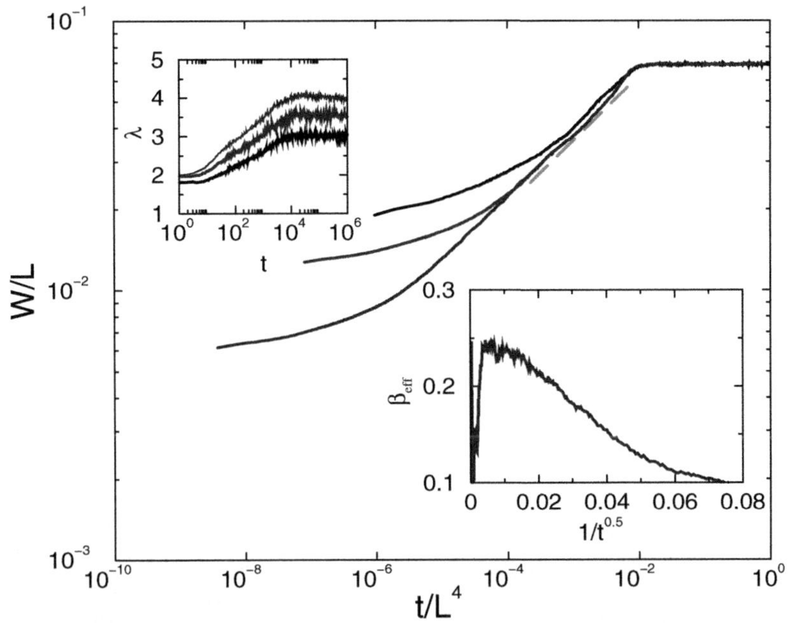

Figure 9.6 Scaling behavior of the isotropic LCOD model for $L = 32, 64, 128$ (top to bottom at the left side). The data collapse has been achieved with the Mullins–Herring class exponents. For the growth exponent fitting (dashed line) results in $\beta = 0.25(1)$. The insert on the right shows the same by local slopes. The insert on the left shows the evolution of λ. Reprinted from reference [67]. See also Color Insert.

9.5 Pattern Generation by Competing Inverse Mullins–Herring and KPZ Processes

As we have shown in the previous section, in the zero noise limit, our RSOS surface diffusion processes generate growth with MBE or Mullins–Herring scaling behavior. In this section, we investigate them in the presence of competing KPZ updates. In our simulations, we initiate hopping with probabilities: p_{+x}, p_{-x}, p_{+y}, p_{-y} alternately with the deposition (with probability p) and removal (with probability q) processes. We follow the surface roughness and pattern formation with the corresponding wavelength growth.

9.5.1 Spatially Anisotropic Surface Diffusion

From the point of pattern formation, the LHOD and LCOD models behave differently. We always start the simulations from a flat surface and watch if stable patterns can arise. In case of an anisotropic inverse LHOD model of diffusion probability $p_{\pm x} = 1$, $p_{\pm y} = 0$, a competing Edwards–Wilkinson process always generates ripple patterns as shown on Fig. 9.7), which is stable for all $p = q \leq 1$. This formation is metastable against KPZ (height anisotropy), but for very large times (in the steady state), the ripples become uneven, blurred, and cut into smaller pieces. The wavelength, defined as (9.26), grows only a little (in a power law manner) and saturates quickly.

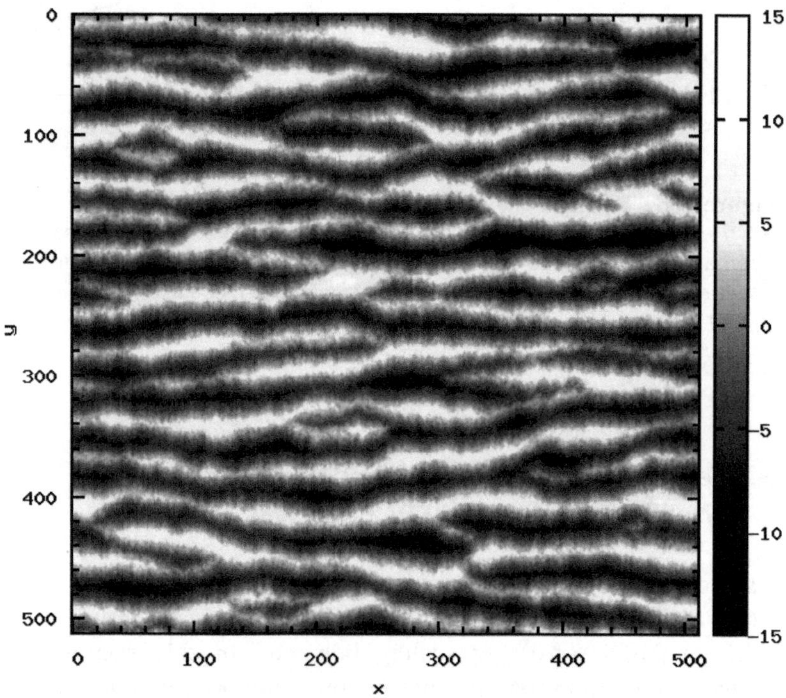

Figure 9.7 Snapshot of surface heights of the ripple patterns generated by the parameters $p_{\pm x} = 1$, $p_{\pm y} = 0$ (anisotropic, inverse Mullins–Herring) and $p = q = 1$ adsorption/desorption at $t = 10^4$ MCs, in the LHOD model of linear size $L = 512$. Reprinted from reference [67].

However, if we create such anisotropy in which a steady, direct current flows, for example when only $p_x = 1$ and all the others are zero, we can find a different behavior. In this case, for weak Edwards–Wilkinson or KPZ, the ripples are not completely straight and exhibit a coarsening as

$$\lambda \propto t^{0.24(1)} . \tag{9.32}$$

The ripples are straighter in the KPZ case than for up-down isotropic deposition/removal. Furthermore, a good data collapse can be obtained with the Mullins–Herring class exponents for sizes $L = 32, 64, 128$ as shown on Fig. 9.8). However, this scaling can be destroyed by increasing the strength of the nonconserved reaction

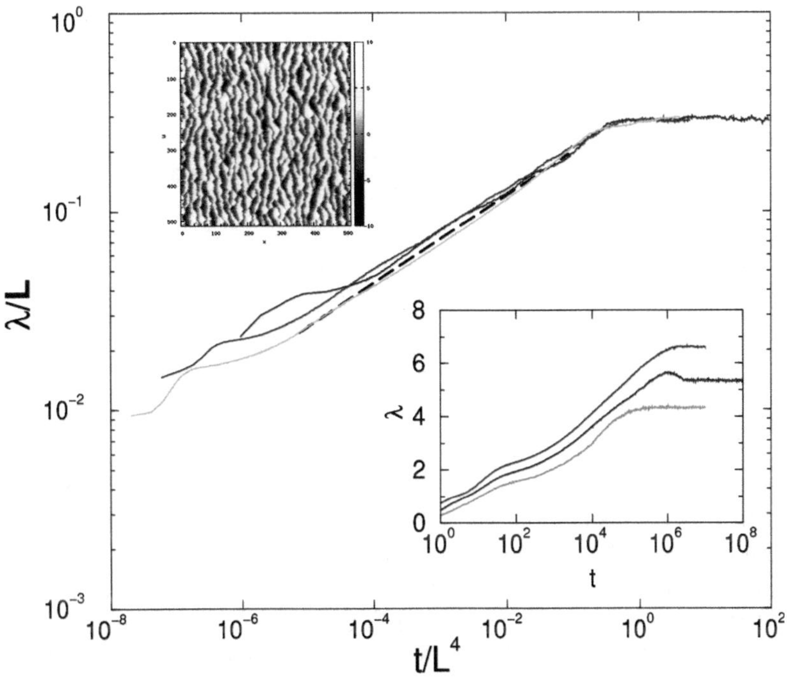

Figure 9.8 The wavelength growth in the LHOD model for anisotropic diffusion with steady direct current $p_y = 1$, $p = q = 0.005$ for sizes $L = 32, 64, 128$ (top to bottom at the beginning). Dashed line: power law fit with the exponent $\beta = 0.24(1)$. The left insert shows the corresponding pattern. The right insert corresponds to the isotropic diffusion case $p_{\pm x} = p_{\pm y} = 1$, where $\lambda(t)$ grows logarithmically. Reprinted from reference [67].

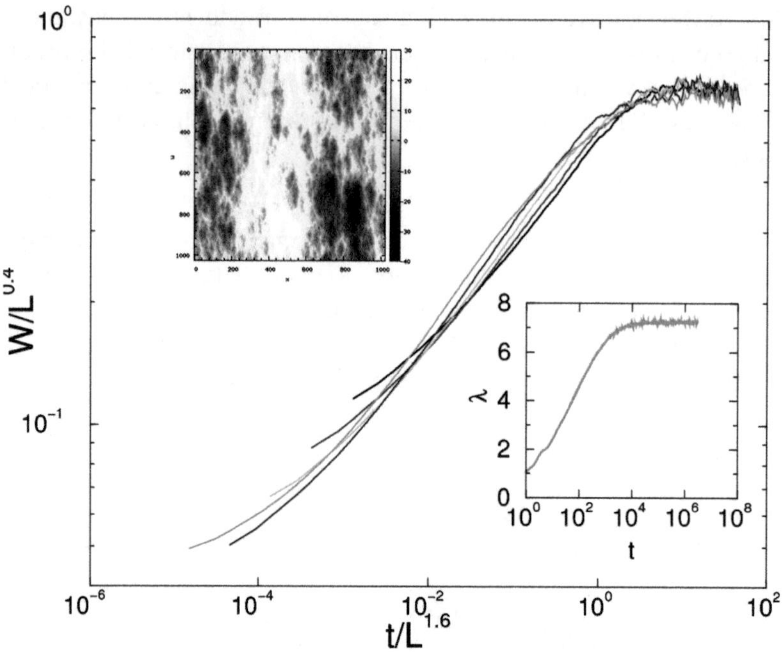

Figure 9.9 Data collapse for deposition ($p = 1, q = 0$) and anisotropic, inverse diffusion $p_{\pm x} = 1$ in the LHOD model with KPZ class exponents for $L = 64, 128, 256, 512, 1024$ (top to bottom curves at the right side). Right insert: $\lambda(r)$ for $L = 1024$, left insert the blurred ripple structure. Reprinted from reference [67]. See also Color Insert.

and the power law crosses over to logarithmic growth of λ. In case of strong KPZ deposition ($p = 1, q = 0$), the asymptotic scaling of LHOD becomes completely KPZ type, the wavelength saturates very quickly (see Fig. 9.9) and the ripple structure smears. In principle, we should expect anisotropic KPZ behavior here, but it is well known that in two dimensions, such spatial anisotropy is irrelevant; hence, the isotropic KPZ class behavior [76] is not surprising. Note that our anisotropic Kuramoto–Sivashinsky model is different from what is called and considered to be the "anisotropic Kuramoto–Sivashinsky" in the literature [77], because in our case, the surface diffusion (corresponding to the ∇^4 term) is anisotropic. Such models are very hard for analytic treatment, and spatial anisotropy is introduced in the ∇^2 terms usually.

The anisotropic LCOD is less effective for ripple formation than the LHOD. In this case, the patterns are smoother, and the λ scales logarithmically in time and by the size. The only exception is when we allow a steady direct current again. In this case, similar power laws as shown in Fig. 9.9 emerge for weak Edwards–Wilkinson or KPZ.

When direct current is not allowed ($p_{\pm x} = 1$, $p_{\pm y} = 0$), only spatial anisotropy in LCOD and we add a KPZ ($p = 1$, $q = 0$) update, we can see the emergence of KPZ scaling. In this case, the wavelength depends logarithmically both on time and the size L. It is important to realize that for short, one decade length, time windows the wavelength growth can also be well fitted with a power law

$$\lambda(t, L \to \infty) \propto t^{0.17(1)}\,, \tag{9.33}$$

which resembles to experimental results, but since the steady state values exhibit a clear logarithmic dependence on the sizes

$$\lambda(t \to \infty, L) \propto \ln(L) \tag{9.34}$$

we do not think that this "power law" fit would correspond to a real asymptotic behavior in the thermodynamic limit.

9.5.2 *Spatially Isotropic Surface Diffusion*

When the isotropic, inverse surface diffusion competes with the (smoothing) Edwards–Wilkinson process, one can observe dot formation both in the LHOD and LCOD models. Figure 9.10 shows a snapshot of the growing dots for LHOD in the presence of weak Edwards–Wilkinson. Here one sees rectangular-shaped patterns corresponding to the lattice symmetry. In case of LCOD, the contrast of the patterns is smoother, roughly circular. However, this pattern coarsening is much slower than in case of ripples.

We shall discuss the LHOD results first, which according to our previous numerical analysis, corresponds to the nonlinear equation (KPZ + MBE):

$$\partial_t h = \sigma \nabla^2 h + \lambda(\nabla h)^2 + \nu_4 \nabla^4 h + \lambda_{22} \nabla^2 [\nabla(h)^2 + \eta \tag{9.35}$$

In case of Edwards–Wilkinson type of deposition/removal, the nonlinear term vanishes ($\lambda = 0$) and in fact we model the CKPZ behavior. The characteristic size (λ) of the dots grows

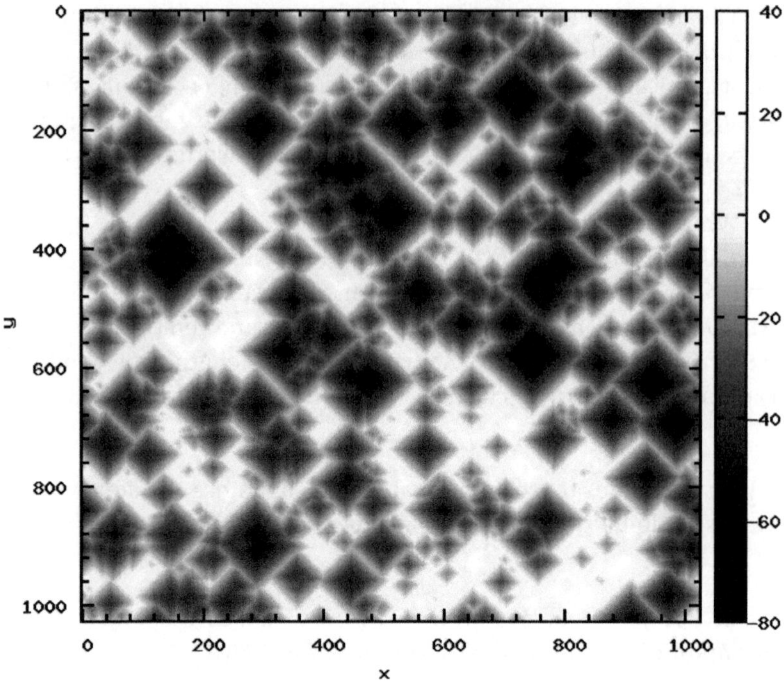

Figure 9.10 Snapshot of surface heights of the dot patterns generated by the LHOD model with parameters: $p_{\pm x} = p_{\pm y} = 1$ (isotropic, inverse Mullins–Herring) and $p = q = 0.1$ at $t = 10^4$ MCs. Reprinted from reference [67].

logarithmically (see insert of Fig. 9.8) in time and $\lambda_{max} \propto \ln(L)$. The pattern formation is more pronounced in the LHOD model than in the LCOD model. We associate it to the up/down anisotropy present in the LHOD.

The same kind of patterns can be also observed in case of strong KPZ anisotropy $p = 1, q = 0$ for short times, but later the dots are smeared. For the LHOD + KPZ case, a very slow crossover to KPZ scaling (see Fig. 9.11) occurs. Although the data collapse with KPZ exponents is rather poor for smaller sizes (it would be better with larger z and α exponents corresponding to the MBE class) for $L \geq 512$, it agrees with KPZ. The wavelength saturates very quickly (see insert of Fig. 9.11) in agreement with KPZ, where no coarsening is expected. Using our knowledge that the LCOD

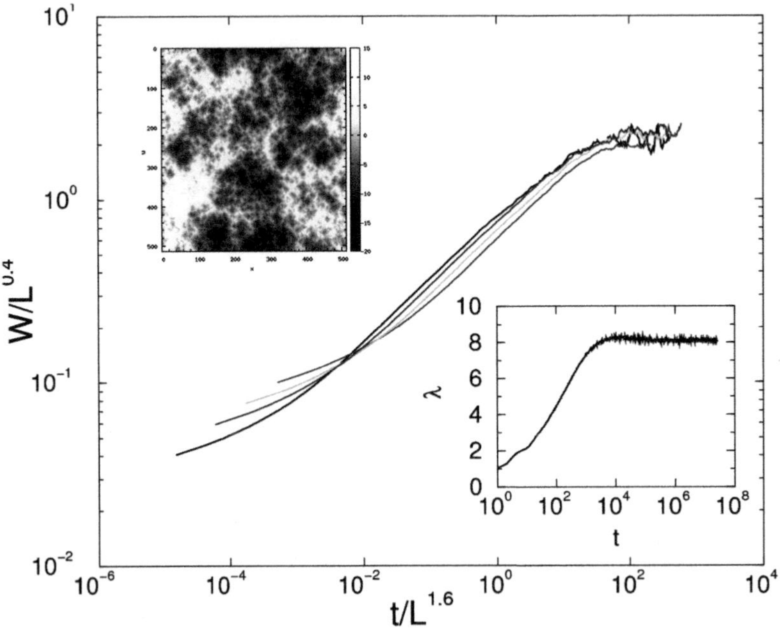

Figure 9.11 Data collapse of the $L = 128, \ldots, 1024$ LHOD model ($p_{\pm x} = p_{\pm y} = 1$) with a competing deposition ($p = 1$) process. One can see a slow crossover toward KPZ scaling. The right insert shows the growth of λ for $L = 512$. The left insert is a snapshot of the steady state, corresponding to the smeared KPZ height distribution. Reprinted from reference [67].

model exhibits Mullins–Herring scaling now, we can investigate the scaling behavior of the (inverse) Kuramoto–Sivashinsky equation (9.7), described by the combination of inverse Mullins–Herring and normal KPZ processes. We have run extensive simulations up to $t = 3 \times 10^6$ MCs (for $L = 128, 256, 512, 1024$) to obtain firm numerical evidence. As Fig. 9.12 shows, the finite size scaling collapse with the KPZ exponents is satisfied and the effective β extrapolates to $1/4$. This value agrees well with our high precision KPZ simulation result [63]. The wavelength grows logarithmically in time (9.33) (see insert of Fig. 9.12) and saturates well before the steady state. In the steady state, it grows slowly with the system size as (9.34). For short, one decade time window, one can fit the data with $\lambda(t, L \to \infty) \propto t^{0.12(2)}$, but due to the clear logarithmic behavior in the

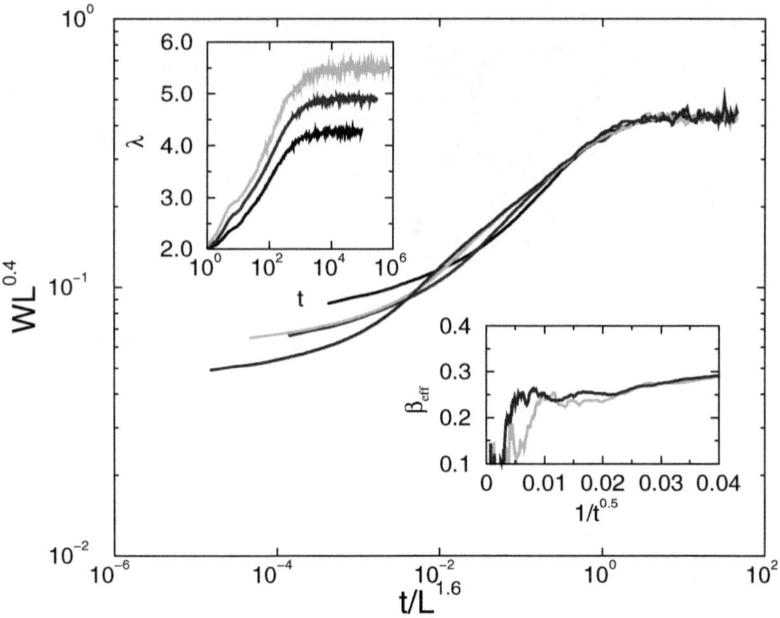

Figure 9.12 Data collapse of the $L = 128, \ldots, 1024$ (top to bottom at the right) LCOD model ($p_{\pm x} = p_{\pm y} = 1$) with competing deposition ($p = 1$). One can see clear KPZ scaling. The left insert shows the logarithmic growth of λ for $L = 128, 256, 512$ (bottom to top). Right insert: β_{eff} as the function of time. Reprinted from reference [67].

steady state (9.34), one should not take such power law fitting very seriously.

For the sake of completeness, we have performed similar analysis for the anisotropic LCOD model ($p_{\pm x} = 1$, $p_{\pm y} = 0$) with KPZ too, for sizes $L \leq 1024$. We have obtained agreement with the KPZ scaling for the width W and logarithmic growth for λ as in case of isotropy.

9.6 KPZ in the Presence of Normal Surface Diffusion

Here we study the scaling behavior of the KPZ process in the presence of normal (smoothing) LHOD, introduced in the previous section. First let us consider the weak isotropic diffusion case: $p_{\pm x} = p_{\pm y} = 0.1$. As Fig. 9.13 shows, the $W(t)$ curves for sizes $L = 64$,

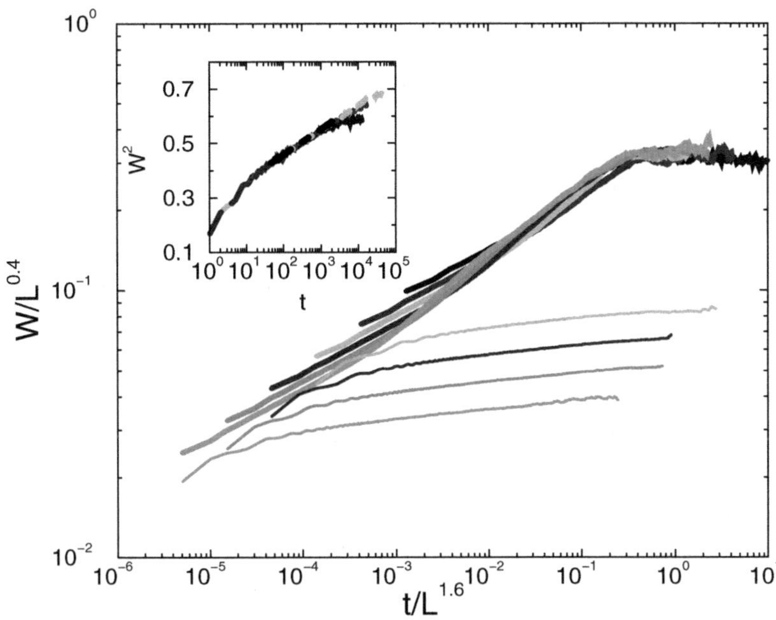

Figure 9.13 Data collapse of KPZ deposition ($p = 1$) and weak, isotropic normal LHOD (higher curves) for $L = 64, 128, \ldots, 2048$ (top to bottom). In case of strong diffusion (lower curves), the KPZ scaling disappears and as the insert shows, logarithmic growth can be observed. Reprinted from reference [67].

$128, \ldots, 2048$ exhibit a good collapse when we rescale them with the 2+1-dimensional KPZ exponents ($\alpha = 0.4, z = 1.6$). This means that the KPZ scaling is stable against of the introduction smoothing, MBE type of surface diffusion.

On the other hand, when we add strong LHOD diffusion $p_{\pm x} = p_{\pm y} = 0.9$ to the KPZ process ($p = 1$), the surface growth slows down and we do not find KPZ scaling any more as can be seen on the lower part of Fig. 9.13. Instead logarithmic surface growth emerges as shown in the insert of Fig. 9.13. A fitting with the form $W^2(t) = A + B\ln(t)$ for the asymptotic regime gives $A = 0.40(1)$ and $B = 0.26(1)$. The amplitude of this growth is different from the exactly known universal value of the Edwards–Wilkinson class in two dimensions: $A_{EW} = 0.151981$ [78].

The wavelength saturates very quickly to the maximal value, which for weak diffusion scales logarithmically with the system size (9.34), while in case of the strong diffusion the characteristic length remains on the order of lattice unit: $\lambda_{max} \simeq 1$, with a very weak size dependence, corresponding to uncorrelated surface heights.

It is well known that in the strong diffusion limit, the relevant fluctuations below d_c can be washed away, hence the logarithmic surface growth we observe should correspond to the mean-field (MF) behavior of the strong-coupled KPZ. This provides us a unique way to study the crossover behavior between the KPZ and KPZ-MF behavior.

9.6.1 *Probability Distribution Results*

We have tested our surface scaling results for KPZ in the presence of diffusion. In case of competing KPZ and LHOD/LCOD processes (i.e., for $p = 1$, $q = 0$, $p_{\pm x} = p_{\pm y} = 1$), we have determined the

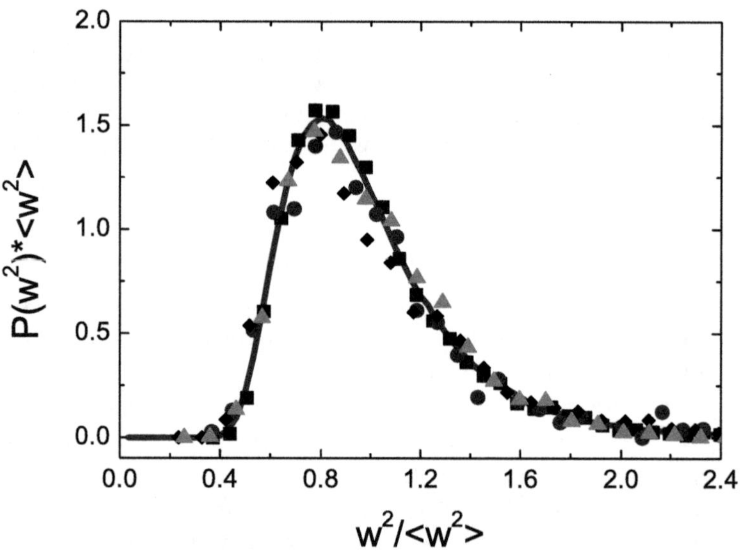

Figure 9.14 Comparison of $P(W^2)$ of the KPZ + LHOD (boxes); KPZ+ inverse LHOD (dots); KPZ+inverse, anisotropic LHOD (pink rhombuses); KPZ+inverse LCOD (triangles) with that of the KPZ from reference [22] (solid line). Reprinted from reference [67].

$P(W^2)$ distributions well in the saturation regime of systems of size $L = 1024$. The steady state could be reached for: $t > \sim 10^5$ MCs in case of KPZ + LHOD and for $t > 5 \times 10^7$ MCs in case of the KPZ + LCOD model. We have generated 100 independent samples and cut out the steady state $W^2(t)$ data. We have determined the scaled steady state probability distributions as shown in Fig. 9.14 for different combinations. The comparison with the histogram of the 2+1-dimensional KPZ from reference [22] shows a good agreement in general, providing further numerical evidence that the Kuramoto–Sivashinsky model asymptotically exhibits the KPZ universality class scaling. Our results complement the one-dimensional simulations results by reference [55], where the equivalence of Kuramoto–Sivashinsky and KPZ scaling was confirmed numerically.

9.7 Conclusions and Outlook

In conclusion we have shown that the mapping of a KPZ surface growth model onto driven lattice gases (DLG) can be extended to higher dimensions. Although the growth of the surfaces exhibits the spatial symmetry of the underlying lattice, one can map it onto an anisotropic DLG of more complex objects. The coarse grained, continuum description of these d-mers is an anisotropic Burgers equation. Still the DLG model is nontrivial because it is just an oriented drift of d-mers with hard-core exclusions. The topological constraint is the consequence of the required surface continuity by the mapping. In two dimensions, we confirmed [67] that the probability distribution $P(W^2)$ matches the universal scaling function determined for another KPZ model [22]. We presented effective bit-coding simulations and high precision results for the exponents α, β, and z independently (see Table 9.1). The sensitive local slope analysis provides numerical agreement with former simulation results, but for larger sizes, which have not been investigated so far, we see a slight growing tendency in the β_{eff} exponents in all dimensions. For $d = 2$, our results marginally overlap with the $\beta = 1/4$ value suggested some time ago by renormalization group. The change toward a trivial behavior in higher dimensions in the DLG language would mean the disappearance of the topological constraints among the extended d-mer objects as they could follow a simple ASEP dynamics of point particles. This is the target of further studies using massively parallel algorithms on graphic cards [65, 66]. We hope that we will be able to obtain a firm estimate for the upper critical dimension using extrapolation techniques.

Using this method, we have revisited some unresolved questions of basic surface growth phenomena. Contrary to earlier studies, we could perform numerical analysis of models with surface diffusion in $2+1$ dimensions, due to the efficient mapping of RSOS models onto binary lattice gases. We have shown that in the zero external noise limit, RSOS models with short-range interactions can be constructed, which exhibit MBE or Mullins–Herring type of surface growth. For inverse (roughening) diffusion, which increases the local curvature, unstable growth resulting in pyramid-like structures emerges. The size of these structures is limited only by L, which is not directly comparable with real materials. We created these MBE or Mullins–Herring type of atomistic models in order to study them in a competition with nonconserved KPZ processes. The simulations provided numerical evidence that strong, smoothing surface diffusion can slow down the KPZ to a logarithmic growth; thus we are able to reach the mean-field behavior of the strongly coupled KPZ fixed point in two dimensions, which is expected to show up in high dimensions only.

The mapping of the surface models onto lattice gases implies that the (anisotropic), oriented diffusion of dimers (KPZ) is stable against the introduction of an attracting force among them, but a strong repulsion can destroy the fluctuations, resulting in a mean-field behavior. We provided strong numerical evidence using surface scaling and probability distribution studies that the Kuramoto–Sivashinsky model exhibits KPZ scaling in $2+1$ dimensions as conjectured by field theory. We summarized the models we considered in case of spatially isotropic surface diffusion and nonconserved noise in Table 9.2.

Further studies of LCOD, with different boundary conditions can set the target of the research of the surface inclination by mapping the surface tilt onto the total particle concentration of the lattice gas. In particular, the angle dependence of the phase transitions

Table 9.2 Overview of the surface models with spatial isotropy and nonconserved noise

Model acronym	σ	λ_2	λ_{22}	ν_4	Properties
EW	+	0	0	0	smooth surface, (logarithmic) growth
KPZ	+	+/−	0	0	rough surface, (power law) growth
MH	0	0	0	+	smoothing diffusion, $z = 4$
iMH	0	0	0	−	roughening diffusion, unstable, $z = 4$
MBE	0	0	+/−	+	normal diffusion, $z = 10/3$
iMBE	0	0	+/−	−	roughening diffusion, $z = 10/3$, pyramids
KS	+	+/−	0	−	KPZ scaling, dots, (logarithmic) growth
iKS	+	+/−	0	+	KPZ scaling, KPZ-MF for strong diffusion

Source: Columns 2–5 show the sign of couplings of the differential equations considered.

among different growth phases can be understood by considering the underlying driven gas. Using our method, one can transform results of disorder or of anomalous diffusion between the surface and lattice gas models.

We introduced a characteristic length scale λ to follow the dynamics of patterns, which occur, if normal (smoothing) KPZ competes with inverse (roughening) diffusion.[d] We investigated this race for Mullins–Herring and MBE process, with and without spatial anisotropies. In case of uni-axial surface diffusion ripple, while for x/y lattice isotropy, dot-like pattern formation could be achieved. The wavelength growth is slow and saturates much before the steady state. Usually, we found logarithmic time and system size dependence of λ, except when steady direct current flows through the system. In this case, the interfaces are more rough, the ripples are bended, and power law scaling is observable. In this case, the scaling of λ agrees with the scaling of the width, i.e., characterized by the Mullins–Herring class exponents. This finding agrees with three-dimensional kinetic Monte Carlo simulations of epitaxial growth and erosion on (110) crystal surfaces [79] and with analytic arguments [80]. Furthermore, in case of ion beams with grazing incidence, the dislocation dynamics results in such athermal, kinetic coarsening of the patterns [81].

[d]Another type of wavelength measure based on the power spectrum density calculations introduced by us in reference [74].

The wavelength behavior can be understood with the help of considering the underlying lattice gas model, since the ripple or dot structures correspond to a phase separation. The coalescing surface pattern dynamics can be mapped onto the generalization of the reaction-diffusion process [75] of extended objects. It was shown that in ASEP type of models, where strong phase separation is present, the domain growth follows slow, logarithmic behavior [82, 83] in case of smooth surfaces. On the other hand, for rough surfaces, power law coarsening of λ has been derived using simple scaling arguments [84].

Finally, we mention that our models enable effective, bit-coded, stochastic cellular automaton type of simulation of surfaces, hence they could be run extremely fast on advanced, graphic cards [65, 66, 85].

Acknowledgments

We thank Z. Rácz and J. Krug for the useful comments. Support from the Hungarian research fund OTKA (Grant No. T77629), OSIRIS FP7, and the bilateral German–Hungarian exchange program DAAD-MÖB (Grant Nos. 50450744, P-MÖB/854) is acknowledged. The authors thank for the access to the HUNGRID, Cluster-grid, and NVIDIA for supporting the project with high-performance graphics cards.

References

1. S. Facsko, T. Dekorsy, C. Koerdt, C. Trappe, H. Kurz, A. Vogt, and H. L. Hartnagel (1999) *Science*, **285**, 1551.
2. M. Makeev, R. Cuerno, and A.-L. Barabási (2002) *Nucl. Instrum. Methods B*, **197**, 185–227.
3. T. Halpin-Healy and Y.-C. Zhang (1995) *Phys. Rep.*, **254**, 215.
4. A. L. Barabási and H. E. Stanley (1995) *Fractal Concepts in Surface Growth*, Cambridge University Press, Cambridge.
5. M. Kardar, G. Parisi, and Y. Zhang (1986) *Phys. Rev. Lett.*, **56**, 889.
6. J. M. Burgers (1974) *The Nonlinear Diffusion Equation*, Riedel, Boston.
7. D. Forster, D. R. Nelson, and M. J. Stephen (1977) *Phys. Rev. A*, **16**, 732.

8. J. Krug (1997) *Adv. Phys.*, **46**, 139.
9. M. Kardar (1987) *Nucl. Phys. B*, **290**, 582.
10. H. van Beijeren, R. Kutner, and H. Spohn (1985) *Phys. Rev. Lett.*, **54**, 2026.
11. H. K. Janssen and B. Schmittmann (1986) *Z. Phys. B*, **63**, 517.
12. T. Hwa (1992) *Phys. Rev. Lett.*, **69**, 1552.
13. S. F. Edwards and D. R. Wilkinson (1982) *Proc. R. Soc.*, **381**, 17.
14. M. A. Moore, T. Blum, J. P. Doherty, M. Marsili, J.-P. Bouchaud, and P. Claudin (1995) *Phys. Rev. Lett.*, **74**, 4257.
15. M. Lässig and H. Kinzelbach (1997) *Phys. Rev. Lett*, **78**, 903.
16. J. K. Bhattacharjee (1998) *J. Phys. A*, **31**, L93.
17. H. C. Fogedby (2005) *Phys. Rev. Lett.*, **94**, 195702.
18. L. Canet, H. C. B. Delamotte, and N. Wschebor (2010) *Phys. Rev. Lett.* **104**, 150601.
19. T. Ala-Nissila, T. Hjelt, J. M. Kosterlitz, and O. Venäläinen (1993) *J. Stat. Phys.*, **72**, 207.
20. T. Ala-Nissila (1998) *Phys. Rev. Lett.*, **80**, 887.
21. E. Marinari, A. Pagnani, and G. Parisi (2000) *J. Phys. A*, **33**, 8181.
22. E. Marinari, A. Pagnani, G. Parisi, and Z. Rácz (2002) *Phys. Rev. E*, **65**, 026136.
23. C. Herring (1950) *J. Appl. Phys.*, **21**, 301.
24. W. W. Mullins (1957) *J. Appl. Phys.* **28**, 333.
25. M. Constantin, *et al.* (2004) *Phys. Rev. E*, **69**, 061608.
26. S. D. Sarma, P. P. Chatraphorn, and Z. Toroczkai (2002) *Phys. Rev. E*, **65**, 036144.
27. A. Bru, J. M. Pastor, I. Fernaud, I. Bru, and S. Melle (1998) *Phys. Rev. Lett.*, **81**, 4008.
28. S. Morel, *et al.* (1998) *Phys. Rev. E*, **58**, 6999.
29. S. D. Sarma, *et al.* (1994) *Phys. Rev. E*, **49**, 122.
30. M. Castro, *et al.* (1998) *Phys. Rev. E*, **57**, R2491.
31. J. M. López and M. A. Rodríguez (1996) *Phys. Rev. E*, **54**, R2189.
32. J. Krug (1991) *Phys. Rev. Lett*, **67**, 1882.
33. J. M. Kim and S. D. Sarma (1994) *Phys. Rev. Lett.*, **72**, 2903.
34. J. Krug (1994) *Phys. Rev. Lett.*, **72**, 2907.
35. M. Siegert and M. Plischke (1994) *Phys. Rev. E*, **50**, 917.
36. F. Family (1986) *J. Phys. A*, **19**, L411.
37. D. Wolf and J. Villain (1990) *Europhys. Lett.*, **13**, 389.

38. S. D. Sarma and P. Tamorenea (1991) *Phys. Rev. Lett.*, **66**, 325.
39. Y. Kim, D. K. Park, and J. M. Kim (1994) *J. Phys. A*, **27**, L533.
40. C. Dasgupta, S. D. Sarma, and J. M. Kim (1996) *Phys. Rev. E*, **54**, R4552.
41. G. Ódor, B. Liedke, and K.-H. Heinig (2009) *Phys. Rev. E*, **79**, 021125.
42. Y. Kuramoto and T. Tsuzuki (1977) *Prog. Theor. Phys.*, **55**, 356.
43. G. I. Sivashinsky (1979) *Acta Astronaut.*, **6**, 569.
44. R. Cuerno and K. B. Lauritsen (1995) *Phys. Rev. E*, **52**, 4853.
45. I. Procaccia, *et al.* (1992) *Phys. Rev. A*, **46**, 3220.
46. C. Jayaprakash, F. Hayot, and R. Pandit (1993) *Phys. Rev. Lett*, **71**, 12.
47. J. T. Drotar, Y.-P. Zho, T.-M. Lu, and G.-C. Wang (1999) *Phys. Rev. E*, **59**, 177.
48. V. S. L'vov and I. Procaccia (1992) *Phys. Rev. Lett.*, **69**, 3543.
49. V. S. L'vov and V. V. Lebedev (1993) *Europhys. Lett.*, **22**, 419.
50. V. S. L'vov and I. Procaccia (1994) *Phys. Rev. Lett.*, **72**, 307.
51. C. Jayaprakash, F. Hayot, and R. Pandit (1994) *Phys. Rev. Lett*, **72**, 308.
52. Z. Rácz, M. Siegert, D. Liu, and M. Plischke (1990) *Phys. Rev. A*, **43**, 5275.
53. J. Krug (1993) *Phys. Rev. Lett*, **70**, 3271.
54. M. Rost and J. Krug (1995) *Physica D*, **88**, 1–13.
55. J. Krug and F. Hontinfinde (1997) *J. Phys. A*, **30**, 7739–7752.
56. H. Hinrichsen and G. Ódor (1999) *Phys. Rev. Lett.*, **82**, 1205.
57. G. Ódor (2004) *Rev. Mod. Phys*, **76**, 663.
58. M. Plischke, Z. Rácz, and D. Liu (1987) *Phys. Rev. B*, **35**, 3485.
59. P. Meakin, P. Ramanlal, L. Sander, and R. Ball (1986) *Phys. Rev. A*, **34**, 5091.
60. T. Ligget (1985) *Interacting Particle Systems*, Springer-Verlag, Berlin.
61. S. Katz, J. L. Lebowitz, and H. Spohn (1984) *J. Stat. Phys.*, **34**, 497.
62. B. Schmittman and R. K. P. Zia (1996) Statistical mechanics of driven diffusive systems, in *Phase Transitions and Critical Phenomena* (C. Domb and J. L. Lebowitz, eds), vol. 17, Academic Press, London.
63. G. Ódor, B. Liedke, and K.-H. Heinig (2010) *Phys. Rev. E*, **81**, 031112.
64. M. Matsumoto, *et al.* (2006) *J. Univ. Comp. Sci.*, **12**, 672–690.
65. H. Schulz, G. Ódor, J. Kelling, K.-H. Heinig, B. Liedke, and N. Schmeisser (2010) Computing the KPZ equation using gpu acceleration, 3rd international workshop on innovation in information technologies, theory and practice.

66. Henrik Schulz, Géza Ódor, Gergely Ódor, Máte Ferenc Nagy (2011) Simulation of 1+1 dimensional surface growth and lattices gases using GPUs, *Comp. Phys. Comm.* **182**, 1467.
67. G. Ódor, B. Liedke, and K.-H. Heinig (2010) *Phys. Rev. E*, **81**, 051114.
68. L. H. Tang, B. M. Forrest, and D. E. Wolf (1992) *Phys. Rev. A*, **45**, 7162.
69. F. Family and T. Vicsek (1985) *J. Phys. A*, **18**, L75.
70. F. D. A. A. Reis (2004) *Phys. Rev. E*, **69**, 021610.
71. S.-V. Ghaisas (2006) *Phys. Rev. E*, **73**, 022601.
72. M. Lässig (1998) *Phys. Rev. Lett*, **80**, 2366.
73. S. Gupta, S. N. Majumdar, C. Godrèche, and M. Barma (2007) *Phys. Rev. E*, **76**, 021112.
74. G. Ódor, B. Liedke, K.-H. Heinig, and J. Kelling (2012) *Applied Surface Science*, **258**, 4186.
75. G. Ódor (2008) *Universality in Nonequilibrium Lattice Systems*, World Scientific.
76. U. C. Täuber and E. Frey (2002) *Eur. Phys. Lett.*, **59**, 655.
77. S. Facsko, T. Bobek, A. Stahl, H. Kurz, and T. Dekors (2004) *Phys. Rev. B*, **69**, 153412.
78. T. Nattermann and L.-H. Tang (1992) *Phys. Rev. A*, **45**, 7156.
79. L. Golubovic, A. Levandovsky, and D. Moldovan (2002) *Phys. Rev. Lett.*, **89**, 266104.
80. L. Golubovic and A. Levandovsky (2008) *Phys. Rev. E*, **77**, 051606.
81. H. Hansen, *et al.* (2009) *Phys. Rev. Lett*, **102**, 146103.
82. M. R. Evans, Y. Kafri, H. M. Koduvely, and D. Mukamel (1998) *Phys. Rev. E*, **58**, 2764.
83. Y. Kafri, D. Biron, M. Evans, and D. Mukamel (2000) *Eur. Phys. J.*, **16**, 669.
84. M. R. Evans, Y. Kafri, E. Levine, and D. Mukamel (2002) *Phys. Rev. E*, **62**, 7619.
85. J. Kelling and G. Ódor (2011) *Phys. Rev. E*, **84**, 061150.

Chapter 10

Applications of Ion Beam Produced Patterned Substrates in Plasmonics

Mukesh Ranjan,[a,*] Thomas W. H. Oates,[b] and Stefan Facsko[c]

[a] *FCIPT, Institute for Plasma Research, Sector-25, Gandhinagar, Gujarat, India*
[b] *Leibniz-Institut für Analytische Wissenschaften — ISAS-Berlin, Albert-Einstein-Str. 9, 12489 Berlin, Germany*
[c] *Institute of Ion Beam Physics and Materials Research, Forschungszentrum Dresden-Rossendorf e.V., P.O. Box 51 01 19, 01314 Dresden, Germany*
*ranjanm@ipr.res.in

Plasmonic properties of self-assembled silver nanoparticles/nanowires arrays on periodically patterned Si(100) substrate are reported. The advantage of this bottom-up approach over other self-assembling and lithographic methods is the flexibility to tune array periodicity down to 20 nm with interparticle gaps as low as 5 nm along the ripple. Ripple patterns have shallow modulation (~2 nm) still particles self-assembly was observed in nonshadow deposition. Therefore, adatoms diffusion and kinetics is important on ripple surface for the self-assembly process. The physical vapor deposition (PVD) e-beam evaporation method used for deposition has proven to be superior to sputter deposition due to lower incident flux and lower atom energy. Self-assembled nanoparticles are optically anisotropic, i.e., they exhibit a direction-dependent shift

Nanofabrication by Ion-Beam Sputtering: Fundamentals and Applications
Edited by Tapobrata Som and Dinakar Kanjilal

ISBN 978-981-4303-75-0 (Hardcover), 978-981-4303-76-7 (eBook)
www.panstanford.com

in localized surface plasmon resonances (LSPR). The reason of the observed anisotropy is a direction-dependent plasmonic coupling. In this way, plasmon resonance tuning using various nanoparticles arrays is shown. Different in plane and out of the plane dielectric coefficients calculated by modeling Jones matrix elements confirms that nanoparticle/nanowire array are biaxial anisotropic ($\varepsilon_x \neq \varepsilon_y \neq \varepsilon_z$). The nanoparticles are predominantly insulating while nanowires are both metallic and insulating depending on the dimension. Application of such nanowire arrays as transparent conductive electrode (TCE) and nanoparticle arrays as metamaterial is shown.

10.1 Introduction

The ongoing miniaturization of electronic circuits will soon reach physical limits [1]. A major bottleneck of modern processors, which consist of multi-cores or several graphical processing units, is the data flow between the cores and between the processor and the memory [1, 2]. Optical interconnects have been proposed to provide a solution to this problem. The size of such optical fibers or waveguides is of the order of the wavelength of visible light and thus much bigger than typical electric interconnects. They can provide higher speed of data transfer by optical fibers; however, they add an additional limit to miniaturization.

A possible solution to this downside effect is offered by plasmonic waveguides. Plasmonics is a rapidly evolving branch of photonics based on phenomena of light interacting with metallic nanostructures. Thus plasmonic circuits offer the potential to carry both optical signals and electric currents, thereby, combining optical and electronic communication on one chip [1–4]. The resonant interaction between charge oscillations near the surface of a metal or metallic nanostructure and the electromagnetic field of the light creates surface plasmons (SP) or localized surface plasmons (LSP), respectively. This unique property of SP and LSP provides the possibility of localization and guiding of light in subwavelength structures, and thus can be used to construct miniaturized optoelectronic circuits with subwavelength components.

Localized surface plasmons in metallic nanoparticles have given birth to several other important applications. Due to the enhanced electromagnetic (EM) near field in the vicinity of metal nanoparticles, the Raman scattering cross section of a nearby molecule is drastically enhanced [5–10]. This phenomenon opened a new direction of study known as surface-enhanced Raman scattering/spectroscopy (SERS) [5, 6]. In between two nanoparticles, coupled SP mode can induce even higher field enhancement [11]. In chains of nanoparticles, the field enhancement can be a few orders of magnitude higher than for single nanoparticles [7–15]. Field coupling and field enhancement are strongly dependent on interparticle gap and are prominent for gaps below 50 nm [16, 17]. Due to this large field enhancement, SERS has achieved single-molecule sensitivity [6, 7]. Already plasmonic sensors to detect biological tissues using SERS are available. Other applications where plasmonics is extensively studied are plasmonic switching [18, 19], biological sensors [20], near-field applications such as SNOM [21], metamaterials [22], superlenses [23], photovoltaics [24, 25], display and recording media [26, 27], etc.

Form a technological point of view, the application of metal nanoparticles with plasmonic properties demands full control of the growth process. Methods to grow ordered particles can be divided into two approaches: top-down, i.e., lithographic methods, and bottom-up, i.e., self-assemble methods. Currently, the production of nanoparticles arrays of 20 nm period with interparticle gap down to 5 nm using lithographic methods is problematic [28]. However, field enhancement and coupling effects start to be prominent in this regime. Another drawback of some of the lithographic methods is their serial and thus time-consuming character. Self-assembling and self-organization methods have been successfully demonstrated using faceted alumina [29], anodized alumina [30, 31], and block copolymers [32, 33] as substrates. Several chemical methods have also been presented [34, 35] with limited success. There are different drawbacks in these approaches, such as limited control over the interparticle gap and array periodicity. In case of anodized alumina substrates, only the top surface of the metal particles or wire are exposed; therefore, interparticle coupling is reduced. Synthesis using colloidal chemistry offers no control over the ordering. All these methods are, therefore, lacking tunability.

A different approach is to use periodic ripple patterns induced by ion irradiation as templates followed by oblique incidence physical vapor deposition (PVD) to grow highly ordered nanoparticles and nanowires [36–40]. Ripple formation of different periodicity and on different substrate materials has been extensively studied [41]. Therefore, a knowledge base to produce them with full control is already available. The scaling laws to tune the ripple periodicity and amplitude have been investigated in detail for Si [42–46], a main material used in electronics devices. Due to the scalability and the universality of the ripple pattern formation induced by ion irradiation, this approach is very promising to develop aligned nanoparticles and nanowires for plasmonic applications. Interesting plasmonic properties of metal films deposited on such rippled templates have already been reported [36, 37, 39, 40, 47, 48].

In this chapter, the method to produce highly ordered nanoparticles and nanowire arrays using pre-patterned ripple templates by ion beam irradiation will be presented first. The important experimental parameters to tune the periodicity, size, and aspect ratio of the particles will be presented. In the following sections, the plasmonic properties of nanoparticles and nanowire array will be discussed. We shall show that ordered nanowires grown on ripple patterned substrates have the potential to be used as transparent conductive electrode (TCE) material for solar cells application [49, 51]. This is due to their bulk metal like dielectric properties along the wires and their high optical transmission for light polarized perpendicular to the wires. In addition, arrays of highly ordered nanoparticles grown on ripple patterned substrates might be used as negative index material within a certain wavelength range [52].

10.2 Scaling Laws of Ripple Pattern Formation

The near-field coupling of localized surface plasmons decays exponentially with interparticle distance [16, 17]. In Fig. 10.1, an example of the shift in plasmon resonance of 50 nm diameter particle with increasing interparticle gap is shown. It can be observed that the

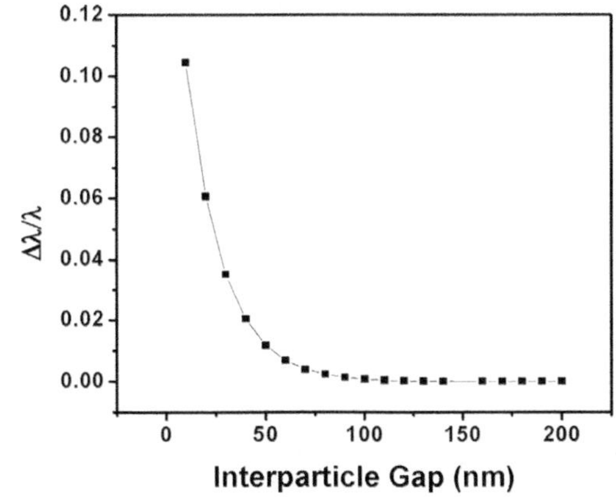

Figure 10.1 Variation in plasmon resonance with interparticle distance both experimentally and simulated as stated in references [16, 17].

shift in plasmon resonance decays exponentially and is negligible above 25 nm particle gap. This is due to the evanescent nature of the plasmonic near field of single nanoparticles. Therefore, closer particles couple much strongly. A shift in plasmon resonance with interparticle gap is usually reported for lithographically produced particles, where the particle size is quite large (~100 nm) and the interparticle gap is also large (minimum 20 nm). With the present approach, nanoparticle arrays aligned along the ripple direction with a much shorter interparticle gap (~5 nm) can be produced. Therefore, the additional control on interparticle gap and array periodicity given by the patterned template make them especially interesting for plasmonic applications. Ripple patterns with a wide range of periodicities from 20 nm to several 100 nm can be produced on a variety of materials using ion beam irradiation. The periodicity depends on several parameters, such as ion energy, ion mass, incidence angle, and fluence [41]. For plasmonic applications, ripple patterns in the range of 20–50 nm are most interesting. In this section, we restrict our discussion to those ion irradiation parameters that fulfill these requirements [42, 44, 46, 52].

The ripple templates presented here are produced by irradiation of Si(100) substrates with 500 eV Ar^+ broad ion beam. The initial root mean square (rms) roughness of the Si wafer is below 0.1 nm.

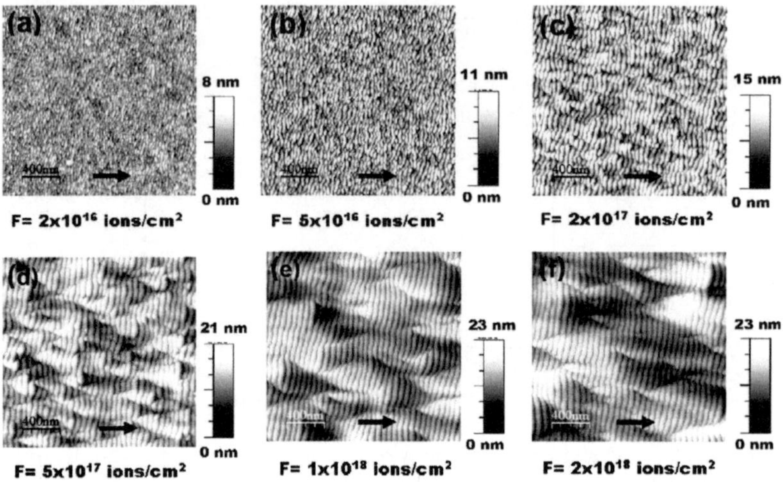

Figure 10.2 AFM images of the surface topography after irradiation with 500 eV Ar^+ ions at 67° angle of incidence and increasing fluence.

The irradiations are performed in a high vacuum chamber at a pressure of 2×10^{-4} mbar. Morphological changes for an ion incidence angle of 67° with respect to the surface normal for different fluences and ion energy of 500 eV are shown in Fig. 10.2. At low fluence of $\sim 10^{16}$ cm^{-2}, shallow ripples appear as shown in Fig. 10.2a. With increasing fluence, wavelength coarsening is observed resulting in larger wavelength ripples as shown in Fig. 10.2b–f. In this way, the ripple wavelength can be tuned from 20 nm to around 45 nm at 500 eV by simply varying the fluence in the range from 10^{16} to 10^{18} cm^{-2}. Figure 10.3 shows the wavelength as a function of ion fluence for two ion energies of 300 eV and 500 eV, respectively. The solid lines represent power law fits, $\lambda \alpha \phi^n$, with a coarsening exponent $n \sim 0.08$ [42, 44]. At higher fluences, the wavelength saturates. In addition, λ increases with ion energy [42, 44, 46, 52, 53].

Atomic force microscopy (AFM) images of ripple patterns with periodicity of 22 nm, 28 nm, 35 nm, 41 nm, 44 nm, and 51 nm along with their rms roughness prepared on Si(100) surfaces are shown in Fig. 10.4. In this case, the ion energy was kept at 500 eV and the time of irradiation was varied. The rms roughness increases with ripple periodicity. Ripples with periodicity in the range of 30–35 nm exhibit

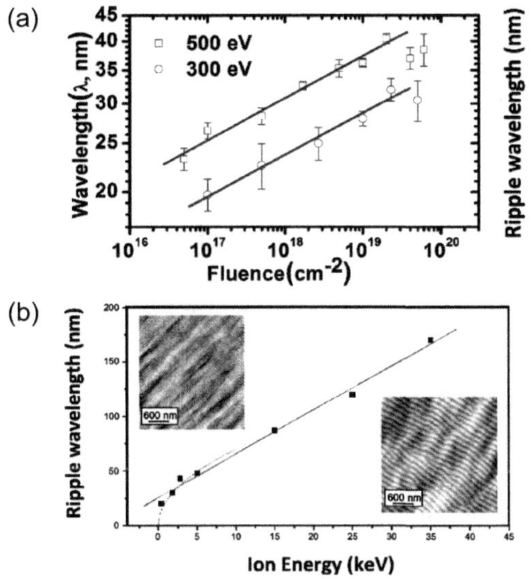

Figure 10.3 (a) Ripple wavelength values as a function of incident ion fluence for two different ion energies of 300 eV and 500 eV, respectively [42, 44]. (b) Ripple wavelength as a function of primary ion energy. The insets show two AFM images for ion erosion with a primary energy of 5 and 25 keV. Both image scan sizes are identical for a better comparison [53].

higher ordering than ripples with 20 nm (disconnected ripples) and with 45–50 nm (slight inhomogeneous). A higher lateral ordering of the 28 nm and 35 nm ripple patterns is reflected by the second-order peak appearing in the FFT (inset of Fig. 10.4). Above 40 nm, a parallel ripple mode with much higher wavelength appears [44], which dominates the roughness at higher fluences. This quasi-periodic pattern is visible in the AFM as high bumps perpendicular to ripple direction. The quality of the ripple pattern directly affects the self-alignment of nanoparticles. Therefore, ripple templates of wavelength less than 40 nm are best suited for good alignment of the nanoparticles and nanowires.

Ripple patterns produced with low-energy (<1 keV) irradiation have a relatively small modulation amplitude, typically in the range of 2–3 nm. An AFM line scan of a well-ordered ripple pattern produced by 500 eV Ar^+ ion irradiation is shown in Fig. 10.5. It can be seen that the peak-to-peak height is around

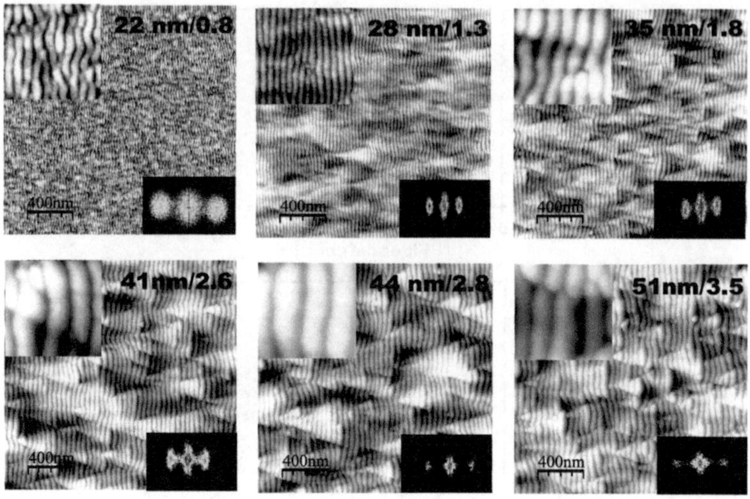

Figure 10.4 Sample of different periodicity ripple used in present study with their rms roughness and FFT (shown in inset) [39]. See also Color Insert.

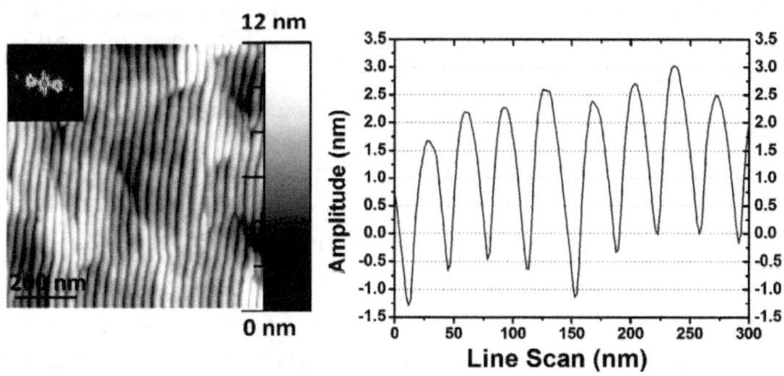

Figure 10.5 AFM topography of ripple pattern produced by 500 eV ion and fluence of 1×10^{17} cm^{-2} and corresponding line scan of ripple modulation is shown. The amplitude is $\sim$ 2.0 nm in this case.

3 nm. However, such small modulations are sufficient for the alignment of nanoparticles on these surfaces without using shadow deposition. During irradiation, a few nanometers of the Si surface are amorphized and oxidized (2–3 nm thick) immediately after exposing to atmosphere. The oxide layer formation is a favorable condition for silver cluster formation due to larger surface free

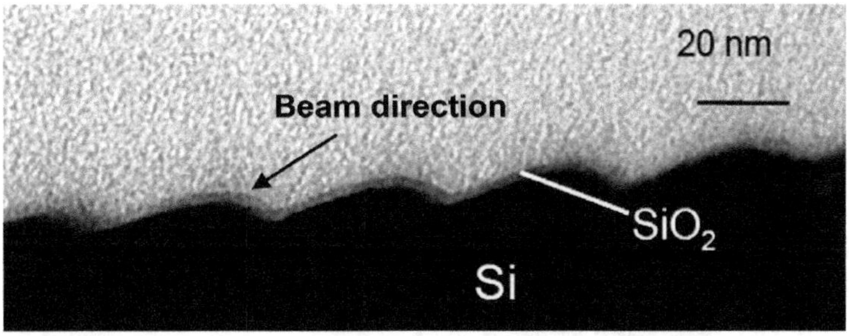

Figure 10.6 Cross-sectional TEM image showing ripple asymmetry and oxide layer on top.

energy differences with silver. The oxide also prevents metal silicide formation. An interesting property of the rippled templates is their asymmetry observed in cross-sectional transmission electron microscopy (TEM), as shown in Fig. 10.6. Ripples are tilted in the direction of the incident ion beam. This is also a favorable condition for the nanocluster alignment providing a larger ridge portion on one side and a smaller steeper portion on the other side. Depending on the direction of deposition with respect to this asymmetry, tilt-ordered nanoparticles and nanowires can be formed, respectively.

10.3 Metal Film Growth on Ripple-Patterned Substrate by PVD

To understand the process of self-alignment of nanoparticles on pre-patterned Si substrates, the deposition parameters, i.e., angle of incidence of silver atoms, deposition time, deposition rate, substrate temperature, and azimuthal direction of deposition, were varied [49, 54]. A sketch of the two-step process is shown in Fig. 10.7. In the first step, the ripple-patterned substrate is produced by low-energy ion irradiation followed by deposition of silver atoms at an oblique angle of incidence and with an azimuthal direction perpendicular to the ripple direction. Self-aligned silver nanoparticles are formed in the ridge portion along the ripples. In this figure, the direction of deposition is opposite to the direction of the incident ion beam used for producing the ripple pattern. As stated before, due to

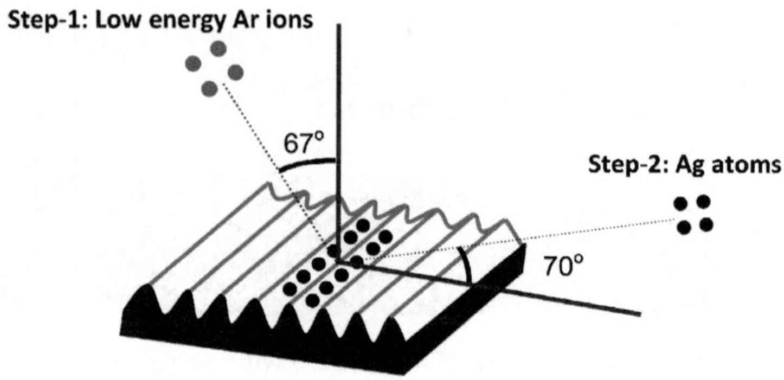

Figure 10.7 Schematic diagram of the two-step process to produce self-aligned Ag nanoparticles. Ripple patterns are produced in the first step followed by Ag deposition at grazing incidence.

the ripple asymmetry, different results are obtained depending on the direction of deposition [49, 54]. All the experimental results shown below were done with the deposition direction as shown in Fig. 10.7. The deposition can be done by electron beam evaporation or magnetron sputter deposition. In the following section, Ag deposition by e-beam evaporation is described.

Examples of silver nanoparticles grown with 0.5 ML min^{-1} deposition rate at an incidence angle of 70° with respect to surface normal after deposition time of 75 min and 105 min, respectively, are shown in Fig. 10.8. The deposition was done at room temperature. Slightly elongated particles are formed first. After prolonged deposition, nanowires are formed aligned along the ripple direction due to particle growth preferentially along the ripples.

Spherical nanoparticles can be formed by post-annealing the silver films prepared at room temperature. For this, particles were deposited at room temperature, exposed to atmosphere, and subsequently annealed at 300°C for 1 h at lower vacuum of 1×10^{-7} mbar. Example of a post-annealed sample is shown in Fig. 10.9a. Well-ordered spherical particles can be identified in the SEM image. Due to the surface free energy minimization, which is lowered by physisorption of oxygen on the silver surface, the particles adopt this spherical shape [55]. An AFM image with corresponding FFT of

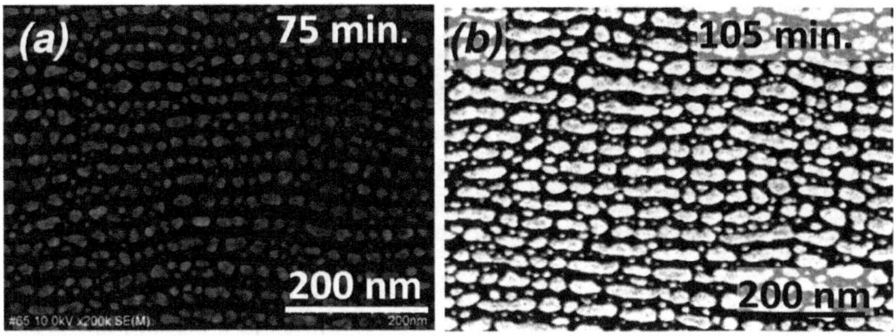

Figure 10.8 SEM images of Ag nanoparticles deposited on rippled SiO_2/Si surface for 75 min (a) and 105 min (b), respectively, with a deposition rate of 0.5 ML min^{-1}.

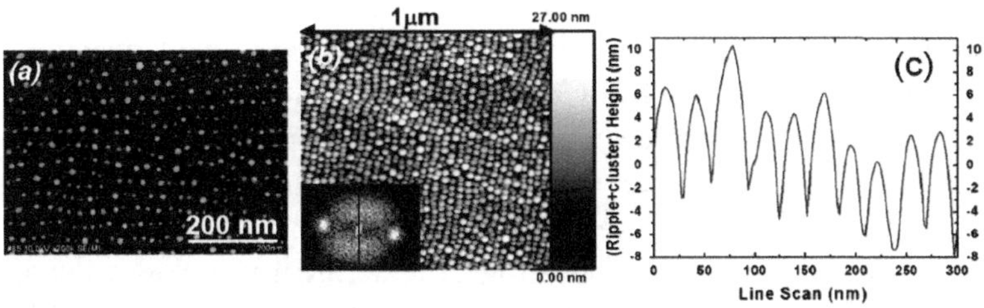

Figure 10.9 (a) SEM images of a sample post-annealed at 300°C after exposing to atmosphere [39]. (b) AFM topography of another sample (inset shows the FFT). (c) Line scan of AFM image at random position showing increased amplitude due to nanoparticles on the ripple pattern.

ordered spherical particles are shown in Fig. 10.9b. The FFT exhibits two pronounced peaks at the same position as the underlying ripple pattern confirming the alignment along the ripple direction and the high degree of ordering of the Ag nanoparticles. Two additional broader peaks also appear in the perpendicular direction indicating a characteristic length in this direction too. This can be attributed to a narrow size distribution of the nanoparticles. A line scan of the AFM image at a random position perpendicular to the ripple direction is shown in Fig. 10.9c. The periodicity of the rippled Si surface is still reproduced; however, the amplitude of the ripple pattern is much higher after deposition due to the presence of silver particles.

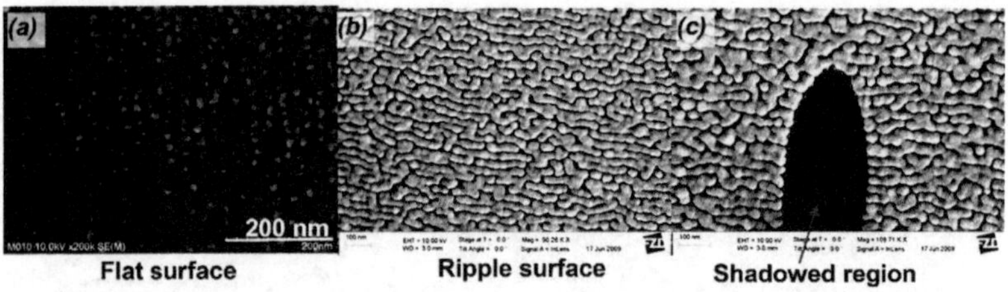

Figure 10.10 SEM image of Ag film on flat (a) and rippled (b and c) Si substrate with large surface coverage. In a shadowed uncoated region, the ripple pattern is visible (c).

A comparison of Ag film grown on flat Si substrate and on rippled Si substrate, respectively, is shown in Fig. 10.10. On the flat Si substrate, the Ag nanoparticles are distributed isotropically without a preferred orientation. The particles grow uniform in all directions resulting in an isotropically structured film at high coverage. In contrast, on the rippled substrate, growth is preferentially along the ripple direction due the modulation of the deposition flux. Therefore, in this case, the structure of the film follows the ripple direction. However, at high coverage, the nanowires will join also in the direction perpendicular to the ripple direction. In Fig. 10.10c, a shadowed region is depicted where the ripple orientation on the substrate is still visible.

A schematic diagram of the nanoparticles growth on rippled substrate is shown in Fig. 10.11a. The growing mechanism can be understood in the following way. First, small particles form on the top ridge region, but since atomic flux comes continuously at certain angle, particle size will become larger with time. Since large particles have lesser surface energy and smaller contact angle [55], increase in particle size reduces interfacial energy; hence, the particles reshape again along the ripple slope to rebalance the energies. Continuous deposition will finally increase the particle diameter toward the valley. Finally, particles become elongated on the ripple slope toward the valley. When the particle diameters are above a certain length ($\sim$ 17.5 nm in the case of 35 nm periodicity), they start tilting in the direction of valley. It is very clear from the SEM and TEM images (Fig. 10.11c,d) that some small particles are still on the hill (<17.5 nm) whereas others are located on the

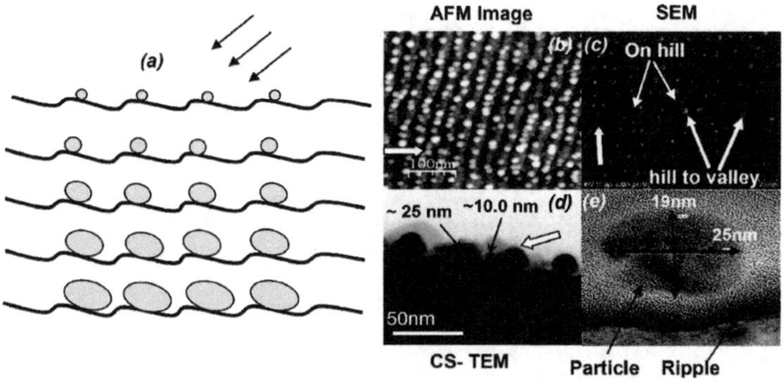

Figure 10.11 (a) Schematic of nanoparticles growth on rippled surface. (b) AFM image and (c) SEM image of Ag particles on the ripples. (d) Cross-sectional TEM (white arrow indicates the direction of the incoming flux). (d) High resolution TEM image of a single particle sitting in a ripple valley. The images demonstrate that smaller particles are on the hills, whereas larger particles expend from the hills into the valleys.

slope from hill to valley ($>$17.5 nm). The condition described here is applicable only for the deposition direction shown in Fig. 10.11a and for 35 nm periodicity ripples. For the case when deposition is performed from another direction, the situation is different; in this case, particles already start growing from the valley region and further movement is restricted.

As has been demonstrated in the last paragraph, nanoparticles growth follows the surface ripple pattern forming aligned nanoparticle arrays. The localized surface plasmon resonances (LSPR) of these nanoparticle arrays depends on the size and shape of the particles, but also on the interparticle gap. Hence, by changing the pattern periodicity and the growth conditions, the LSPR can be tuned accordingly. In Fig. 10.12a–c, aligned nanoparticles arrays are shown with periodicities of 20 nm, 35 nm, and 45 nm, respectively. Annealing at different temperature and for different time can also change the mean interparticle gap along the ripple. Similarly, parallel nanowires of different width can be formed by controlling deposition time and rate as demonstrated in Fig. 10.12d–e. In conclusion, with e-beam evaporation, the characteristics of the nanoparticles arrays, such as size, shape, interparticle distance, and

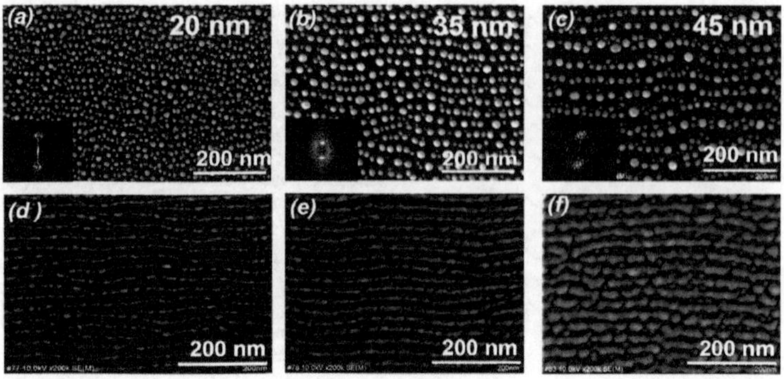

Figure 10.12 Ordered nanoparticle of 20 nm, 35 nm, and 45 nm periodicity, respectively, after annealing at 300°C for 1 h (top SEM images). Nanowires of different width produced on 35 nm ripple substrate. Deposition rate of 0.5 ML min^{-1} was used for different times 75 min, 105 min, and 125 min, respectively.

periodicity, can be determined with high flexibility by the ripple template and by the deposition conditions.

In addition to deposition by e-beam evaporation, also magnetron or ion beam deposition can be used to grow nanoparticles on ripple patterned surfaces [36, 37, 38, 40]. In this section, we compare results of Ag deposition by e-beam evaporation and by magnetron sputtering deposition. Magnetron sputtering deposition works at very high working pressures, typically in the range of 10^{-2} to 10^{-3} mbar. The deposition rate with magnetron sputtering is almost two orders of magnitude higher than by e-beam evaporation. In addition, the energy of deposited atoms is much higher (1–10 eV) compared to the thermal energy of evaporated atoms (few 100 meV). Therefore, the adatom mobility and the nucleation rate are higher leading to a high density of nanoparticles. Hence, nucleation and growth kinetics with magnetron sputtering deposition is very different to deposition by e-beam evaporation.

Figure 10.13 shows SEM images of Ag nanoparticle films deposited on flat Si substrates using e-beam evaporation and magnetron sputtering at room temperature as well as at 300°C substrate temperature with same deposition rate of 0.5 ML min^{-1} and for 25 min, respectively. Image (b) and (c) look similar, which means in magnetron sputtering more surface is covered at room

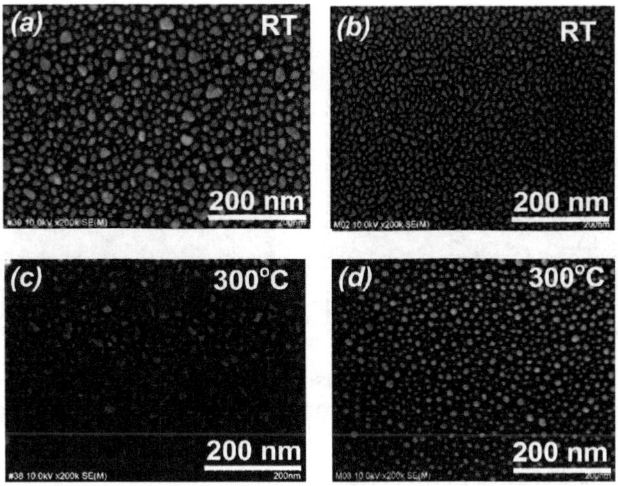

Figure 10.13 SEM images of Ag nanoparticles deposited on flat, oxidized Si substrate using e-beam evaporation (a, c) and magnetron sputtering (b, d) at room temperature and 300°C, respectively.

temperature due to higher flux and high mobility of adatoms, which is the case in e-beam evaporation at high temperature. In image (d), particles shown are grown at heated substrate of 300°C, which look similar to annealing results of evaporated particles. Thus due to high pressure growth during magnetron sputtering, the oxygen content is increased, resulting in spherical particles similar to those obtained by e-beam evaporation after annealing.

In order to produce aligned spherical Ag nanoparticles using magnetron sputtering, the ripple-patterned substrate must be heated. Examples of Ag deposition using magnetron sputtering with an angle of incidence of 70° from the surface normal and deposition from both sides of ripples are shown in Fig. 10.14. In case of perpendicular deposition onto the long ridge side, no alignment was observed (Fig. 10.14a), whereas in case of deposition onto the short, steeper ridge side, very few but aligned particles appear (Fig. 10.14b). The second case gives the possibility for particle alignment if the process runs for sufficiently long time. However, deposition by e-beam evaporation gives better alignment due to the lower nucleation density and higher mean free path of the arriving atoms.

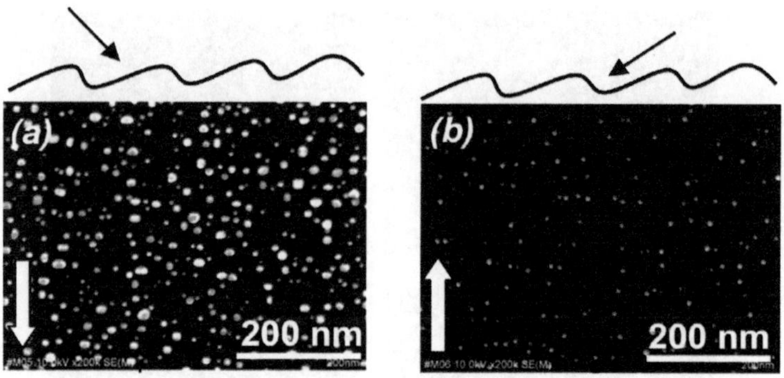

Figure 10.14 Cluster growth using magnetron sputtering from two different directions at 300°C; white arrow shows the direction of deposition.

10.4 Plasmonic Properties

10.4.1 *Optical Anisotropy in Self-Aligned Nanoparticles*

A comparison of reflection measurements conducted for bulk silver and for randomly distributed nanoparticles is shown in Fig. 10.15a,b, respectively. For bulk silver, reflectance is almost constant over the entire visible range, while for nanoparticles, it exhibits a peak at plasmon resonance wavelength. The combined effects of the conduction electrons and the bound d-electrons influence the reflectance properties of the metal film in the visible range. The conduction electrons result in strong reflection from visible wavelengths, right out to virtually DC. Interband absorption from the 4d–5d orbitals occurs in the UV region at around 320 nm (3.9 eV) (Fig. 10.15a) [56, 57].

The Drude theory models the conduction electrons as plasma, with a positive ionic background, resulting in a plasma frequency, ω_p, dependent primarily on the free electron density within the metal. For silver, this is predicted at around 9.1 eV. However, the highly polarizable d-band electrons contribute a background polarizability, which shifts the observed plasma frequency to just below the interband transition at 3.8 eV. This is referred to as the screened plasma frequency, ω_{sp}. Plasma oscillations are observed at the screened plasma frequency in thin silver films (ca. 20 nm) at oblique incidence under p-polarized illumination. They are also

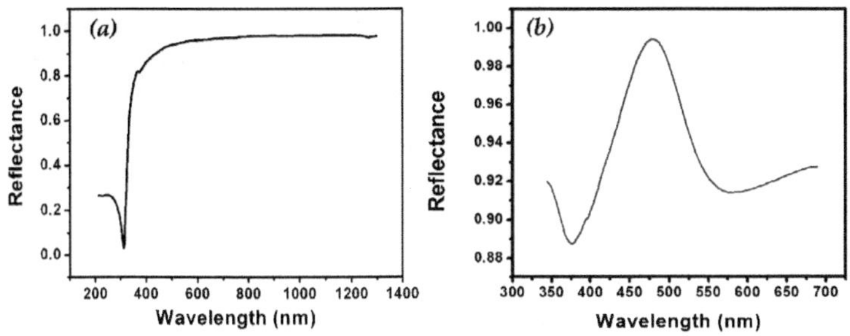

Figure 10.15 Reflectance of bulk silver (a) and nanoparticles (b).

observed in spherical nanoparticles, where they referred to as LSPR. The spherical geometry of the nanoparticles shifts the observed resonance wavelength from ω_{sp} to $\omega_{sp}/\sqrt{3}$ for a spherical particle in vacuum. For particles in a dielectric medium or deposited on a surface, the resonance wavelength will be further red shifted. The reflection curve in Fig. 10.15b shows a typical resonance for silver nanoparticles on a glass substrate.

A comparison of reflection measurements performed for both nonordered and ordered nanoparticles by incident p-polarized (along the ripple, $E_{\parallel}$) and s-polarized (across the ripple, $E_{\perp}$) light is shown in Fig. 10.16a–c. In Fig. 10.16a,b, particles are nonordered and deposited on flat Si substrate, rippled Si substrate, respectively. In both cases, the plasmon resonance occurs at the same wavelength for both p and s polarizations. Only for the case of ordered particles shown in Fig. 10.15c, a red shift in wavelength occurs for light with electric field vector parallel to the ripples [36, 37], demonstrating an optical anisotropy.

Particles of random shape and size show a broader resonance (Fig. 10.16a, b), while annealed spherical particles exhibit a much sharper resonance peak (Fig. 10.16c). It can be concluded that by annealing, aligned particles with narrower size distribution and tunable optical anisotropy are produced [39]. For nonordered particles, the LSPR is not polarization dependent. Still, near-field coupling is expected for these sense nanoparticle films. However, the coupling is isotropic and does not cause any optical anisotropy. The anisotropy due to single-particle shape is also averaged out for randomly oriented particles in the far field.

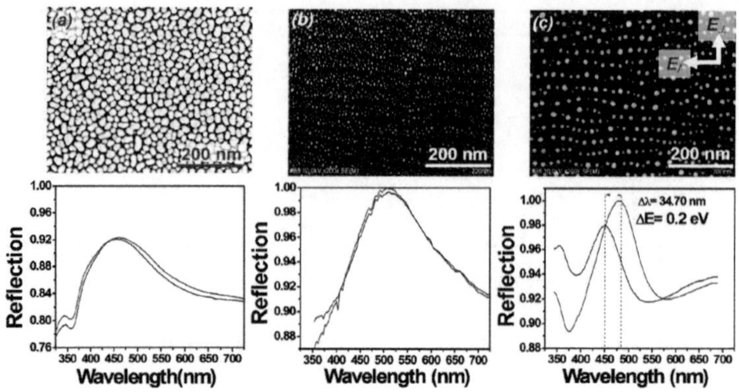

Figure 10.16 Reflection measurements corresponding to nonordered particles on normal Si, rippled Si substrate, and for ordered nanoparticles are shown in plot (a), (b), and (c), respectively. Red and blue lines are corresponding to p-polarization and s-polarization, respectively. See also Color Insert.

For the aligned spherical Ag particles in Fig. 10.16c, no significant shape anisotropy is expected. Nevertheless, a red shift in the LSPR of 0.2 eV is observed for the light polarized in direction of the ripples. We attributed this optical anisotropy to a stronger near-field coupling along the direction of the ripples. Along the ripples, particles are densely arranged with ~10 nm interparticle distance, while parallel to the ripples, they are separated by the ripple wavelength of ~35 nm. It was mentioned earlier that particle coupling falls off exponentially with the interparticle gap [16, 17]. It is also visible in Fig. 10.16c that the resonance strength perpendicular to the ripple direction is weaker than parallel to it. Correspondingly, the number of aligned particles is higher along the ripples compared to the perpendicular direction. The higher density of particles corresponds to a higher average free-electron density, resulting in a stronger LSPR signal.

In the following, we present measurement of the LSPR for aligned nanoparticle arrays on pre-patterned ripple surface with different periodicity and interparticle distance. In Fig. 10.17a, SEM images of aligned particle chains with periodicity of 20 nm, 30 nm, 35 nm, and 45 nm, respectively, are presented. By annealing, the shape and the interparticle gap can be adjusted. For every ripple periodicity,

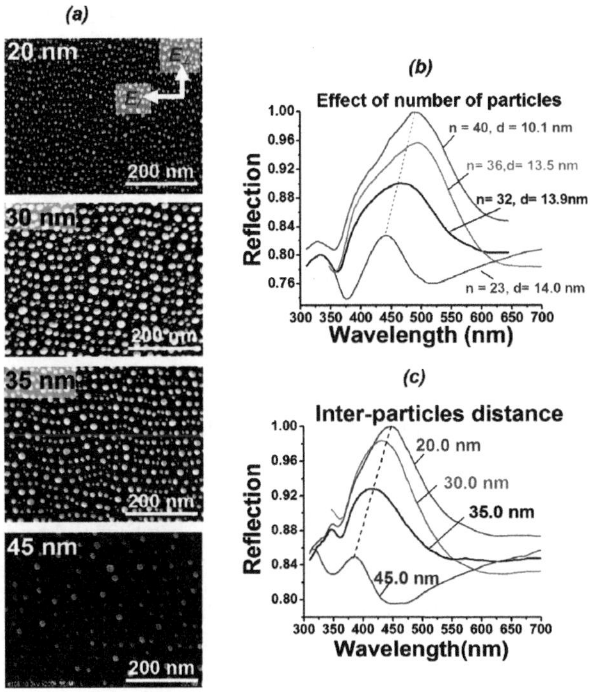

Figure 10.17 (a) SEM images of nanoparticles arrays produced on 20 nm, 30 nm, 35 nm, and 45 nm wavelength ripple substrate, respectively. Along the ripple, particle density and interparticle gap are higher than across the ripple. SPR tunability by varying the number of particles and with ripple periodicity is shown in (b) and (c), respectively. The number of particles shown in the SEM image is denoted by *n*.

samples were annealed for different times to adjust the interparticle gap along the ripples. However, the interparticle gap change is accompanied by a change of particle numbers. Perpendicular to the ripple direction, the interparticle gap is determined by the periodicity of the pattern. The averaged particle size and density along the ripple direction were extracted from the SEM image (Fig. 10.17b,c). In this image, it can be seen that the number of particles changes along the ripples and the interparticle gap changes across the ripple.

In Fig. 10.17b, the influence of the particle density in the chain on the LSPR for light polarized parallel to the ripple direction is shown.

The position of the LSPR is red shifted with increasing particle density in the chain [58]. This is attributed to the near-field coupling of particles, which is stronger along the chain [58]. In the direction perpendicular to the ripples, the interparticle distance changes with the ripple periodicity. Therefore, with increasing periodicity, a blue shift in the LSPR is observed (Fig. 10.17c). These two examples clearly demonstrate the capability to tune the LSPR with the help of ripple pattern templates.

10.4.2 *Dielectric Function of Nanowire Array of 35 nm Period*

Reflection measurements clearly show that self-aligned nanoparticles are optically anisotropic. To further investigate the symmetry of this anisotropy, i.e., whether it is uniaxial or biaxial, the dielectric tensors were investigated. Assuming that field coupling of nanoparticles is responsible for the shift in the plasmon resonance, the symmetry of the dielectric tensor should reflect the symmetry of the ripple pattern. Therefore, we investigated an array of nanowires with fixed periodicity of 35 nm and three different widths of 10 nm, 12.5 nm, and 15 nm, respectively. SEM images of the three nanowire arrays grown on ripple-patterned templates are shown in Fig. 10.18a–c [49]. Along the wire length, the optical properties are similar to bulk silver. However, perpendicular to the nanowires, they exhibit an LSPR. The Jones matrix elements measured using generalized ellipsometry at an azimuthal angle of 45° and incident angle of 75° are shown in Fig. 10.18. Nonzero values of the off-diagonal Jones matrix elements $(\Psi)_{ps,sp}$ are due to the anisotropic nature of the samples [59, 60]. The dielectric functions were calculated using a biaxial layer model with a Drude function along the wire long axis (x-axis) and perpendicular to substrate plane (z-axis), and a Lorentz function across the wires (y-axis) [49, 59, 60, 72]. Measured and modeled Jones matrix elements are shown in Fig. 10.18. This simple model neglects the ripple modulation; however, it works well and the calculated Jones matrix elements are in fair agreement to the measured values.

The dielectric tensor elements for the three orthogonal directions ε_x, ε_y, and ε_z are shown in Fig. 10.19. The elements are

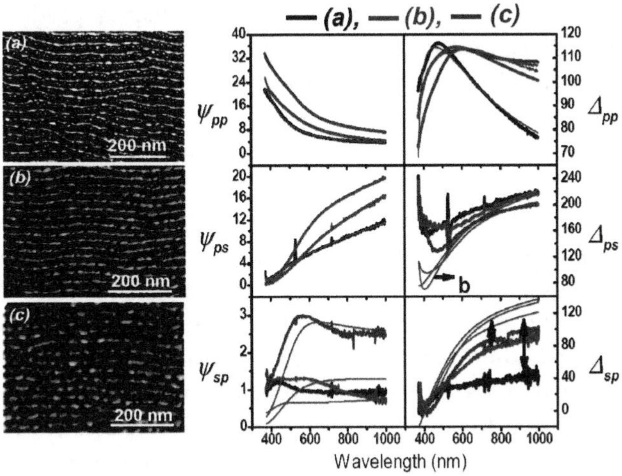

Figure 10.18 (a–c) Silver nanowires of different width decorating the ridges of the ripple patterns. Measured (thick lines) values of the GE parameters, corresponding to (a) black, (b) red, and (c) blue. The thin red lines represent the respective model results. See also Color Insert.

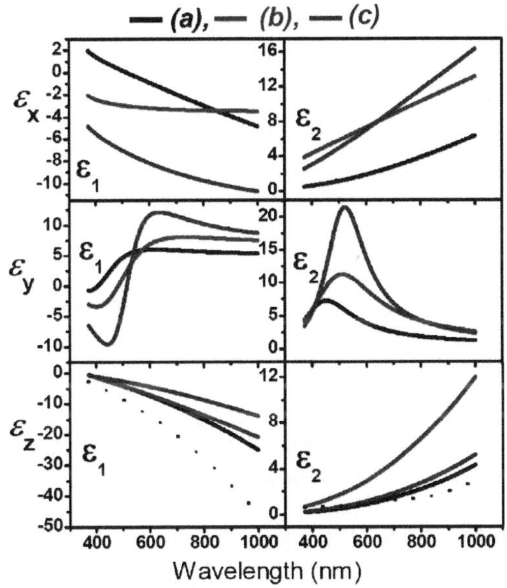

Figure 10.19 Biaxial dielectric functions of the arrays shown in (a) black line, (b) red line, and (c) blue line determined by fitting generalized ellipsometry measurements. See also Color Insert.

complex such that $\varepsilon = \varepsilon_1 + i\varepsilon_2$. ε_x and ε_z display predominantly metallic behavior, although ε_x of sample shown in Fig. 10.18a is positive in the UV range indicating an insulating behavior, while at wavelengths higher than 550 nm, ε_x is negative. ε_z is very close to the values of bulk silver. Note the strong resonance absorption in the imaginary parts of ε_y due to the LSPR. The amplitude and wavelength maximum of the resonance increase with increasing wire thickness, both due to the increased proportion of silver in the volume content of the layer [49, 61, 71]. Note also that the maximum values of the imaginary part of ε_y, located at 520 nm, are consistently higher than those of the imaginary part of ε_x across the measurement range. This illustrates the strength of the absorption cross section of the LSPR. Different values of dielectric functions indicate the biaxial anisotropic nature of the nanowires array.

Based on the modeled dielectric tensor, reflection and transmission properties of the nanowire film on different substrates can be predicted. A transmission spectrum measured for Ag nanowires on ripple-patterned SiO_2 substrate is shown in Fig. 10.20ii along with the calculated transmission values using the model described above. Light polarized parallel to the wire long axis is transmitted, while light polarized perpendicular to the wires is absorbed due to the LSPR. Figure 10.20i shows the simulated transmission of the three arrays shown in Fig. 10.18a–c on a 0.1 mm-thick silicon dioxide

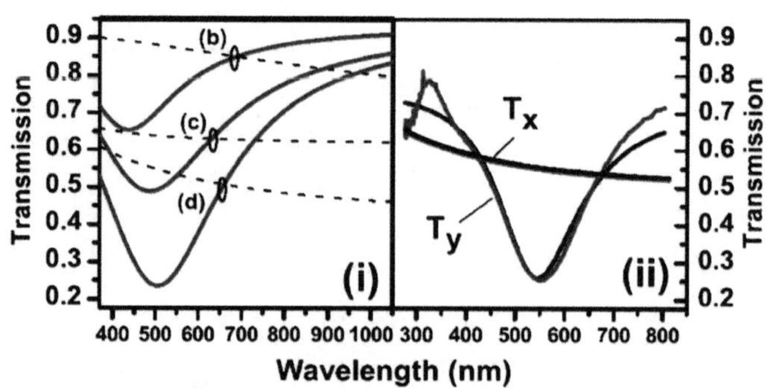

Figure 10.20 (i) Simulated transmission spectra using the dielectric functions of three wires shown above on a SiO_2 layer (b, c, and d correspond to a, b, c, respectively). (ii) Comparison of modeled (black) and measured (red) transmission spectra of silver wires grown on rippled SiO_2 substrate. See also Color Insert.

substrate. Light is incident normal to the substrate surface, with electric field vectors parallel (dashed black line) and perpendicular (solid red line) to the direction of the wires, respectively. The main message from the curves is that for all three samples, the extinction ratio is highly wavelength dependent. Transmission of parallel polarization is relatively constant across the range, the difference in magnitude being primarily dependent on the layer thickness of the silver. Light polarized perpendicular to the wires is weakly transmitted in the visible near the LSPR maximum. However, in the IR region, transmission is above 80% for all three materials.

The unusual plasmonic properties of such dense nanowire array lead us to speculate on their application in photovoltaics. A desirable goal in this field is to utilize ultrathin metal films as the TCE. The materials considered here would function not only as the TCE with transparencies above 80% over most of the measured spectral range (sample a) with bulk silver like dielectric functions (high electrical conductivity), but also as a plasmonic scatterer to improve the light trapping in the absorption layer [62]. By combining the two elements into a single layer, the compactness can be increased.

10.4.3 *Nanoparticles Array as Negative Refraction Material*

Negative refraction has been demonstrated in highly anisotropic materials [63]. These materials — termed indefinite materials — have one orthogonal effective tensor element different in sign to the other two. The dispersion relation thus describes a hyperbola, and the Poynting vector is negatively refracted for all incident angles [64]. Arrays of aligned metallic nanowires fulfill the requirements of indefinite materials and have been used to demonstrate negative refraction [63, 65, 66]. The nanowire arrays described above also fulfill the requirements of an indefinite material. In Fig. 10.19, one observes a positive real ε_y, but negative real ε_x and ε_z. Even for materials where the resonance is weak, materials with an anisotropic plasmonic response are still an interesting occurrence [67]. Such materials may arise from an anisotropic arrangement of spherical particles, or from an isotropic ordered arrangement of nonspherical particles. In this section, we will show that

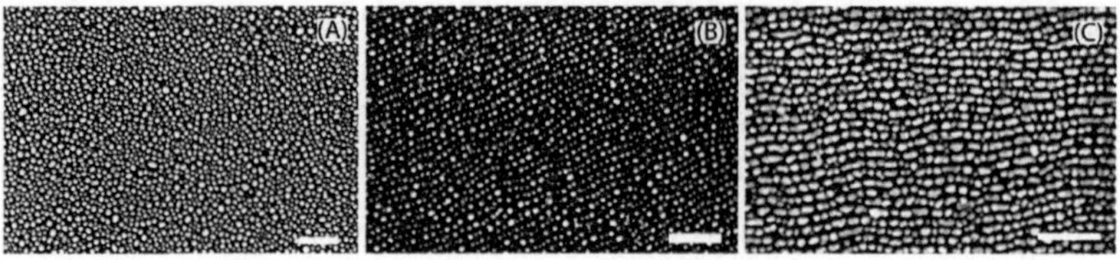

Figure 10.21 SEM images of the samples A, B, and C. Sample A shows a random distribution of silver particles, while samples B and C show aligned nanospheres and nanorods, respectively. The scale bar is 200 nm in all images. The x direction is defined as parallel to the alignment, y is perpendicular, and the z direction is normal to the substrate plane.

nanoparticles arrays produced on rippled templates also fulfill the condition of indefinite materials within certain wavelength range.

SEM images of the three samples used are shown in Fig. 10.21. Sample A is a typical randomly nucleated silver island film with particle diameters around 20 nm. In the templated samples B and C, the particles are also randomly nucleated parallel to the ripple direction (x direction); however, they are aligned along the ripple ridges in the perpendicular plane (y direction). Sample C shows aligned nanorods with lengths up to 50 nm [51].

The DFs of sample A determined by fitting a uniaxial model is shown in Fig. 10.22. The in-plane DF shows the characteristic plasmon resonance around 2.7 eV and the interband transition at 3.8 eV. The out-of-plane DF is similar to the bulk DF, but the broadening of the Drude component is increased to account for surface scattering, while the Drude amplitude is decreased to account for the reduced effective electron density. Note that in the out-of-plane direction, although not explicit in the dielectric function, a volume plasmon resonance is still observed in transmission and reflection at around 3.5 eV [51]. Note that we observe an absorption peak in the z direction at the *effective* plasma frequency ($\varepsilon_1 = 0$) in thin silver island films. In a solid film, this resonance is known as a Ferrell mode [68], the plasmonic analogue of the phononic Berreman mode [69], and is extensively discussed in basic texts [70]. We should expect that as the metal volume content of the film increases, the out-of-plane resonance will blue shift to the bulk plasma resonance frequency and the in-plane resonance will red shift to zero (i.e.,

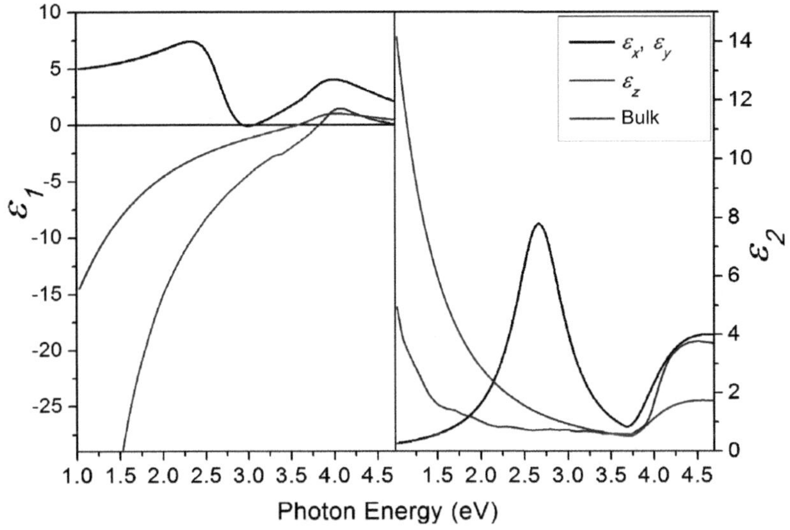

Figure 10.22 Uniaxial DFs of sample A compared to those of bulk silver from reference [71].

Drude behavior) [70]. Therefore, we observe that nonordered nanoparticles are uniaxially anisotropic, i.e., homogeneous in-plane dielectric tensor elements and a different out-of-plane dielectric tensor element.

The effective dielectric functions of samples B and C are shown in Fig. 10.23 [51]. As observed for the uniaxial case above, the out-of-plane tensor element ε_z is similar to that of bulk silver, with the characteristic Drude tail at low energies. The two peaks in the in-plane components of the imaginary parts ε_2 of the dielectric tensor are due to the particle plasmon resonances, occurring at 2.60 eV and 2.71 eV for sample B and 1.83 eV and 2.52 eV for sample C. The latter resonances are considerably stronger. ε_1 shows a region of anomalous dispersion around the resonances and drops below zero on the high frequency side of the resonance in sample C. In the region between the resonance peaks, the signs of ε_1 are opposite in the in-plane directions. This leads us to conjecture as to the possibility of creating an indefinite material from multilayers of elongated particles on rippled templates.

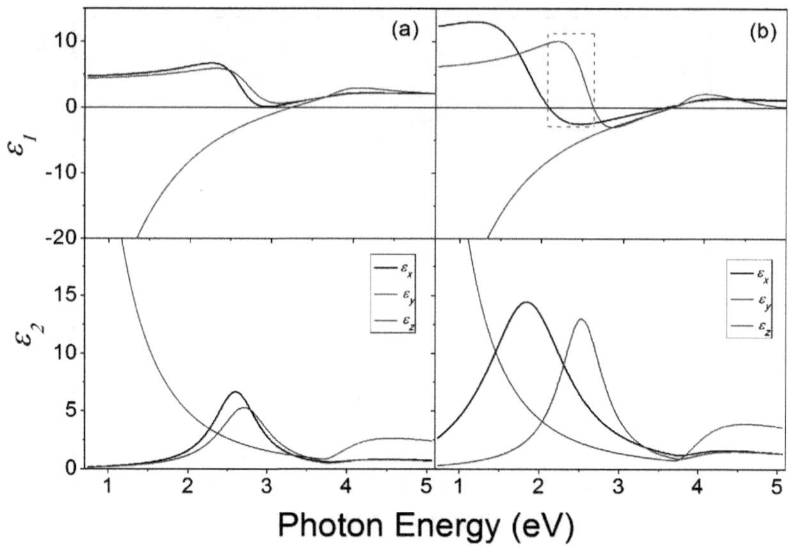

Figure 10.23 Orthogonal elements of the dielectric tensor for samples (a) B and (b) C. The box in (b) shows a region where the signs of the real part of the in-plane tensors are opposite, suggesting the potential to fabricate indefinite materials with tunable properties.

An indefinite material is one whose dielectric tensor elements are of opposite signs [73]. In an anisotropic material, a TM plane wave with wavevector $\boldsymbol{k}$ in the x–y plane will be dispersed according to

$$\frac{k_x^2}{\varepsilon_y} + \frac{k_y^2}{\varepsilon_x} = \frac{\omega^2}{c^2} \quad (10.1)$$

where ω is the angular frequency and c is the speed of light in vacuum. In an anisotropic material with all tensor elements of the same sign, Eq. 10.1 describes an ellipse for a given frequency (equifrequency curve). If the signs of the tensor elements are opposite in the propagation plane, the equifrequency curve will be hyperbolic. The group velocity is perpendicular to the equifrequency surface and is, therefore, negative in an indefinite material, described as a negative group-index [65]. Since the phase velocity direction is still positive, the group and phase velocities are antiparallel, resulting in negative refraction.

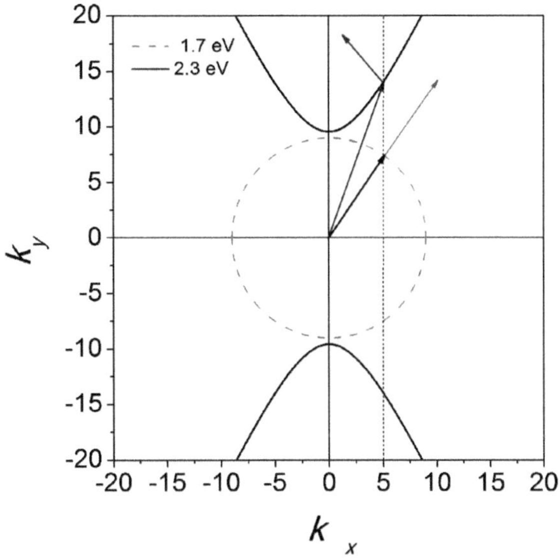

Figure 10.24 Equifrequency curves for a plane wave in the x–y plane propagating in sample C at $\hbar\omega = 1.7$ eV (positive isotropic ε, dashed curve) and 2.3 eV (indefinite ε, solid curve). At 1.7 eV, the refracted wavevector (black arrow) and group velocity (magenta arrow) are parallel, while at 2.3 eV the x component of the refracted wavevector (blue arrow) and group velocity (red arrow) are antiparallel.

The equifrequency curves for sample C at $\hbar\omega = 1.7$ eV and 2.3 eV are shown in Fig. 10.24 for a wavevector in the x–y plane. At 1.7 eV, the material is positive isotropic in the x–y plane and is positively refracting (parallel group and phase velocities). At 2.3 eV, the group and phase velocities are antiparallel, with the x component of the group velocity being negative. Such negative refraction has been observed in nanowire arrays [63]. The difference with our material is that the range of negative refraction is limited to a region between the plasmon resonances, which may be desirable in designing photonic devices. We predict that by the addition of more identical layers, separated by a thin dielectric spacer layer, one could construct a multilayer material composed of layers of sample C, which would allow light coupling into a waveguide structure. Thus one could design a material where for a certain energy range ε_1 is positive for ε_y and ε_z and negative for ε_x. Our simple two-step self-organized patterning procedure is conducive to large-scale fabrication. Optical components such as wedges and prisms may then be stamped or etched into such a multilayer.

10.5 Conclusion

Ripple patterns produced by ion irradiation are powerful templates providing a great flexibility to grow aligned nanoparticles and nanowires arrays. The pattern periodicity and thus the array periodicity can be tuned by the ion energy and fluence, while deposition parameters determine particle size, interparticle distance along the ripples, and aspect ratio of the nanoparticles. Any PVD method capable of providing low atomic flux and ion energy should be suitable for growing aligned nanoparticles on ripple-patterned surfaces in a controlled manner. Silver nanoparticle arrays exhibit a strong anisotropy in their optical properties due to the direction-dependent LSPR. Both nanoparticle and nanowire arrays produced on ripple-patterned templates are biaxially anisotropic. Our template-based approach can be exploited for diverse applications demanding controlled growth of nanoparticles and nanowires, such as in photovoltaics, sensors, and many others.

References

1. E. Ozbay, *et al.* (2006) *Science*, **311,** 189.
2. R. Zia, J. A. Schuller, A. Chandran, and M. L. Brongersma (2006) *Materials Today*, **9**, 20.
3. W. L. Barnes, A. Dereux, and T. W. Ebbesen (2003) *Nature*, **424**, 824.
4. D. K. Gramotnev and S. I. Bozhevolnyi (2010) *Nature Photonics*, **4**, 83.
5. M. Moskovits (1985) *Reviews of Modern Physics*, **57**, 783.
6. S. Nie and S. R. Emory (1997) *Science*, **275**, 1102.
7. K. Kneipp, Y. Wang, H. Kneipp, L. T. Perelman, I. Itzkan, R. R. Dasari, and M. S. Feld (1997) *Physical Review Letters*, **781**, 667.
8. K. D. Alexander, M. J. Hampton, S. Zhang, A. Dhaw-an, H. Xuc, and R. Lopeza (2009) *Journal of Raman Spectroscopy*, **40**, 2171.

9. F. J. Garcı'a-Vidal and J. B. Pendry (1996) *Physical Review Letters*, 77, 1663.
10. J. Jiang, K. Bosnick, M. Maillard, and L. Brus (2003) *Journal of Physical Chemistry B*, **107**, 9964.
11. E. Hao and G. C. Schatz (2004) *Journal of Chemical Physics*, **375**, 120.
12. F. Song, T. Wang, X. Wang, C. Xu, L. He, J. Wan, C. V. Haesendonck, S.P. Ringer, M. Han, Z. Liu, and G. Wang (2010) *Small*, **6**(3), 446.
13. R. L. Chern, X. X. Liu, and C. C. Chang (2007) *Physics Review E*, **76**, 016609.
14. S. Enoch, R. Quidant, and G. Badenes (2004) *Optics Express*, **12**(15), 3422.
15. Z. B. Wang, B. S. Luk'yanchuk, W. Guo, S. P. Edwardson, D. J. Whitehead, L. Li, Z. Liu, and K. G. Watkins (2008) *Journal of Chemical Physics*, **128**, 094705.
16. K.-H. Su, Q.-H. Wei, X. Zhang (2003) *Nano Letters*, **3**, 1087.
17. P.·K. Jain, W. Huang, and M. A. El-Sayed (2007) *Nano Letters*, **7**, 2080.
18. N. Large, M. Abb, J. Aizpurua, and L. M. Otto (2010) *Nano Letters* **10**, 1741.
19. V. K. S. Hsiao, Y. B. Zheng, B. K. Juluri, and T. J. Huang (2008) *Advanced Materials*, **20**, 3528.
20. S. S. Yee (1999) *Sensors and Actuators B*, **54**, 1.
21. G. P. Wiederrecht (2004) *Euro Physics Journal of Applied Physics*, **28**, 3.
22. P. Nagpal (2009) *Science*, **325**, 594.
23. T. Taubner, D. Korobkin, Y. Urzhumov, and R. Hillenbrand (2006) *Science*, **313**, 1595.
24. H. A. Atwater and A. Polman (2010) *Nature Materials*, **9**, 205.
25. V. E. Ferry, L. A. Sweatlock, D. Pacifici, and H. A. Atwater (2008) *Nano Letters*, **8**, 12.
26. S. Park and J. W. Hahn (2009) *Optics Express*, **17**, 20203.
27. C. Z. Xin and W. U. L. (2008) *Nan Chinese Science Bulletin*, **53**,2257.
28. R. Lloyd (2001) *Proceedings of The IEEE*, **89**, 366.
29. S. Camelio, D. Babonneau, D. Lantiat, and L. Simonot (2007) *Euro Physics Letters*, **79**, 47002.
30. G. Riveros, S. Green, A. Cortes, H. Gómez, R. E. Marotti, and E. A. Dalchiele (2006) *Nanotechnology*, **17**, 561.
31. C. T. Wu, F. H. Ko, H. Yen, and H.-Y. Hwang (2006) *Microeletronic Engineering*, **83**, 1567.

32. C. Mendoza, T. Pietsch, J. S. Gutmann, D. Jehnichen, N. Gindy, and A. Fahmi (2009) *Macromolecules*, **42**, 1203.
33. J. Y. Cheng, H. I. Smith, and E. L. Thomas (2006) *Advanced Material*, **18**, 2505.
34. H. Schmidt (2001) *Applied Organic Chemistry*, **15**, 331.
35. Q. Liu, Y. Cui, D. Gardner, X. Li, S. He, and I. I. Smalyukh (2010) *Nano Letters*, **10**, 1347.
36. T. W. H. Oates, A. Keller, S. Facsko, and A. Mücklich (2007) *Plasmonics*, **2**, 47.
37. A. Toma, D. Chiappe, D. Massabo, C. Boragno, F. B. Mongeot (2008) *Applied Physics Letters*, **93**, 163104.
38. T. W. H. Oates, A. Keller, S. Noda, and S. Facsko (2008) *Applied Physics Letters*, **93**, 063106.
39. M. Ranjan, S. Facsko, M. Fritzsche, and S. Mukherjee (2012) *Microelectronic Engineering*, doi:10.1016/j.mee.2012.02.018.
40. S. Camelio, D. Babonneau, and L. Simonot (2007) *Euro Physics Letters*, **79**, 47002.
41. W. L. Chan and E. Chason (2007) *Journal of Applied Physics*, **101**, 121301.
42. A. Keller, S. Facsko, and W. Möller (2008) *New Journal of Physics*, **10**, 063004.
43. W. L. Chan and E. Chason (2007) *Applied Physics Letters*, **101**, 121301.
44. A. Keller, S. Roßbach, S. Facsko, and W. Möller (2008) *Nanotechnology*, **19**, 135303.
45. A. D. Brown, J. Erlebacher, W.-L. Chan, and E. Chason (2005) *Physical Review Letters*, **95**, 056101.
46. B. Ziberi, F. Frost, T. Höche, and B. Rauschenbach (2005) *Physical Review B*, **72**, 235310.
47. S. Camelio, D. Babonneau, D. Lantiat, L. Simonot, and F. Pailloux (2009) *Physical Review B*, **80**, 155434.
48. O. M. Liedke, B. Liedke, A. Keller, B. Hillebrands, A. Mucklich, S. Facsko, and J. Fassbender (2007) *Physical Review B*, **75**, 220407.
49. M. Ranjan, T. W. H. Oates, S. Facsko, and W. Möller (2010) *Optics Letters*, **35**, 2576.
50. A. Kumar and C. Zhou (2010) *Nano*, **4**, 11.
51. T. W. H. Oates, M. Ranjan, S. Facsko, and H. Arwin (2010) *Optics Express*, **19**, 2014.
52. A. Keller, R. Cuerno, S. Facsko, and W. Möller (2009) *Physical Review B*, **79**, 115437.

53. J. Fassbender, T. Strache, M. O. Liedke, D. Markó, S. Wintz, K. Lenz, A. Keller, S. Facsko, I. Mönch, and J. McCord (2009) *New Journal of Physics*, **11**, 125002.
54. S. Numazawa, M. Ranjan, K.-H. Heinig, S. Facsko, and R. Smith (2011) *Journal of Physics: Condensed Matter*, **23**, 222203.
55. K. K. Nanda, A. Maisels, and F. E. Kruis (2008) *Journal of Physics C*, **112**, 13488.
56. U. Kreibig and M. Vollmer (1993) *Optical Properties of Metal Clusters*, Springer, Michigan; *Springer Series in Material Science*, vol 25.
57. S. A. Maier (2007) *Plasmonics: Fundamental and Applications*, Springer, Bath, UK.
58. R. Chern, X. Liu, and C. Chang (2007) *Physical Review E*, **76**, 016609.
59. H. Fujiwara (2007) *Spectroscopic Ellipsometry: Principles and Applications*, John Wiley and Sons, UK.
60. M. Schubert (1996) *Physical Review B*, **53**, 4265.
61. R. Doremus (1998) *Thin Solid Films*, **326**, 205.
62. V. E. Ferry, L. A. Sweatlock, D. Pacifici, and H. A. Atwater (2008) *Nano Letters*, **8**, 4391.
63. J. Yao, Z. W. Liu, Y. M. Liu, Y. Wang, C. Sun, G. Bartal, A. M. Stacy, and X. Zhang (2008) *Science*, **321**, 930.
64. M. Notomi (2000) *Physical Review B*, **62**, 10696.
65. Y. M. Liu, G. Bartal, and X. Zhang (2008) *Optics Express*, **16**, 15439.
66. V. A. Podolskiy, A. K. Sarychev, and V. M. Shalaev (2003) *Optics Express*, **11**, 735.
67. I. Romero and F. J. G. de Abajo (2009) *Optics Express*, **17**, 22012.
68. R. A. Ferrell (1958) *Physical Review*, **111**, 1214.
69. D. W. Berreman (1963) *Physical Review*, **130**, 2193.
70. C. Kittel (1996) *Introduction to Solid State Physics*, John Wiley and Sons, New York.
71. T. W. H. Oates and A. Mucklich (2005) *Nanotechnology*, **16**, 2606.
72. D. W. Lynch and W. R. Hunter (1985) In *Handbook of Optical Constants of Solids* (E. D. Palik, ed), Academic Press, New York.
73. D. R. Smith and D. Schurig (2003) *Physical Review Letters*, **90**, 077405.

Index

Color Insert

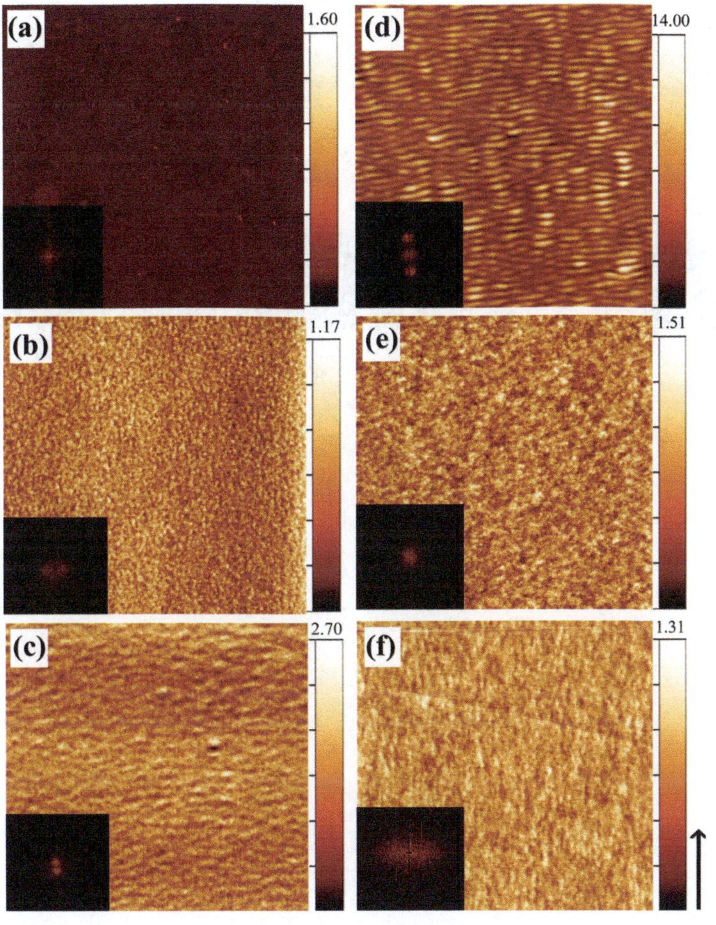

Figure 2.4

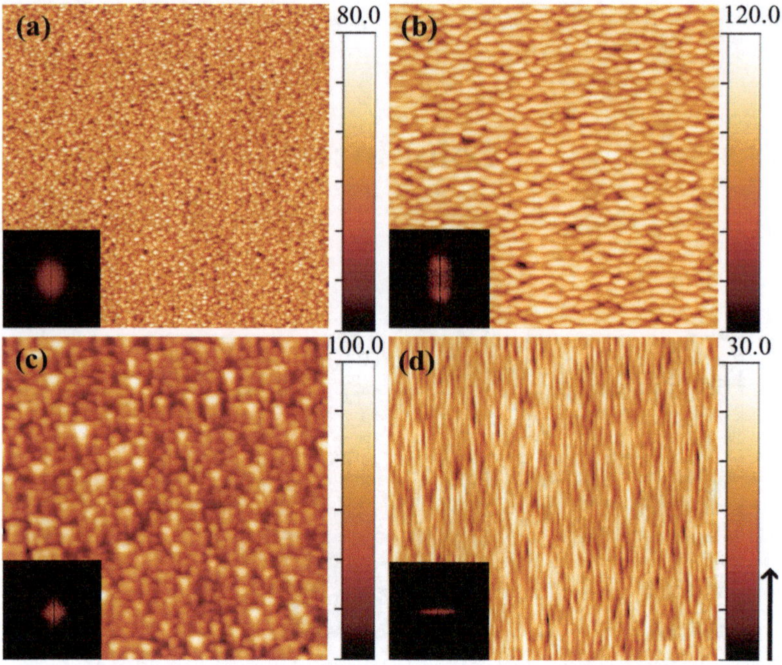

Figure 2.8

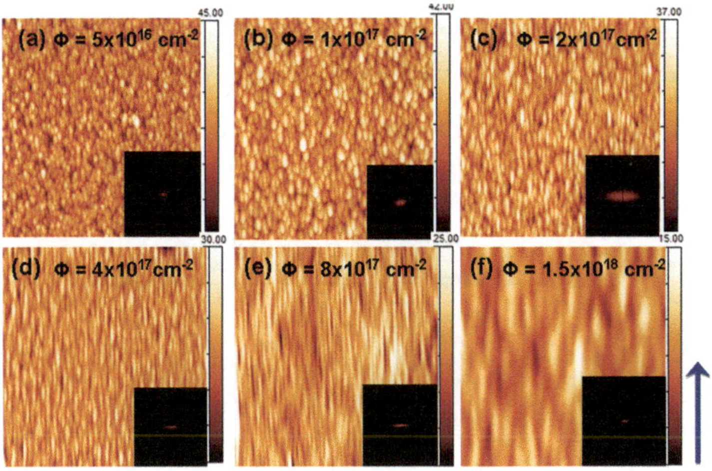

Figure 2.16

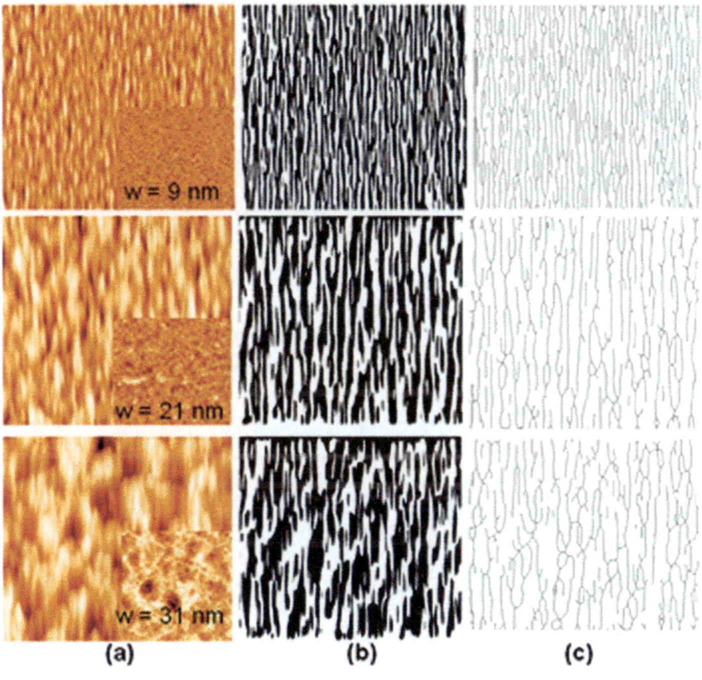

Figure 2.19

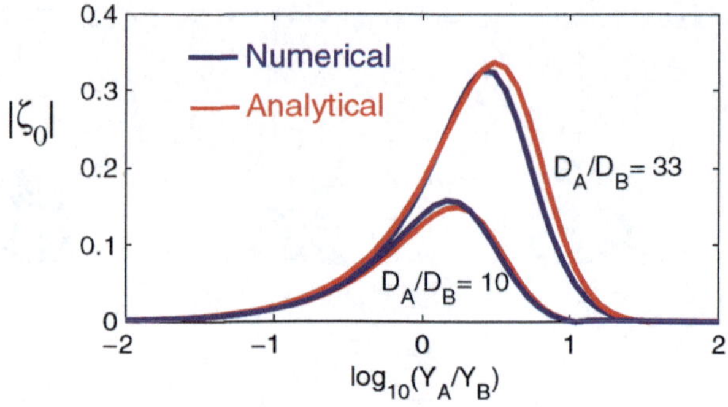

Figure 3.2

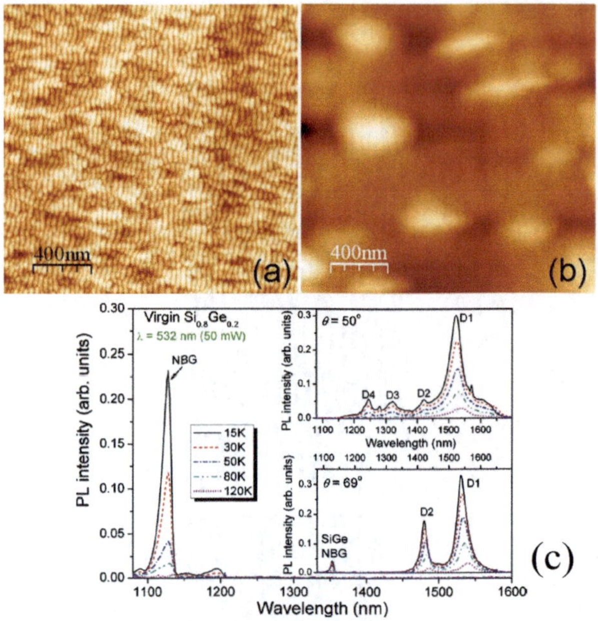

Figure 3.5

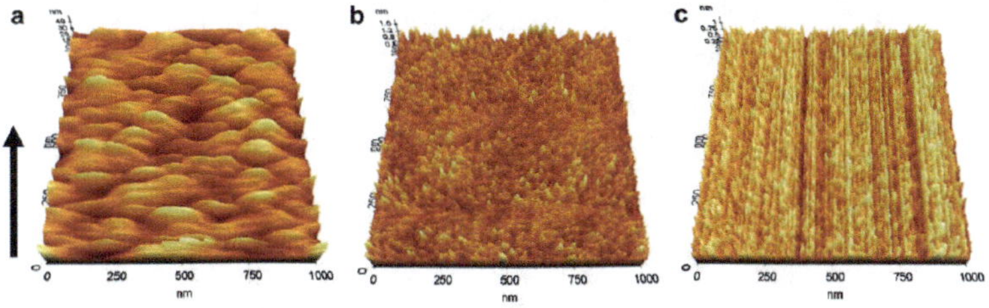

Figure 3.6

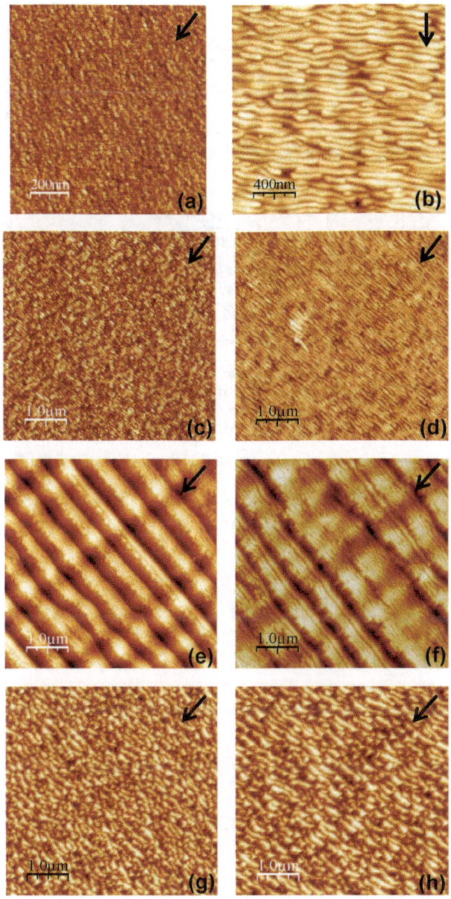

Figure 3.9

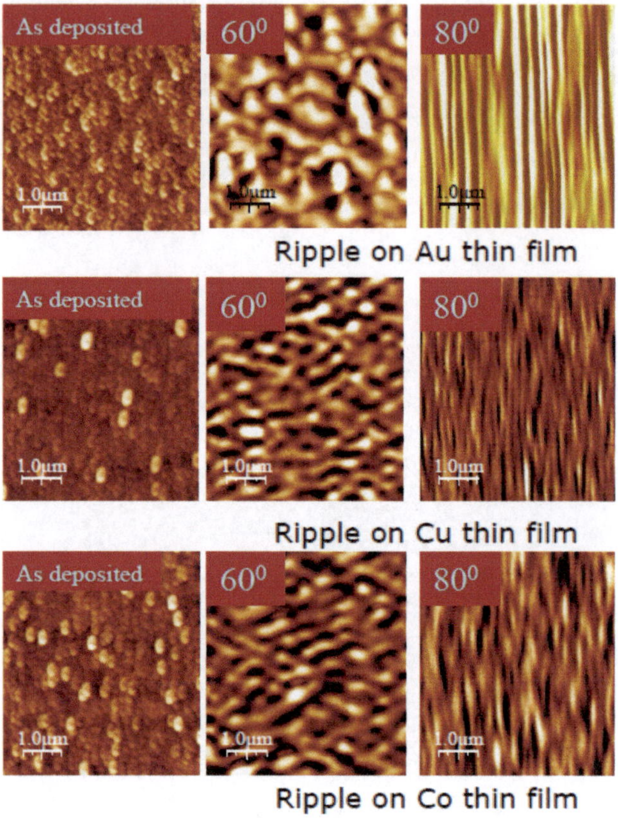
As deposited
60⁰
80⁰
1.0μm
Ripple on Au thin film
Ripple on Cu thin film
Ripple on Co thin film

Figure 5.3

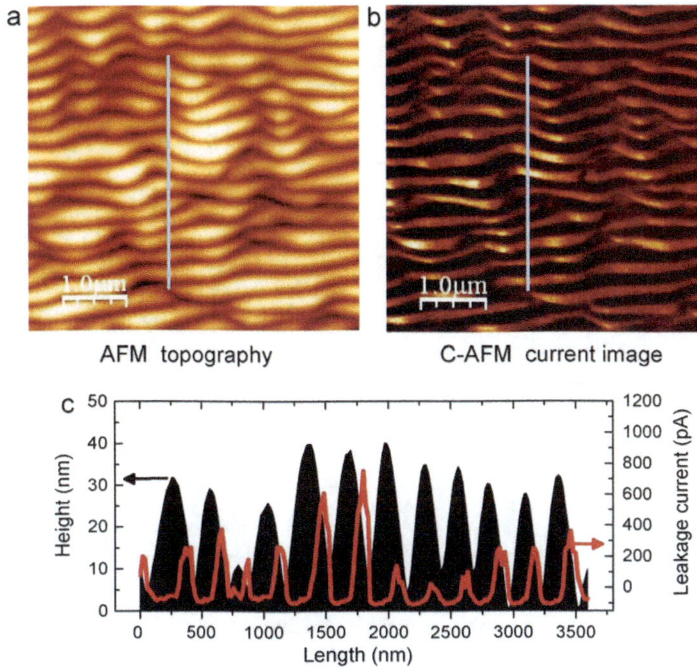

Figure 5.13

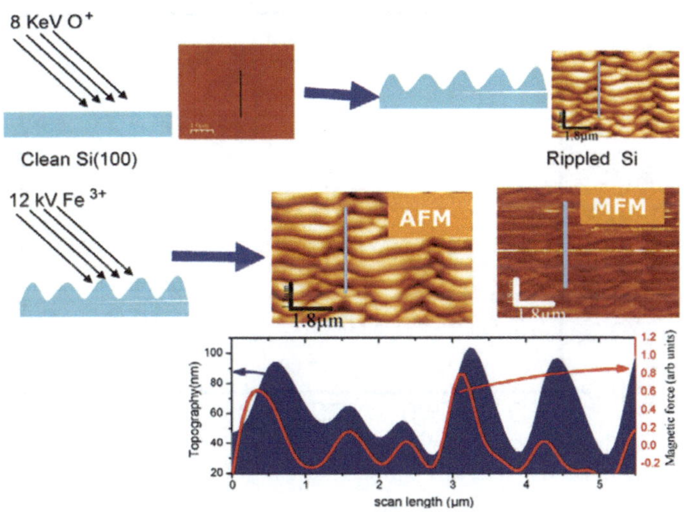

Figure 5.14

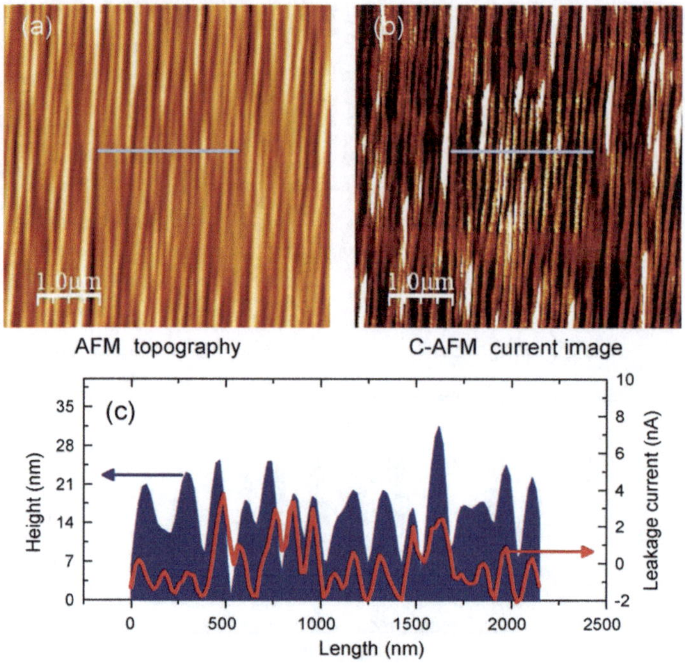

Figure 5.15

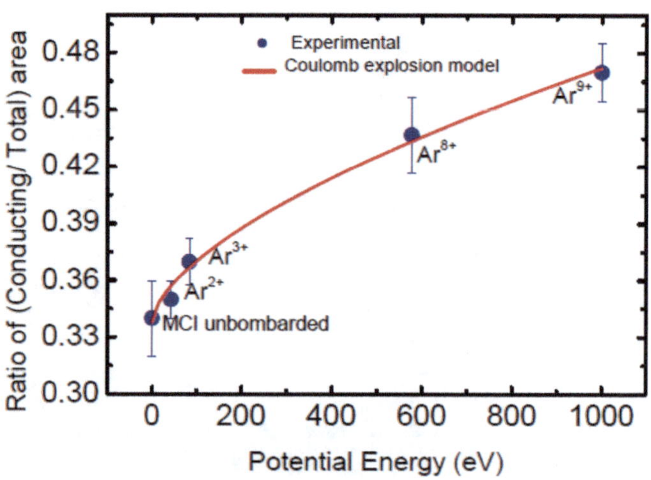

Figure 5.16

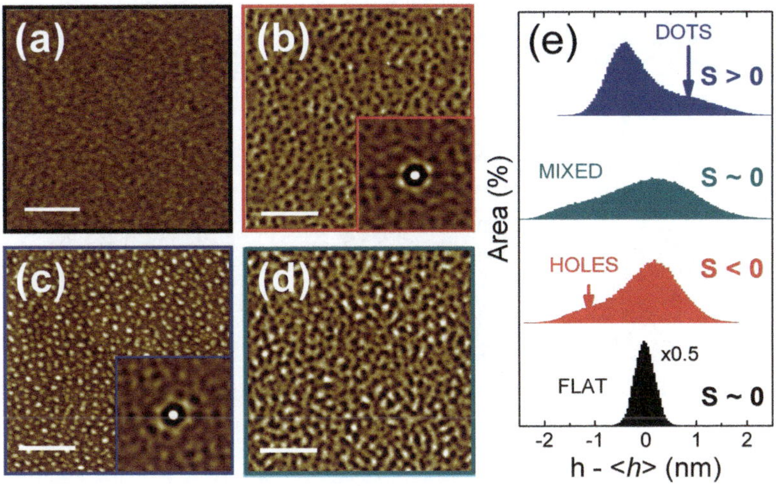

Figure 6.1

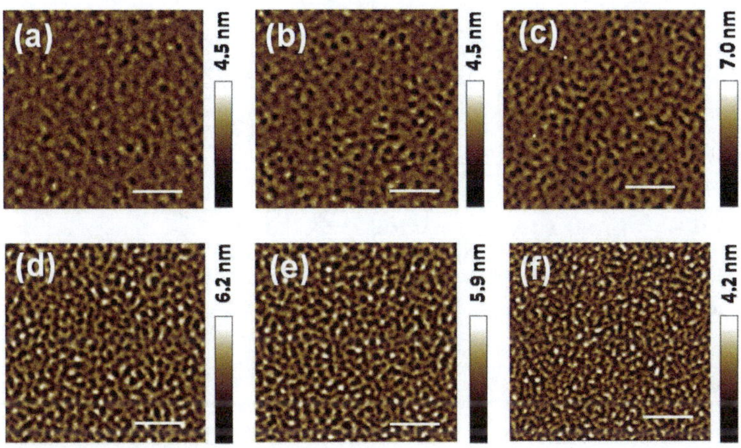

Figure 6.4

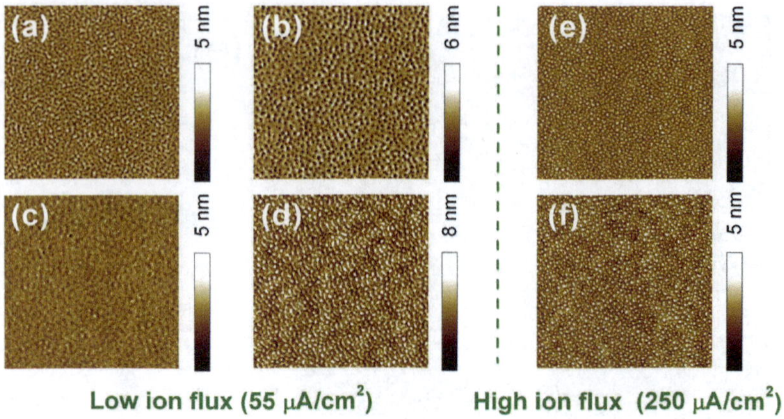

Figure 6.6

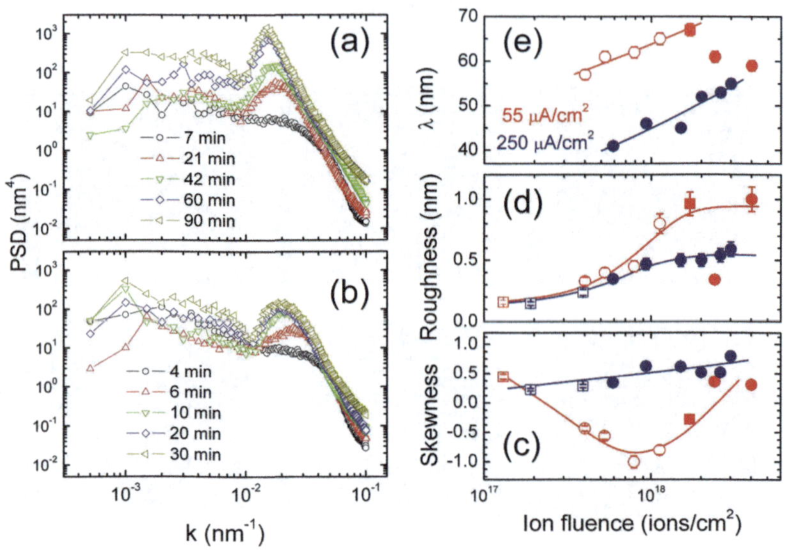

Figure 6.7

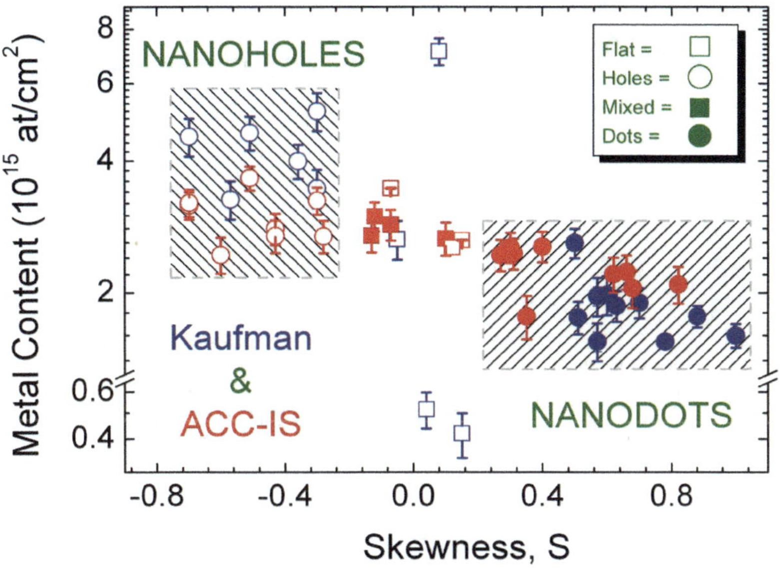

Figure 6.10

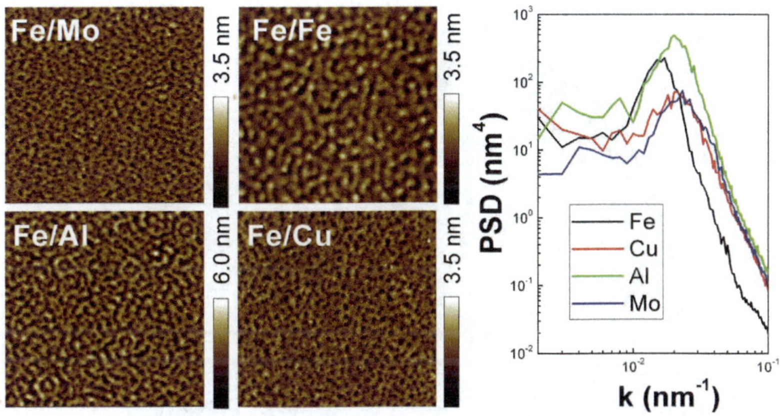

Figure 6.11

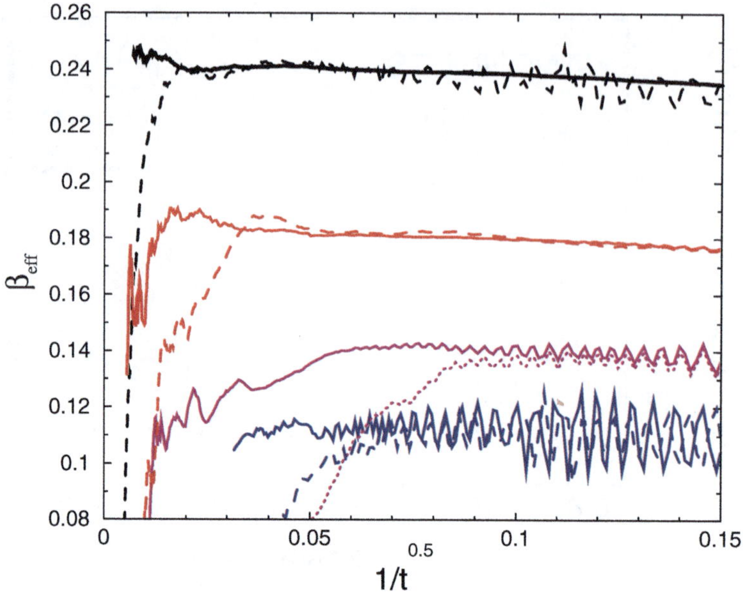

Figure 9.3

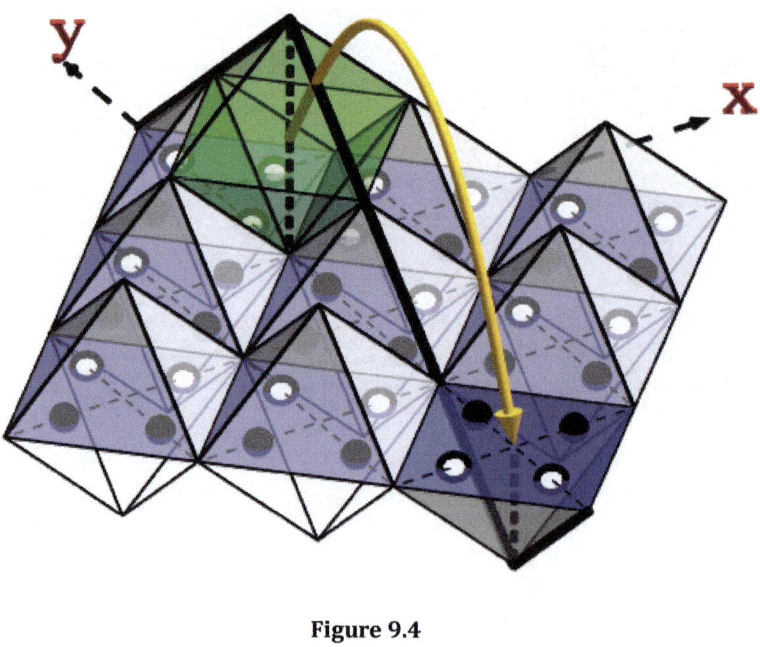

Figure 9.4

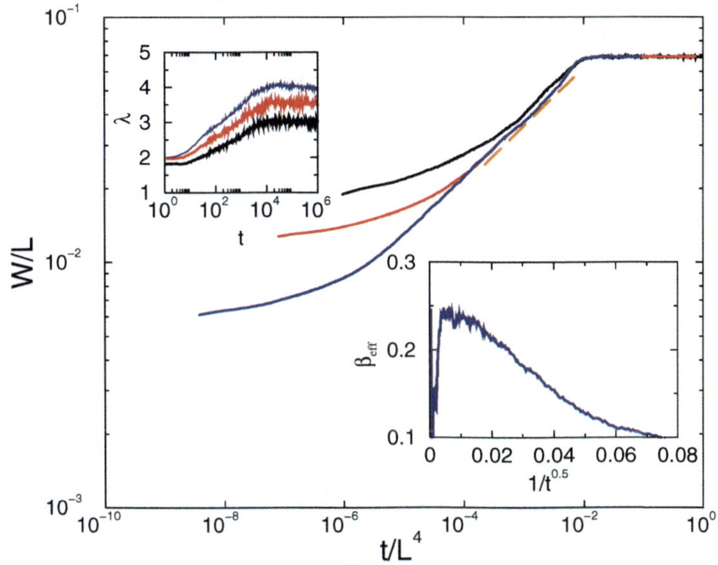

Figure 9.6

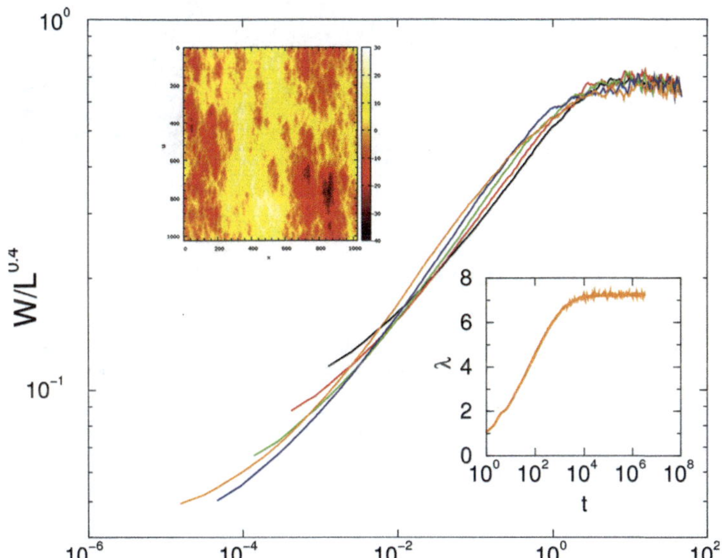

Figure 9.9

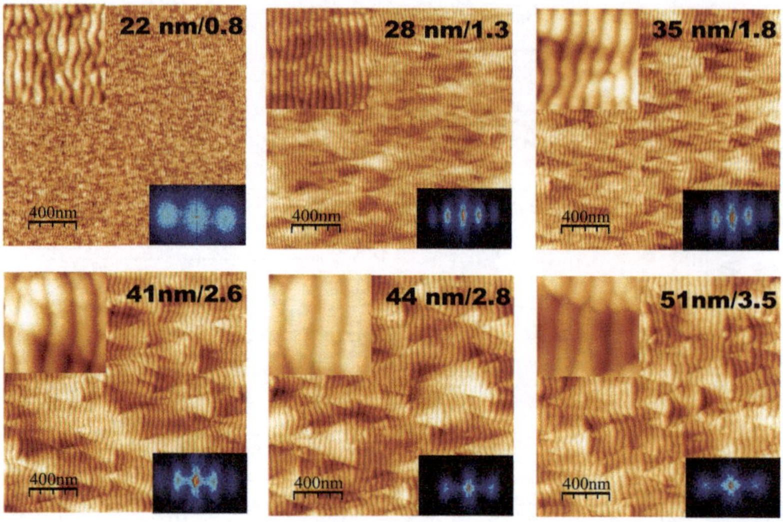

Figure 10.4

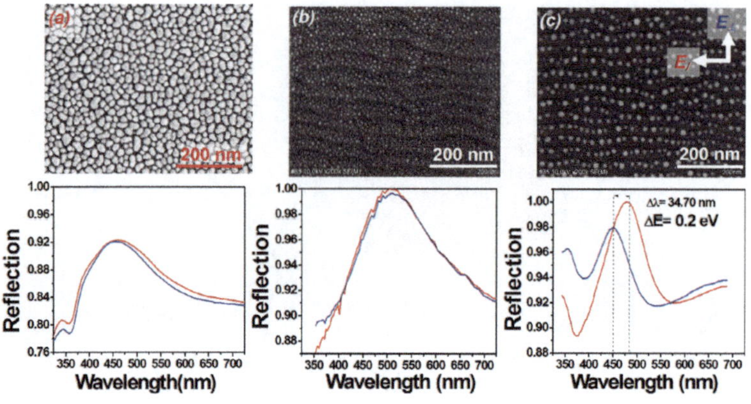

Figure 10.16

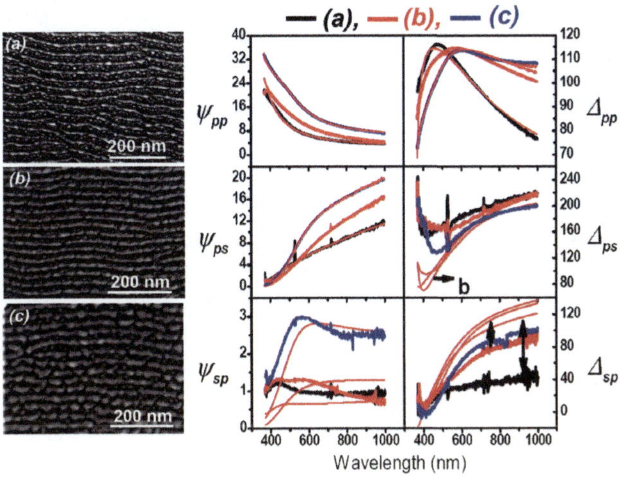

Figure 10.18

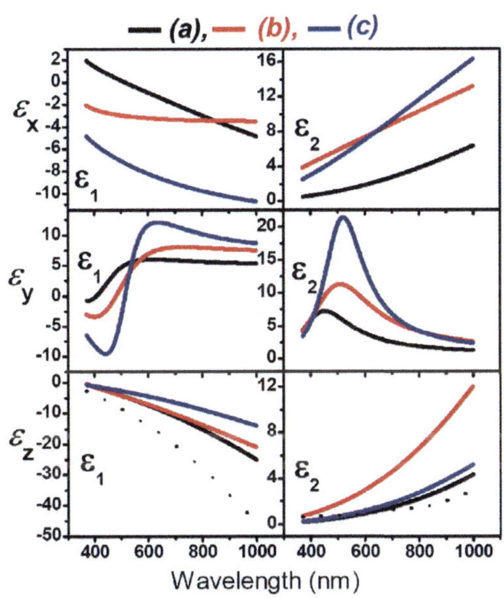

Figure 10.19

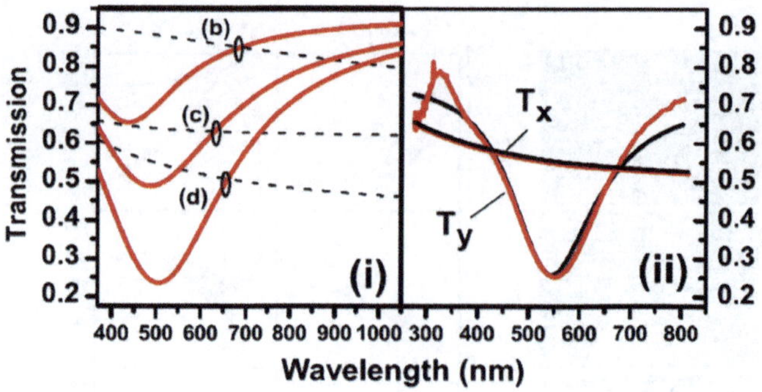
Transmission
0.9
0.8
0.7
0.6
0.5
0.4
0.3
0.2
(b)
(c)
(d)
(i)
400 500 600 700 800 900 1000
T
x
T
y
(ii)
300 400 500 600 700 800
Wavelength (nm)
Transmission

Figure 10.20

For Product Safety Concerns and Information please contact our EU representative GPSR@taylorandfrancis.com Taylor & Francis Verlag GmbH, Kaufingerstraße 24, 80331 München, Germany

Batch number: 10725935

Printed by Printforce, the Netherlands